International Scientific Practical Conference "Materials Science, Shape-Generating Technologies and Equipment" (ICMSSTE 2020)

AA003533

MATEC Web of Conferences Volume 315 (2020)

Yalta, Russia
25 - 29 May 2020

Editor:

Vladimir V. Belov

ISBN: 978-1-7138-1443-6

Printed from e-media with permission by:

Curran Associates, Inc.
57 Morehouse Lane
Red Hook, NY 12571

CURRAN ASSOCIATES INC. proceedings.com

Some format issues inherent in the e-media version may also appear in this print version.

For additional information, please contact EDP Sciences – Web of Conferences at the address below.

EDP Sciences – Web of Conferences
17, Avenue du Hoggar
Parc d'Activité de Courtabœuf
BP 112
F-91944 Les Ulis Cedex A
France

Phone: +33 (0) 1 69 18 75 75
Fax: +33 (0) 1 69 28 84 91

contact-edps@webofconferences.org

Additional copies of this publication are available from:

Curran Associates, Inc.
57 Morehouse Lane
Red Hook, NY 12571 USA
Phone: 845-758-0400
Fax: 845-758-2633
Email: curran@proceedings.com
Web: www.proceedings.com

TABLE OF CONTENTS

Preface

Valery Dyadichev[1,*]
[1]V.I. Vernadsky Crimean Federal University, Simferopol, Crimea, Russia

International Scientific Practical Conference «Materials science, shape-generating technologies and equipment 2020» (ICMSSTE 2020) was held in Yalta, Russia from 25 to 29 May 2020.

ICMSSTE 2020 is organized by the: Federal State Autonomous Educational Institution of Higher Education «V. I. Vernadsky Crimean Federal University»; Federal State Autonomous Educational Institution of Higher Education «Sevastopol State University»; Federal State Autonomous Educational Institution of Higher Education «Polzunov Altai State Technical University»; Limited Liability Company "Engineering Centre The Best Available Technologies".

The Conference provides scientists, industry experts and business representatives from various countries with the opportunity to summarize and share innovative and latest scientific research developments in materials science, material properties, new materials and products manufacturing technologies, industrial equipment and tools, to discuss the development of nanomaterials and nanotechnologies, new equipment and technologies.

In addition, the Conference aim is to establish new scientific relations between conference participants with the purpose to integrate science and business of the various countries, in order to conduct further joint researches, to implement some new joint research projects and introduce the scientific results into the real production industry and business.

ICMSSTE 2020 proceedings cover these topics: Materials science and physical chemistry of materials; Metallic materials: structure, properties, equipment; Non-metal materials: structure, properties, equipment; Composite and powder materials; Coating; Nanomaterials; Building materials, properties, technologies, equipment; Biological materials: structure, properties, technologies, equipment; Medical materials: structure, properties, technologies, equipment; Modeling the structure, properties of materials; Material processing; Theory and technology of welding production; Additive technologies.

This year the organizing committee was forced to lead the conference online, in the virtual form, in connection with the current situation with the epidemiological situation in the world and the impossibility of the personal presence of participants from Russia and other countries in the city of Yalta, because of travel restrictions. The organizing committee decided to hold a conference with online reports of the participants, and not to cancel or a postpone the holding of conference in order to enable scientists to present the obtained topical scientific results and share them with the colleagues, since the science development is important and especially urgent in times of general isolation and problems with personal international communication.

So, the Conference was held on the online platform – "Zoom", with the online reports of authors, with the video and visual presentation of reports, with the opportunity to ask the questions to each other and discuss the scientific results.

The organizers was located in Yalta and some of them n Simferopol, where is located the main unit of V.I. Vernadsky Crimean Federal University. Some of organizers and participant was located at home, because of remote work at home.

The Conference model was as for full-time participation, there were three speeches in the plenary session, all reports were held in accordance with the sections of the conference, each author had from 10 to 15 minutes for his report and the answer to questions from other participants. Also there was time at the ending dates of the conference for general discussion and consideration of poster presentations of participants who were unable to personally submit a video report.

Overall, the 206 authors submitted papers for ICMSSTE 2020, from 6 countries: Russia, Ukraine, Chech Republic, Iraq, Vietnam, Kazakhstan. The Conference Program, video from online reports and presentations of the research articles are available on the Conference website.

Virtual form of reports did not prevent authors from actively participating and establishing new relations and experience exchange. Authors and researches from different countries and organizations, including companies of real sector of the economy, showed the great interest and took an active part in the conference. We look forward to

* Corresponding author: mr.dyadichev@mail.ru

further personal meetings and face-to-face reports by our fellow researchers during the next Conference ICMSSTE 2020.

Editors want to thank the international scientific committee and the reviewers for their great contribution to the work and organization of the Conference. After the peer review process, the submitted papers were selected on the basis of originality, significance, and clarity for the purpose of the Conference.

The organizing committee is very grateful to all the participants and authors of the conference, who presented the results of their research in various fields of materials science, technologies and equipment on our Conference.

Prof., Conference Chair, Valery Dyadichev, V.I. Vernadsky Crimean Federal University, Russia

Conference foto – online reports of conference participants:

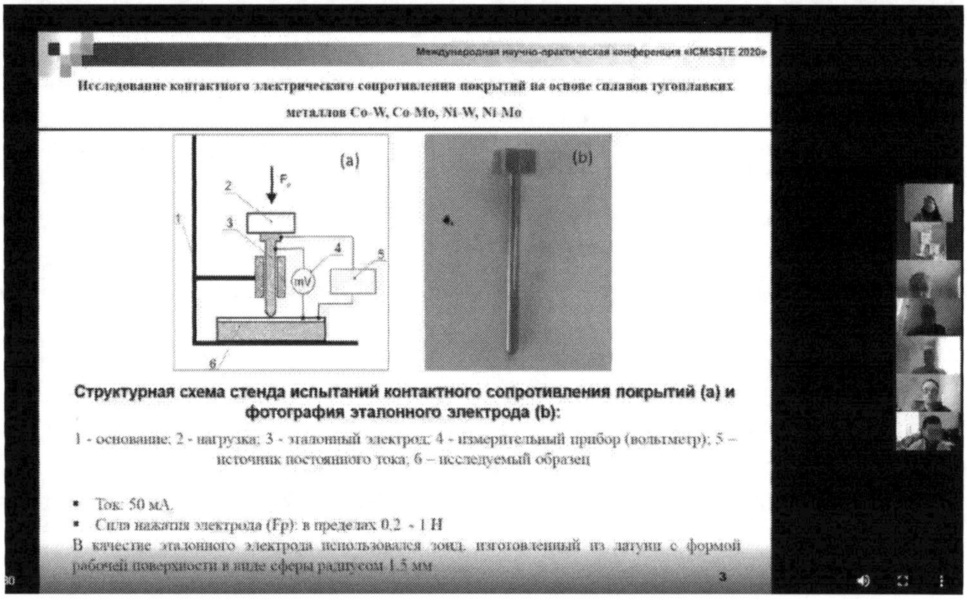

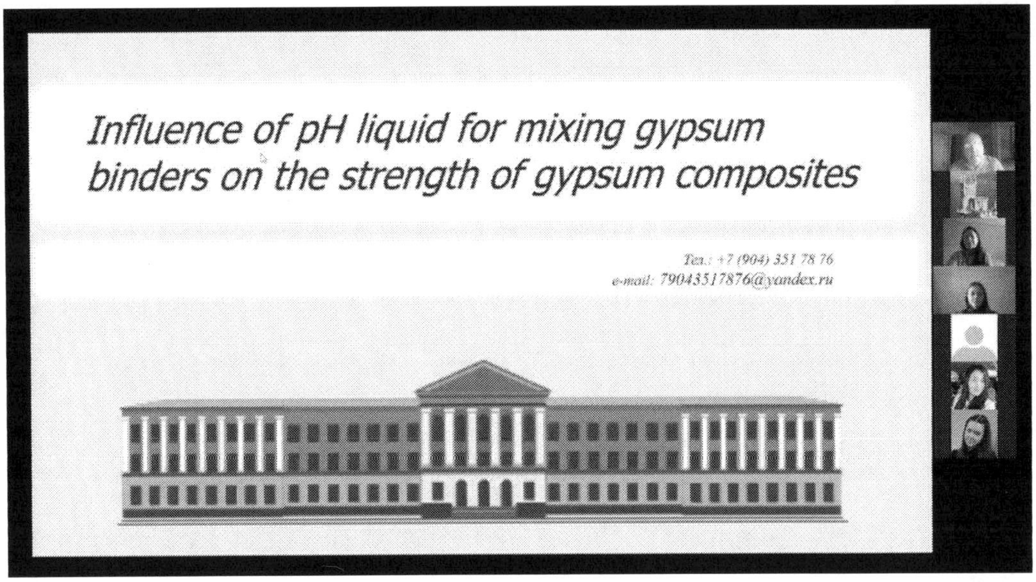
Influence of pH liquid for mixing gypsum binders on the strength of gypsum composites
Тел.: +7 (904) 351 78 76
e-mail: 79043517876@yandex.ru

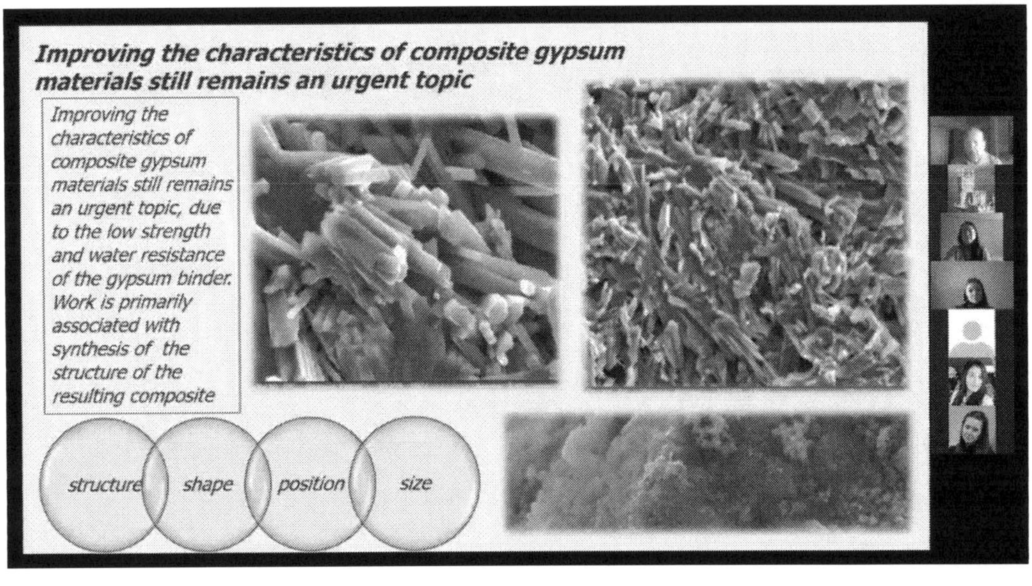
Improving the characteristics of composite gypsum materials still remains an urgent topic
Improving the characteristics of composite gypsum materials still remains an urgent topic, due to the low strength and water resistance of the gypsum binder. Work is primarily associated with synthesis of the structure of the resulting composite
structure
shape
position
size

International Scientific Practical Conference «Materials science, shape-generating technologies and equipment 2020» (ICMSSTE 2020): Yalta, Russia, 25-29 May, 2020.

Scientific Editors:

- Prof., Dr., Vladimir V. Belov - Vice-Rector for Research, Head of the Department of Production of Building Products and Structures of Tver State Technical University, Counselor of Russian Academy of Architecture and Building Sciences
- Limited Liability Company "Engineering Centre The Best Available Technologies".

ICMSSTE 2020 is organized by the: Federal State Autonomous Educational Institution of Higher Education «V. I. Vernadsky Crimean Federal University»; Federal State Autonomous Educational Institution of Higher Education «Sevastopol State University»; Federal State Autonomous Educational Institution of Higher Education «Polzunov Altai State Technical University»; Limited Liability Company "Engineering Centre The Best Available Technologies".

edp sciences

Web of Conferences

Statement of Peer review

In submitting conference proceedings to *Web of Conferences*, the editors of the proceedings certify to the Publisher that

1. They adhere to its **Policy on Publishing Integrity** in order to safeguard good scientific practice in publishing.
2. All articles have been subjected to peer review administered by the proceedings editors.
3. Reviews have been conducted by expert referees, who have been requested to provide unbiased and constructive comments aimed, whenever possible, at improving the work.
4. Proceedings editors have taken all reasonable steps to ensure the quality of the materials they publish and their decision to accept or reject a paper for publication has been based only on the merits of the work and the relevance to the journal.

Title, date and place of the conference

International Scientific Practical Conference „Materials science, shape-generating technologies and equipment 2020" (ICMSSTE 2020)
May 25-29, 2020
Yalta, Russia

Proceedings editor(s): Limited Liability Company "Engineering Centre The Best Available Technologies" - Sergey Menyak

Date and editor's signature 18 June,

17, avenue du Hoggar - PA de Courtabœuf – BP 112 - 91944 Les Ulis Cedex A (France)
Tél. : 33 (0)1 69 18 75 75 - Fax : 33(0)1 69 07 45 17 – www.edpsciences.org

Conducting a microscopic analysis of the powders obtained by electroerosion dispersion of waste of steel X13

E.V. Ageeva[1], *S.V.* Khardikov[2*], and *A. N.* Novikov[3]

[1]Southwest state University, 305040, Kursk, Russian Federation
[2]Southwest state University, 305040, Kursk, Russian Federation
[3]OSU named after I. S. Turgenev, 302026, Orel, Russian Federation

Abstract. Alloyed steels and alloys containing a significant amount of non-ferrous metals are used for the manufacture of critical parts for various purposes, since it is the alloyed steels and alloys that possess a complex of high physicomechanical and other properties. Chromium is the most common alloying element. Chromium alloyed steel has increased strength, hardness, wear resistance, corrosion resistance in aggressive and oxidizing environments. The presence of chromium increases the hardenability of steel. The process of obtaining powder materials by the method of electroerosion dispersion favorably differs by the possibility of obtaining powders even from very hard and refractory materials. In the case of grinding material with this method, the only important criterion is the electrical conductivity of the material being ground. Worn parts and other scrap can be used as feedstock during the process. The purpose of the study is to investigate the powder, obtained from the wastes of chromium-containing steels. Thanks to the study of the samples of the used powder, on modern devices, results were obtained, indicating that the chromium-containing powder, obtained by the method of electroerosive dispersion, can be reused to restore and strengthen machine parts.

1. Introduction

Powder technology is a broad area for the production of dispersed bodies, used in various branches of production - powder metallurgy, the ceramic industry, the production of food and medicinal products, fertilizers, fuel, building materials, etc. Due to some external similarity of the technology of powder metallurgy with the technology of ceramic production, products manufactured by powder metallurgy are also widely known as metal-ceramic [1–3].

Along with the advantages of powder metallurgy, it should be noted that there are disadvantages that impede and limit its wide distribution. The main disadvantages include the high cost of metal powders and the lack of mastered methods for producing alloy powders - steels, bronzes, brass, etc. One of the solutions to these problems is the use of

* Corresponding author: hardikov1990@mail.ru

electric erosion. Currently, alloyed steels and alloys containing a significant amount of non-ferrous metals are used for the manufacture of critical parts for various purposes, since it is the alloyed steels and alloys that possess a complex of high physicomechanical and other properties.

Carbon steels do not have sufficiently high mechanical and physicochemical properties and do not meet the requirements for metals in modern engineering. One of the main ways to improve the properties of steel is the introduction of various elements into it. The process of introducing various elements into steel to enhance its properties is called alloying, and the resulting steels are called alloyed [4–6]

The introduction of alloying elements leads to the formation of new phases and structural components, changes the kinetics of transformations, causes new phenomena and changes the technology of heat treatment. The alloying elements introduced into carbon steel, change its mechanical and physical properties. Doping significantly increases the strength and hardness while maintaining good toughness of steel, increases its hardenability, allows for quenching in moderate coolers. Doping gives the steels a number of special properties: heat resistance, corrosion resistance, scaling resistance, magnetoresistance and many other properties.

Chromium is the most common alloying element. Chromium alloyed steel has increased strength, hardness, wear resistance, corrosion resistance in aggressive and oxidizing environments. The presence of chromium increases the hardenability of steel. With 13% chromium in steel, it becomes stainless. A further increase in the chromium content gives the steel anti-corrosion at high temperatures, as well as magnetoresistance.

Due to the widespread use of chromium-containing steels in industry, there is currently an acute problem of recycling their waste. The most promising method of processing such wastes is the electro-erosion dispersion method (EED), which is distinguished by the ecological purity of the process and relatively low energy costs [7–9].

For the development of technologies for the practical application of powder, obtained from wastes, and evaluating the effectiveness of its use, complex theoretical and experimental studies are required.

The aim of the work was to study the particle size distribution of the powder particles, obtained by electro-erosion dispersion of chromium-containing wastes in lighting kerosene.

2. Materials and methods

To carry out the planned researches, chromium-containing steel wastes were selected. Lighting kerosene was chosen as a working fluid. To obtain a powder by the method of electroerosive dispersion, an installation for EED of conductive materials was used. Wastes were loaded into a reactor filled with a working fluid - lighting kerosene; the process was carried out with the following electrical parameters: discharge capacitor capacity 45 μF, voltage 100 ... 110 V, pulse repetition rate 55 ... 65 Hz. As a result of the local impact of short-term electrical discharges between the electrodes, the waste material was destroyed with the formation of dispersed powder particles.

Using the QUANTA 600 FEG scanning electron microscope, a direct analysis of powder particles with a sufficiently high resolution was carried out. The QUANTA 600 FEG - (manufactured by FEI (Holland)) is an electron-ion scanning microscope with an electron beam column equipped with a tungsten cathode. Its accelerating voltage is from 200 eV to 30 kV and its resolution is 3.5 nm at 35 kV at optimum WD, 3.5 nm at 30 kV in natural mode and less than 15 nm at 1 kV in low vacuum mode. The ion column with a gallium liquid-metal ion source Magnum has an accelerating voltage of 5 kV to 30 kV and a resolution of 20 nm. The system is equipped with a five axial motorized 50x50x25 mm table, gas injection systems for spraying conductors and dielectrics, as well as for etching samples.

Powders obtained from electro-erosion dispersion in kerosene from chrome-containing steel wastes were analyzed using an Analysette 22 NanoTec laser particle size analyzer to determine the size distribution of the obtained powder particles.

3. The study of the microstructure of powders

Using the method of scanning electron microscopy (SEM), particles of a powder sample were investigated using a secondary electron detector (Fig. 1).

Fig. 1. Photograph of powder particles, obtained by EED

It can be seen that the shape of the powder particles is determined by the form in which the material is ejected from the well in the EED process. It is also seen that particles with a regular spherical or elliptical shape prevail in the powder. They are obtained by crystallization of the molten material (liquid phase). Particles formed during the crystallization of a boiling material (vapor phase) have an irregular shape, an order of magnitude smaller than particles formed from the liquid phase, and usually agglomerate with each other and on the surface of other particles. In the process of EED such particles are most susceptible to chemical and phase changes.

The results of measuring the particle size of the obtained powder are presented in Figure 1.

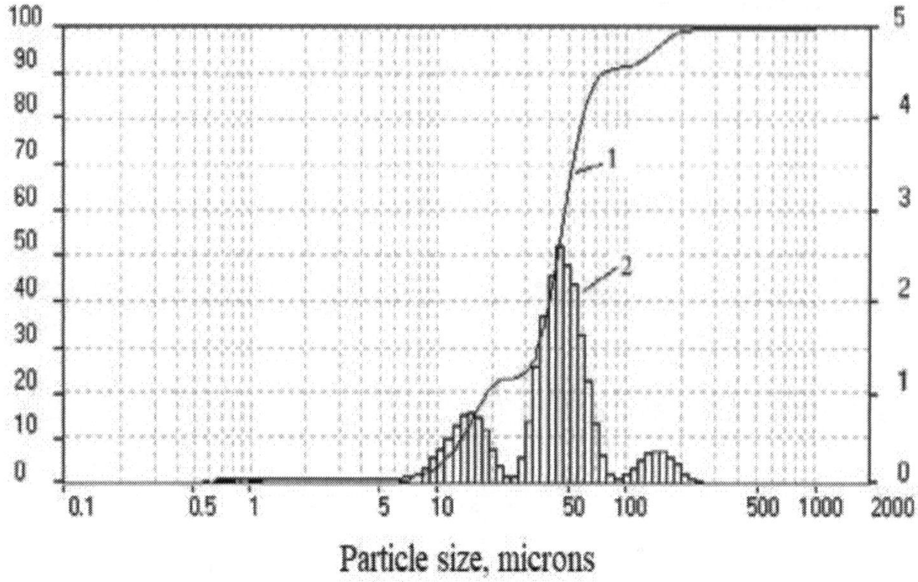

Fig. 2. The size distribution of powder microparticles: 1 - Integral curve, 2 - Histogram.

It was established that the average particle size is 49.33 μm, the arithmetic value is 49.328 μm, and the specific surface area is 2732.76 cm^2/cm^3.

4. Conclusion

In the course of the experimental research aimed at the study of the particle size of the chromium powder, which was obtained by electroerosion dispersion of wastes in lighting kerosene with the following electrical parameters: the capacity of the discharge capacitors is 45 μF, the voltage is from 100 V to 110 V, the pulse frequency is from 55 Hz to 65 Hz, the average particle size and the specific surface area were determined. It allows to identify the rational area of its practical application.

On the basis of the obtained results, it can be concluded that the chromium powder, obtained by the method of electroerosive dispersion, can be reused when restoring and hardening machine parts.

The work was supported by a scholarship of the President of the Russian Federation to young scientists and graduate students (SP-945.2019.1).

References

1. Mal'tsev I.M.,. *International Science and Technology Magazine* (2003)
2. Chandler K.M., Mitchell M.D., Pikuz S.A., Shelkovenko T.A., Hammer D.A., Shlyaptseva A.S., Ouart N.D., Hansen S.B., Kantsyrev V.L., Fedin D.A. Rev. Sci. Instrum. **75**, 10-2, 3702. (2004)
3. Wang Z., Qiu K. Electrochem. Commun. **8 (7),** 1075. (2006)
4. Wang Z., Gao G., Zhu H., Sun Z., Liu H., Zhao X., Int. J. Hydrogen. Energ. **34 (23),** 9334 (2009)

5. Kazi I.H., Wild P.M., Moore T.N., Sayer M. Thin Solid Films. **515 (4)**, 2602 (2006)
6. Chen T.J., Lee R.T., Chiou Y.C. Proceedings of the Institution of Mechanical Engineers, Part B: Journal of Engineering Manufacture. T. **227 (11)**, 1596-1606. (2013)
7. Huang Y., Wu H., Shao T., Zhao X., Peng H., Gong Y., Wan H. Chemical Engineering Journal. T. **339**. 322-333. (2018)
8. Sassi W., Dhouibi L., Triki E., Berçot P., Rezrazi M. Electrochimica Acta. T. **117**, 43-452. (2014)
9. Ageeva E.V., Ageev E.V., Horyakova N.M. Russ. Eng. Res. **35** (1), 33. (2015)

X-ray diffraction analysis of electroerosive powder materials, obtained from waste of tungsten-free hard alloy grade KNT16

Vladimir Serebrovsky [1], and *Boris* Sabel'nikov [2*]

[1]Kursk State Agricultural Academy, 305021 Kursk, Russian Federation

[2]Southwest State University, 305040 Kursk, Russian Federation

Abstract. This article presents a study of the phase composition of powder electroerosive materials of TFHA grade KNT16, obtained in ethyl alcohol. It was established that the main phases in the test sample are TiC, $MoNi_3$, Ni, Mo.

1 Introduction

Tungsten-free hard alloys (TFHA) or cermets are materials based not on tungsten carbide, but on another refractory compound (usually titanium carbide or carbonitride); the cementing phase is metals - nickel and / or cobalt, with hardening additives of transition metals 4, 5, 6 groups (Mo, W, Cr, Nb, etc.). These materials combine both ceramic properties (exceptional hardness and wear resistance, oxidation resistance, low adhesion to the material being processed) and metal properties (strength, impact resistance). KNT16 alloy is intended to replace the standard T15K6 alloy, and in some cases instead of VK8, T14K8, T30K4 and T5K10 alloys for semi-finishing and semi-rough turning with continuous cutting of carbon, low alloy and some structural steels.

The widespread use of TFHA KNT16 brand in various industries leads to a large accumulation of its waste requiring processing. Currently, there are methods that allow the processing of solid alloy wastes with a view to their reuse. However, the disadvantages of the known methods are increased energy consumption, multi-operation process [1-3].

The most promising method for processing any conductive material is the method of electroerosive dispersion (EED), which is distinguished by the ecological purity of the process and relatively low energy costs.

Carrying out the planned measures will solve the problem of waste disposal and their further use and thereby reduce the cost of production of the final product.

The aim of the work was to carry out X-ray diffraction analysis of electroerosive powder materials, obtained from wastes of tungsten-free hard alloy grade KNT16 in ethyl alcohol.

2 Materials and Methods

To obtain the powder from TFHA waste KNT16 grade (GOST 19052-80), an EED device, developed by the authors, was used [4]. The waste was loaded into a reactor filled with a working fluid - ethyl alcohol (GOST R 56389-2015), the process was carried out with the

* Corresponding author: sabelnikovboris1@mail.ru

following electrical parameters: capacitance of discharge capacitors 25,5 μF, voltage 195–205 V, pulse repetition rate 95–105 Hz.

The phase composition of the sample was studied by X-ray diffraction on a Rigaku Ultima IV diffractometer in Cu-$K_α$ radiation (wavelength λ = 0,154178 nm) using Soller slits (Fig. 1).

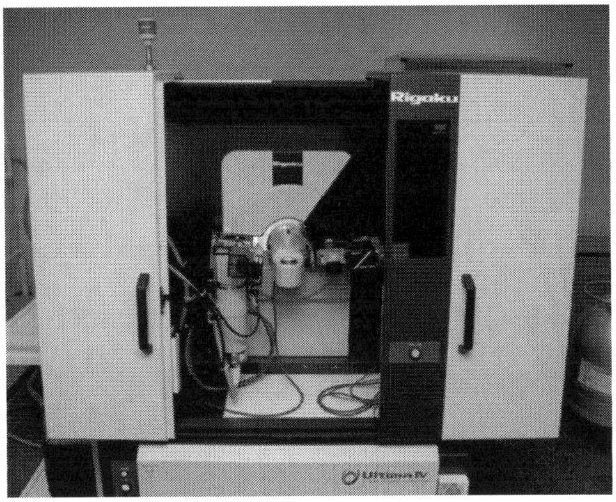

Fig.1. X-ray diffractometer Rigaku Ultima IV.

The diffraction spectrum for phase analysis is taken according to the θ-2Θ scanning scheme with focusing according to Bregt-Brentano in the range of angles of 5 ... 100 degrees. 2 Θ. The shooting is carried out in dotted mode with a scanning step Δ (2θ) = 0.02 deg, speed 0.6 deg / min, operating voltage 45 kV, current 200 mA [5-11].

3 Results

To refine the profile of experimental X-ray diffraction patterns, the PDXL RIGAKU software package was used. The background was subtracted using the Sonneveld – Wisser method, the experimental profile was smoothed using the Savitsky – Naked method, and the components kα1 and kα2 were separated by the Rachinger method. To describe the diffraction maxima, a superposition of the Gauss function and the Lorentz function was used. The phase composition was determined using the ICCD PDF-2 database (2014).

The diffraction pattern and phase composition of the test sample are presented in Fig. 2 and Table 1.

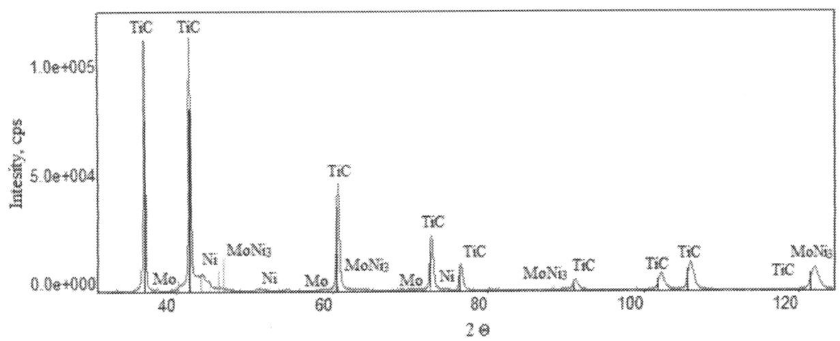

Fig.3. X-ray diffraction pattern of the test sample.

Table 1. The phase composition of the test sample.

Chemical formula	Type of crystal lattice	Lattice parameters
TiC	225:Fm-3m cubic crystal lattice	a = b = c = 4.323534 Å
$MoNi_3$	59:Pmmn,choice-2 orthorhombic crystal lattice	a = 5.167347 Å, b = 4.176675 Å, c = 4.360784 Å
Ni	225:Fm-3m cubic crystal lattice	a = b = c = 3.533749 Å
Mo	229:Im-3m cubic crystal lattice	a = b = c = 3.135084 Å

4 Conclusion

Based on the performed x-ray microanalysis, it was found that the main phases in electroerosive powder materials of the tungsten-free hard alloy KNT16 grade, obtained in ethyl alcohol, are TiC, MoNi3, Ni, Mo. The study will determine the most relevant area of application of the obtained samples and improve the quality of scientific and technological developments.

References

1. E.V. Ageev, *Hardening technologies and coatings* **6**, 8-14, (2011)
2. R.A. Latypov, G.R. Latypova, E.V. Ageev, P.I. Burak, Int. Sci. J., **5**, 80-86 (2013)
3. E.V. Ageev, Electromet., **10**, 24–27 (2011)
4. E.V. Ageev, A. p. 2449859, Russian Federation, C2, B22F9/14 *Installation for producing nanodispersed powders from conductive materials*. - No. 2010104316/02; application 02/08/2010; publ. 05/10/2012
5. E.V. Ageeva, E.V. Ageev, N.M. Horyakova, Russ. Eng. Res., **35**, 33-35 (2015)
6. E. V. Ageev, B. A. Semenikhin, R. A. Latypov, Fund. Prikl. Probl. Tekhn. Tekhnolog, **5**, 39-42 (2010)
7. A. Pereverzev, E. Ageev, MATEC Web of Conf., **298**, 00037 (2019)
8. E.V. Ageev, S.V. Khardikov, E.A. Vorobyev, A.A. Sysoev, MATEC Web of Conf., **298**, 00127 (2019)
9. R.A. Latypov, E.V. Ageeva, G.R. Latypova, MATEC Web of Conf., **298**, 00125 (2019)
10. E.V. Ageev, A.Yu. Altukhov, S.V. Khardikov, S.S. Gulidin, A.N. Novikov, J. Nano-Electron. Phys., **7**, 04080 (2015)
11. E.V. Ageev, A.Y. Altukhov, S.V. Khardikov, A.V. Kirichek, J. Nano- Electron. Phys., **8**, 04055 (2016)

The porosity study of sintered products from electro-erosive materials of alloy Cr17, obtained in lighting kerosene

E.V. Ageev [1*], *A.S.* Pereverzev [1], and *A. A.* Sysoev[1]

[1]Southwest State University, 305040 Kursk, Russian Federation

Abstract. The article presents the results of porosity study of sintered products from electro-erosive materials of alloy Cr17, obtained in lighting kerosene. It was shown that during the consolidation of electro-erosion materials from Cr 17 alloy by the method of spark plasma sintering, the porosity was 0.27%.

1 Introduction

Steel Cr17 in accordance with GOST 5632-72 is defined as nickel-free, chromium heat-resistant corrosion-resistant steel of the ferrite group. In the Cr17 alloy, there is practically no nickel (less than 0.01%); and molybdenum and titanium are completely absent. But its chemical composition provides steel with resistance to corrosion in environments with moderate aggressiveness and resistance to oxidation at high temperatures. A high percentage of chromium alloy provides strength and corrosion resistance. In addition, it contributes to an increase in the melting temperature, and gives the surfaces of products gloss. And the combination of a large amount of chromium with a low carbon number increases ductility.

The widespread use of Cr17 steel in various industries leads to a large accumulation of its waste requiring processing. Currently, there are many ways to recycle metal waste in order to reuse it. However, the disadvantages of the known methods are increased energy consumption, multi-operation process [1-7].

The most promising method of processing metal waste is the method of electro-erosive dispersion (EED), which is distinguished by the ecological purity of the process and relatively low energy costs.

To develop technologies for the practical use of powder materials, obtained from waste alloy Cr17, and to evaluate the effectiveness of their use, complex theoretical and experimental studies are required. The aim of the work was to study the porosity of sintered samples from electroerosive materials of Cr17 alloy, obtained in lighting kerosene.

2 Materials and Methods

* Corresponding author: ageev_ev@mail.ru

To implement the planned studies, the waste of the Cr17 alloy (GOST 5632-72) was loaded into the reactor of the electro-erosive dispersion unit (EED) [7]. Lighting kerosene was used as the working fluid (GOST 4753-68). The process was carried out with the following electrical parameters: capacity of discharge capacitors 55 μF, voltage 120 ... 130 V, pulse repetition rate 95 ... 100 Hz.

As a result, particles were obtained, whose average size is 28 microns. The final porosity of the sintered products is influenced by the pressing and sintering modes.

The powder was consolidated by spark plasma sintering using the SPS 25-10 spark plasma sintering system (Thermal Technology, USA) according to the scheme shown in (Fig. 1).

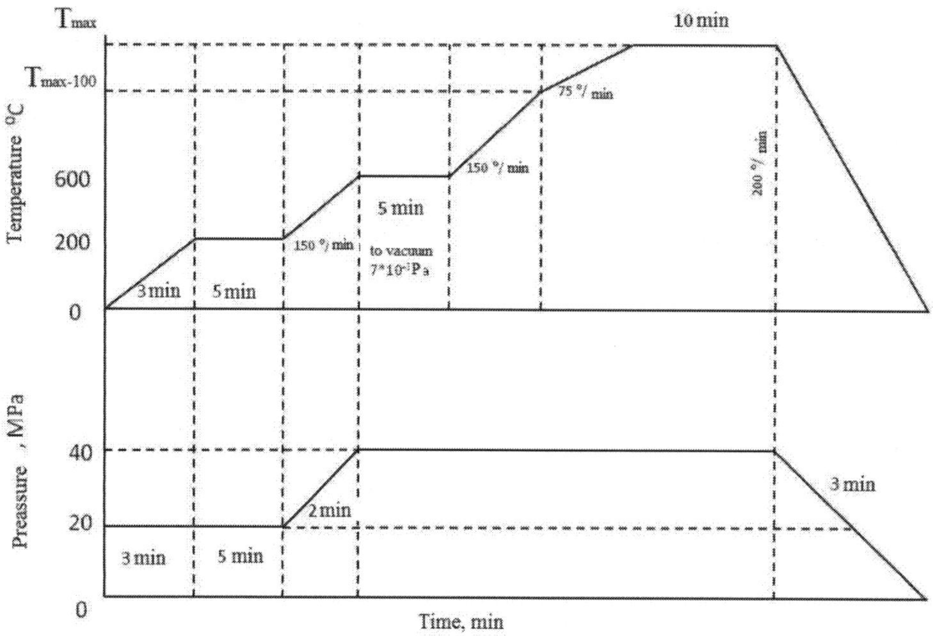

Fig. 1. Scheme of powder consolidation by spark plasma sintering.

The starting material was placed in a graphite matrix, placed under a press in a vacuum chamber. The electrodes, integrated into the mechanical part of the press, fed an electric current to the matrix and created spark discharges between the sintered particles of the material, providing intense interaction.

The advantages of spark plasma sintering technology are the uniform distribution of heat throughout the sample; high density and controlled porosity; lack of need for binding materials; uniform sintering of homogeneous and dissimilar materials; short cycle time; manufacturing of the part immediately in the final form and obtaining a profile close to the specified [8-15].

Porosity was determined using an Olympus GX51 inverted optical microscope with software for quantitative image analysis (Fig.2).

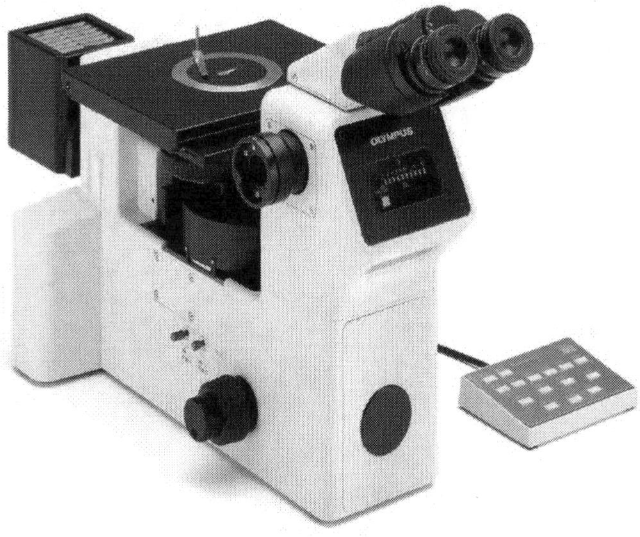

Fig.2. Olympus GX51 Optical Inverted Microscope.

One of the main methods for determining porosity is the metallographic method with elements of qualitative and quantitative analyses of pore geometry (stereoscopic metallography). The use of stereoscopic metallography techniques allows to calculate the specific surface area of large pores, the number of spherical pores per unit volume, the average distance between pores, and the average real diameter of spherical pores [16-20].

3 Results

The prepared samples had no traces of grinding, polishing or chipping of structural components. A thin section was made over the cross section (kink) of the whole product or part of it with an area of less than 2 cm^2. The SIAMS Photolab software, which the microscope is equipped with, has been developed taking into account the specifics of using digital microscopy and image analysis methods for metallographic analysis of compounds.

The results of the sample porosity study by metallographic method are shown in Table 1.

Table 1. Porosity (metallographic method).

Area of analysis, sq. μm	Porosity, %	Dmin	Dmax	Dmed
3788639.6	0.27	3.2	27.1	4.1

Figure 3 shows the microstructure of the obtained sintered samples with a magnification of the microscope 1000 times.

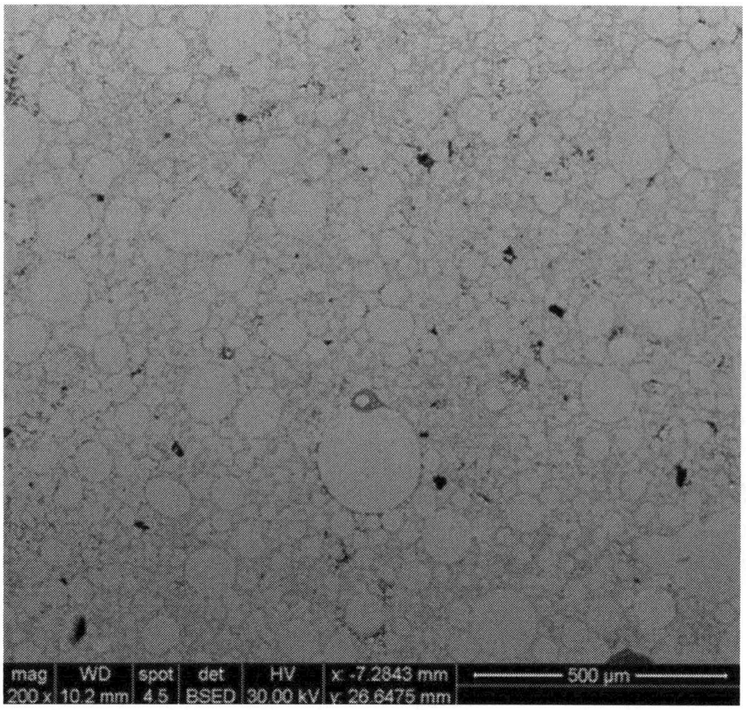

Fig.3. The microstructure of the studied sintered sample.

A histogram of pore size distribution is shown in Fig. 4.

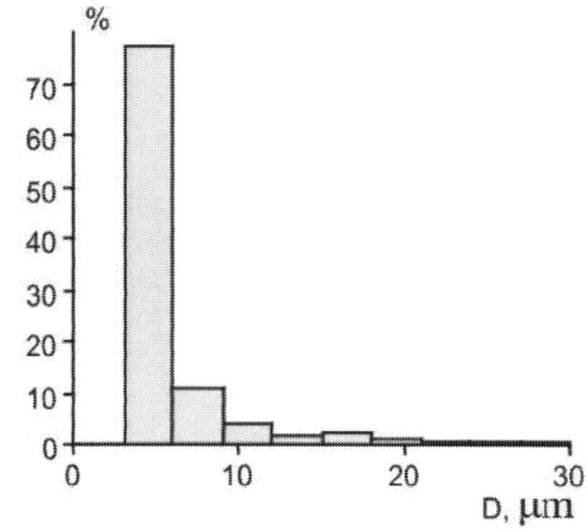

Fig.4. Pore size distribution histogram

According to the presented histogram, more than 90% of the pores have a size of less than 10 microns.

4 Conclusion

On the basis of experimental studies, aimed at studying the porosity of sintered samples by the metallographic method, it was found that when consolidating by spark plasma of sintering electro-erosive materials from Cr17 alloy wastes, obtained in lighting kerosene, the porosity was 0.27%. It is noted that more than 90% of the pores have a size of less than 10 microns. The study will determine the most relevant area of application of the obtained samples and improve the quality of scientific and technological developments.

Acknowledgments: The reported study was funded by RFBR, project number 19-33-90053.

References

1. A.A. Lipatov, RER, **33**, 3, 144-149 (2013)
2. A.M. Adaskin, A.A. Vereshchaka, A.S. Vereshchaka, J. of Frict. and W., **34**, 208-213 (2013)
3. V.L. Bibik, MSF, **762**, 777-781 (2013)
4. E.V. Azarova, E.A. Levashov, V.G. Ralchenko, A.P. Bolshakov, E.E. Ashkinazi, Metall., **54**, 523-529 (2010)
5. Z. Qiao, X. Ma, W. Zhao, H. Tang, B. Zhao, J. of All. and Comp., **462**, 416-420 (2008)
6. K. Maruyama, T. Nonaka, H.Y. Kim, Intermet., **13**, 1116-1121 (2005)
7. E.V. Ageev Patent 2449859, Russian Federation, C2, B22F9 / 14. Installation for producing nanodispersed powders from conductive materials, applicant and patent holder Southwestern State University. - No. 2010104316/02; application 02/08/2010; publ. 05/10/2012, 4.
8. E.V. Ageev, E.V. Ageeva, Bull. of mech.eng., **11**, 51–57 (2013)
9. E.V. Ageev, B. A. Semenikhin, R. A. Latypov, Fund. Prikl. Probl. Tekhn. Tekhn., **5**, 39-42 (2010)
10. T.N. Oskolkova, E.A. Budovskikh, Metal. Sci. Heat Treat, **55**, 96-99 (2013)
11. J. Karlsson, A. Snis, H. Engqvist, J. Lausmaa, J. of Mat. Proc. Techn., **213**, 2109-2118 (2013)
12. D.D. Gu, W. Meiners, K. Wissenbach, R. Poprawe, IMR, **57**, 133-164 (2012)
13. N. Radek, Maint. and Rel., **4**, 10-16 (2009)
14. A.V. Ribalko, O. Sahin, Surf. & Coat.Techn., **168**, 129-135 (2003)
15. Z. Chen, Y. Zhou, Surf. & Coat.Techn, **201**, 1503-1510 (2006)
16. I.V. Galinov, R.B. Luban, Surf. & Coat.Techn, **79**, 9-18 (1996)
17. E.V. Azarova, E.A. Levashov, V.G. Ralchenko, Transl. from Metall. **8**, 50-55 (2010)
18. A. Pereverzev., E. Ageev., MATEC Web of Conf., **298**, 00037 (2019)
19. E.V. Ageev, S.V. Khardikov, E.A. Vorobyev, A.A. Sysoev, MATEC Web of Conf., **298**, 00127 (2019)
20. R.A. Latypov, E.V. Ageeva, G.R. Latypova, MATEC Web of Conf., **298**, 00125 (2019)

X-ray microanalysis of electro-erosive powder materials, obtained from tungsten-nickel-iron (TNI) alloy wastes in water

E.V. Ageev [1*], *V.L.* Selyutin[1], and *S.V.* Pikalov [1]

[1]Southwest State University, 305040 Kursk, Russian Federation

Abstract. This article presents the results of elemental composition study of the electro-erosion powders, obtained from waste alloy TNI in water. It is established that the main elements are tungsten, nickel, iron and oxygen.

1 Introduction

Heavy alloys have a number of very valuable properties, due to which they are effectively used in many industries. Currently, one of the main problems of using these alloys is the presence in it of a significant amount of expensive tungsten [1-12]. This problem can be solved by grinding their waste and reuse. Existing industrial grinding technologies are distinguished by large tonnage, high energy costs and environmental problems. One of the promising and industrially inapplicable methods of grinding any electrically conductive material is the electroerosion method [13-19].

To develop technologies for the reuse of electro-erosion powders obtained from TNI alloy, and to evaluate the effectiveness of their use, complex theoretical and experimental studies are required.

The aim of the work was to study the elemental composition of electroerosive powders, obtained from TNI alloy waste in wate.

2 Materials and Methods

Powders were obtained at the installation of electroerosive dispersion protected by RF Patent No. 2449859, from TNI alloy shavings. Electronic probe microanalysis (EPM) was carried out with the use of EDAX firm manufacture energy-dispersive x-ray emission analyzer installed into Nova NanoSEM 450 scanning-electron microscope. Using the method of scanning electron microscopy by means of a secondary electrons detector powders sample particles were analyzed.

With the use of EDAX firm manufacture energy-dispersive x-ray emission analyzer installed into Nova NanoSEM 450 scanning-electron microscope characteristic X-ray radiation spectra in different points on the surface of powders sample were obtained.

* Corresponding author: ageev_ev@mail.ru

Determination of the elemental composition of microscopic objects from characteristic X-ray radiation being excited in them is meant by electron probe microanalysis. Spectrometers of two types (crystal-free spectrometer or analyzing crystal-type spectrometer) are used in electron probe microanalysis for characteristic spectrum analysis (CSA). The electron-optical system of a scanning-electron microscope is the base for CSA. During the interaction of the electron probe with the sample (Fig. 1 and 1) X-ray radiation is one of the signals being excited. The X-ray radiation can be divided in characteristic and deceleration radiation.

Deceleration X-ray radiation occurs in consequence of primary electrons deceleration in the electrical (Coulomb) field of the atoms of material being analyzed. In this case the kinetic energy of primary electrons is partially or completely transformed into X-ray radiation energy. Consequently, the radiation has a continuous spectrum with the energy from zero level to the energy of a bombarding electron. That is why the radiation is also called continuous X-ray radiation. When carrying out electron probe microanalysis deceleration radiation is undesirable, because it makes the main contribution to increasing background radiation intensity and cannot be excluded.

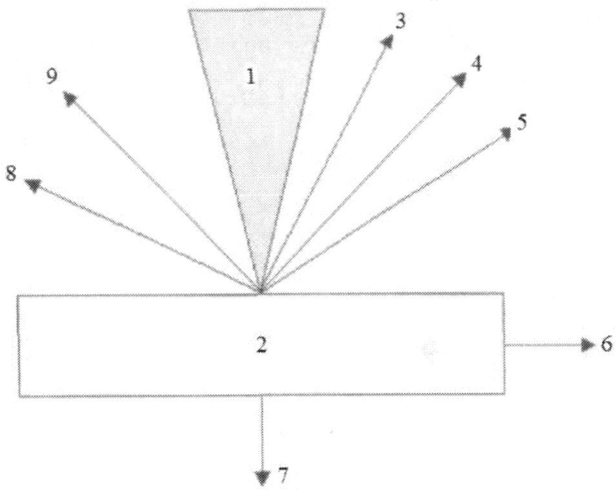

Fig. 1. Effects of the interaction of an electron ray with the object: 1 – electron ray, 2 – object, 3 – reflected electrons, 4 – secondary electrons, 5 – Auger electrons, 6 – absorbed electrons current 7 – passed electrons, 8 – cathodoluminescent radiation, 9 – X-ray radiation.

When primary electrons penetrate a sample, they are decelerated not only by the electrical field of atoms, but also in consequence of immediate collision with the electrons of the material atoms. As a result primary electrons can knock out electrons from the internal K-, L-, or M-shells, leaving an atom of the sample in excited state. Vacancy sites are filled through passage of electrons from higher energy levels. The atom is passed to the main state, and excess energy is released in the form of an X-ray quantum. As far as the energy of the quantum formed depends only on the energy of electron levels involved in the process, and such electron levels are typical for every element, characteristic X-ray radiation occurs. As every atom has a well-defined finite number of levels, between which not only certain type passages are possible, characteristic X-ray radiation forms a discrete linear spectrum.

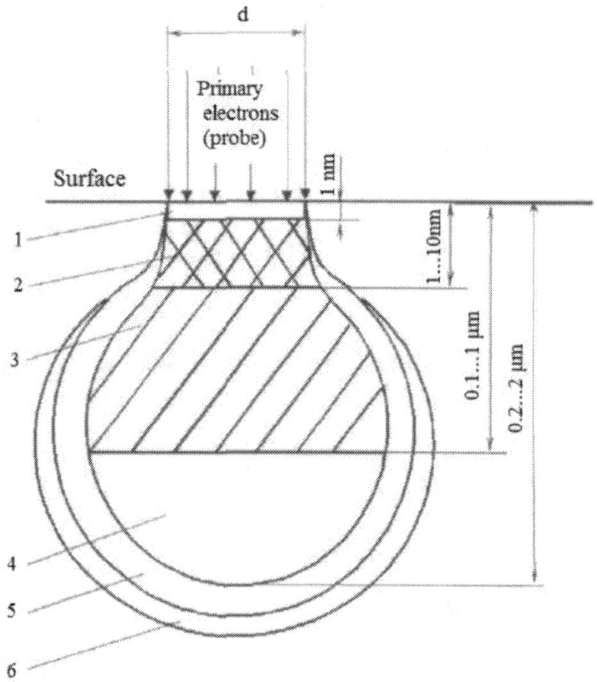

Fig.2. . Signal areas and space resolution under irradiation of the object surface with a flow of electrons (probe). Areas of generation of: 1 – auger electrons, 2 – secondary electrons, 3 – reflected electrons, 4 – characteristic X-ray radiation, 5 – deceleration X-ray radiation, 6 – fluorescence.

It does not succeed to determine light elements with an atomic number less than 4 within an alloy by means of electron probe microanalysis. The same difficulties with elements determination occur, when the lines of L- or M-series of one element are overlaid by the lines of K-series of another element. An important characteristic of CSA is its locality, i. e. the volume of material, in which characteristic X-ray radiation is excited. The volume is determined primarily by the diameter of an electron probe on a sample and depends on accelerating voltage and elemental composition of the material (Figure 1).

Elements distribution analysis can be carried out in qualitative, semi-quantitative, and quantitative form. Qualitative analysis allows determining the type of elements composing an area of a sample being analyzed. If a sample has several phases (regions), the elemental composition of which in unknown, qualitative analysis of each phase is to be carried out. Usually qualitative analysis is used for determining the character of elements distribution along the area of a polished section. After qualitative analysis quantitative analysis is often carried out in separate points chosen. According to the information the obtained software allows determining a phase type on the basis of its elemental composition. Semi-quantitative analysis is carried out, if it is necessary to determine elements distribution along the lines (linear analysis). Linear analysis is carried out by the method of step scanning, i.e. by consequent carrying out analysis in certain points. Thus, quantitative determination of elements concentration is carried out with a specified degree of accuracy.

3 Results

To refine the profile of experimental X-ray diffraction patterns, the PDXL RIGAKU software package was used. The background was subtracted by the Sonneveld – Wisser method, the experimental profile was smoothed by the Savitsky – Golaya method, and the components $k_{\alpha 1}$ and $k_{\alpha 2}$ were separated by the Rachinger method. To describe the diffraction maxima, a superposition of the Gauss function and the Lorentz function was used. The phase composition was determined using the ICCD PDF-2 database (2014).

The results of X-ray microanalysis are shown in Figure 3.

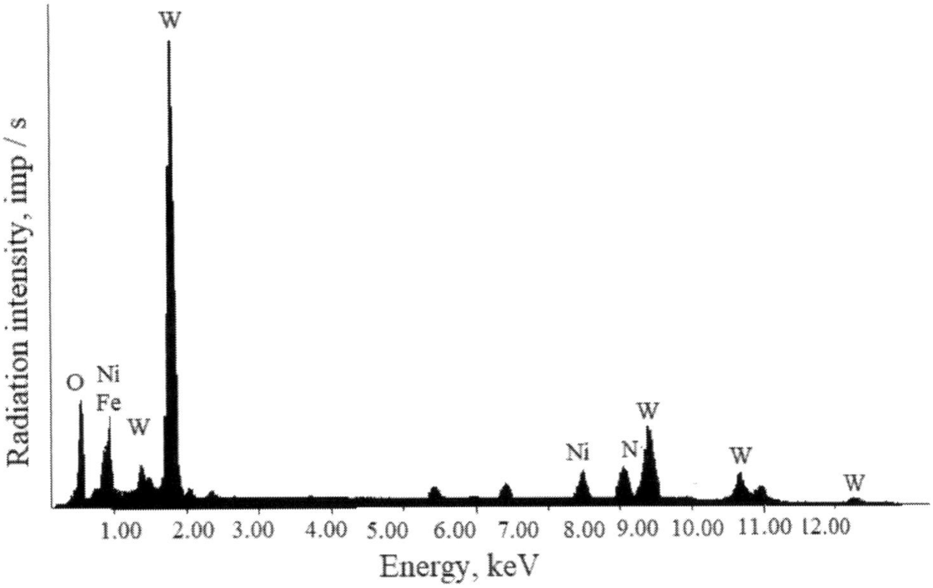

Fig.3. X-ray microanalysis on the surface of electro-erosion TNI alloy particles in distilled water.

4 Conclusion

Based on the elemental composition study of the electro-erosion powders, obtained from waste of a TNI alloy in water, it was found that the main elements are tungsten, nickel, iron and oxygen. The study will determine the most relevant application area of the obtained samples and improve the quality of scientific and technological developments.

This work was supported by a grant from the President of the Russian Federation (NSh-2564.2020.8).

References

1. M.I. Janitor, *Development of physico-chemical and technological fundamentals of the processing of tungsten-cobalt hard alloy by electroerosive dispersion: diss. ... cand. tech. Sciences* / Dvornik Maxim Ivanovich, 116 (Khabarovsk, 2006)
2. M. Sarwar, M. Persson, H. Hellbergh, Wear, **259**, 1144-1150 (2005)
3. A. Basu, J. Chakraborty, S.M. Shariff, G. Padmanabham, S.V. Joshi, G. Sundararajan, J. Dutta Majumdar, I. Manna, Scr. Mat., **56**, 887-890 (2007)
4. V.A. Pozdnyakov, V.T. Borisov, V.T. Molotilov, Russ. Met., **6**, 583-586 (2002)
5. J. Epp, H. Surm, O. Kessler, T. Hirsch, Act. Mat., **55**, 5959-5967 (2007)
6. E.A. Deulin, R.A. Nevshupa, Appl. Surf. Sc., **144-145**, 283-286 (1999)
7. H. Överstam, J. of Mat. Proc. Tech., **171**, 446-450 (2006)
8. J. Zhang, Z. Zhang, W. Liu, Q. Xue, Wear, **219**, 184-187 (1998)
9. S. Jiang, H. Mao, J. of Trib., **133**, 2101 (2011)

10. P. Harlin, P. Carlsson, U. Bexell, M. Olsson, Surf. and Coat. Tech., **201**, 4253-4259 (2006)
11. B. Podgornik, J. Vižintin, Mat. Scien. and Engin., **315**, 28-34 (2001)
12. B.R. Lazarenko, N.I. Lazarenko, A. p. 70000 USSR, B 22f, 09/00 *A method for producing powders and a device for its implementation*. - No. 1371/321510; declared 04/01/1943; publ. 09/23/1964, Bull. 22, 2.
13. V.N. Gusev, *Anodic-mechanical processing of metals*, 321 (1952)
14. E.V. Ageev, Hard.techn.and coat., **6**, 8-14 (2011)
15. R.A. Latypov, G.R. Latypova, E.V. Ageev, P.I. Burak, ISJ, **5**, 80-86 (2013)
16. E.V. Ageev, Electromet., **10**, 24–27 (2011)
17. A. Pereverzev, E. Ageev, MATEC Web of Conf., **298**, 00037 (2019)
18. E.V. Ageev, S.V. Khardikov, E.A. Vorobyev, A.A. Sysoev, MATEC Web of Conf., **298**, 00127 (2019)
19. R.A. Latypov, E.V. Ageeva, G.R. Latypova, MATEC Web of Conf., **298**, 00125 (2019)

Managing the crack energy dissipation as a way to increase the durability of fine-grained limestone concrete

Vladimir Belov[1,*], *Pavel* Kuliaev[1], and *Temur* Barkaya[1]

[1]Tver State Technical University, Civil Construction Department, 22, Af. Nikitin Quay, Tver, 170026, Russia

Abstract. The durability of fine-grained concrete is associated with the process of cracking. The propagation of cracks depends on the energy dissipation rate. The results of the tests presented show the effectiveness of a limestone microdispersed filler and superplasticizer for increasing the durability of fine-grained carbonate concrete.

1 Experiment

1.1 Introduction

The use of limestone crushed stone screenings together with microdispersed limestone filler and superplasticizer to improve the performance and technological properties of fine-grained concrete, and increase the level of durability, is currently relevant for the construction industry. Fine-grained limestone concrete is a heterogeneous solid composite consisting of two or more components: reinforcing elements and a binding matrix [1]. From the very beginning, attempts were made to obtain analytical relationships between the durability, crack resistance of such concrete and the distribution of particles in the mixtures.

Limestone powder is a good modifier of the physical and chemical properties of fine-grained carbonate concrete and a substitute for the main binder. This is due to the ability of limestone powder to work together with a superplasticizer and other ingredients in a concrete body that consists of a matrix, water, binder, various water-reducing additives, and an internal structure capable of organizing the dissipation of internal energy for the formation and growth of cracks [2,3]. The type of dissipation studied in this paper is associated with internal friction, and has an important role in civil construction, and in particular in cases of thin-walled structures, and structures that work on dynamic loads.

Energy dissipation can be considered as an internal change in the structure, or so-called distributed dissipation, due to the characteristic of the material and internal friction, which plays an important role in the case of modification of the dynamic viscosity of the material.

Fine-grained carbonate concrete is usually considered as an elastic-visco-plastic material [4], and its internal matrix as an elastic-plastic medium [5].

[*] Corresponding author: vladim-bel@ya.ru

Distributed dissipation is associated with concrete damage, friction of macro-crack faces caused by constant deformation of plasticity-creep. But internal friction takes a special place, since it is not associated with irreversible changes in the microstructure of concrete. Dissipation associated with internal damage and permanent elastic-plastic deformations can be considered as a critical event, but it also has a positive effect on slowing the growth of main cracks [6].

The destruction of concrete occurs as a result of sprouting of one of the cracks or a cluster of branched cracks [7] through its structure through fillers (under dynamic influence) or through the matrix, skirting the filler grains (under static loading). Tver state technical university conducted a study on improving the durability and crack resistance of fine-grained carbonate concrete depending on the ratio of cement to limestone and the amount of superplasticizer. This study discusses the synergy effect of the combined use of a limestone microfiller and a superplasticizer, which increase the rate of dissipation of crack propagation energy. The rate of dissipation of the crack propagation energy was estimated by the values of the crack formation stresses and the stress intensity coefficient for ten samples of fine-grained carbonate concrete with and without a limestone microfiller.

1.2 Materials

Ten concrete blocks measuring 7x7x7 cm were tested for compression. Limestone filler is a by-product of crushing limestone rocks. Limestone grains ranging in size from 5 to 0.05 mm were used as a filler.

1.3 Technics

The strength, crack resistance, and instantaneous deformations of fine-grained limestone concrete were analyzed on a special stand (Fig. 1.).

Fig. 1. Test unit and sample with electrical sensors on two sides

The appearance of cracks was recorded using electric strain gauges. The accuracy of the measurements was 10-7. To determine the crack resistance, the given stress intensity coefficient was used, taking into account the features of concrete as a non-uniform composite material with a complex conglomerate structure consisting of a cluster of microcracks and a large main crack.

$$K_r = (1-\mu^2) \times K^2{}_1 / E , \tag{1}$$

$$K_1 = \sigma \times \sqrt{\pi \times l}, \quad (2)$$

where l is the half -width of the main crack , μ is the Poisson's ratio, assumed to be 0.3, and σ is the crack -forming stress, E-modulus of elasticity.

2 Test results

Table 1 below shows the composition of the mixture.

Table 1. Composition of the mixture, kg / 1 м^3

№ mixtures	Binder, кg	Aggregates, кg	Cement, кg	Limestone filler, кg	SP-1, кg
1	672	1528	672	0	0
2	672	1528	336	336	0
3	672	1528	672	0	5,04
4	672	1528	672	0	6,72
5	672	1528	672	0	3
6	672	1528	472	200	0
7	672	1528	472	200	2,11
8	672	1528	553	119	4,15
9	672	1528	336	336	0,89
10	672	1528	336	336	2,52

Table 2. Below shows the results of the tests

№ mixture	Crack-forming stress and compressive strength		Stress intensity factor K_r, N × м
	σ, МПа	*R*, МПа	
1	34,4	44,1	0,00099
2	23,3	27,8	0,00372
3	31,5	42,6	0,00075
4	36,7	48,5	0,00073
5	31,9	43	0,00238
6	28,2	35,6	0,00085
7	34,1	41,2	0,00106
8	40,1	46,8	0,00092
9	30,3	37,3	0,00137
10	30	37,1	0,00115

The values of the reduced stress intensity factor K_r for samples containing limestone filler and superplasticizer are higher than for samples without them. In this case, plastic zones are formed at the main crack and at the microcracks in the basin of the latter, which deprive the main crack of energy for further growth, increasing the dynamic viscosity and internal friction. This is the essence of the phenomenon of limiting crack formation through the mechanism of energy dissipation using limestone and superplasticizer SP-1.

The following figures show the relationship between the stress intensity factor K_r (×1000), the content of the superplasticizer, and the content of the limestone filler and the water-cement ratio.

https://doi.org/10.1051/matecconf/202031501005

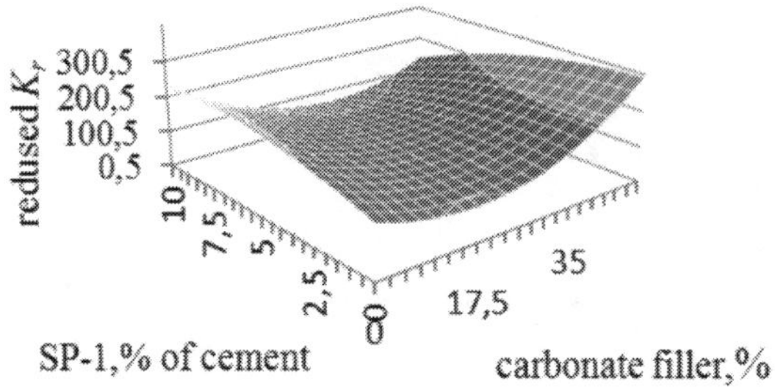

Fig. 2 The ratio between the stress intensity coefficient Kr, the amount of superplasticizer SP-1 and the limestone filler.

The optimal amount of limestone filler as a partial replacement for the binder is 18 %.

In tests, the superplasticizer SP-1 was used in an amount of 0.75% of the volume of cement. Below is a graph of the dependence of the compressive strength on the amount of limestone filler in the concrete test.

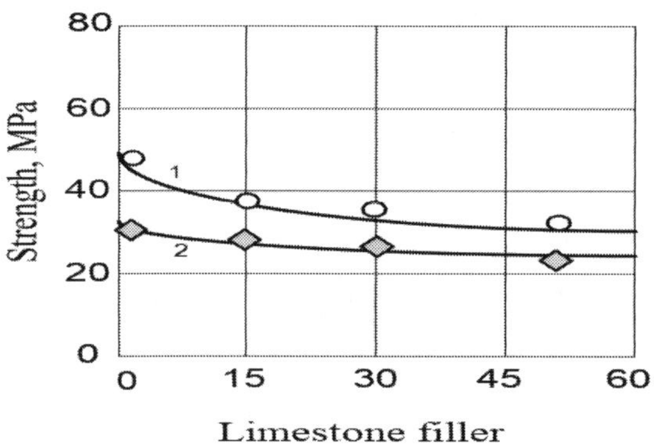

Fig. 3. Compressive strength-curve 1 and crack- forming stresses -curve 2 depending on the content of limestone filler in carbonate fine -grained concrete.

It can be seen that for samples containing limestone filler, the gaps between the crack-forming stresses and the strength values are narrower. This is due to better distribution of cement particles, micro-filler and superplasticizer and filling of air and capillary pores with them. The optimal amount of superplasticizer should compensate for the excess amount of ultrafine limestone particles, which increase the air-trapping capacity of the cement dough (the puffing effect).

It was found that the limestone filler and superplasticizer SP-1 play a positive role in relation to the strength characteristics of limestone fine-grained concrete.

3 Summary and prospects

Limestone aggregate, being manufactured in local quarries, provides good efficiency and a cheap replacement for cement. Easily crushed limestone powder usually has a wide range of granulometric composition, which contributes to the process of filling ultra-voids between cement particles, thereby reducing the need for water, improving the uniformity of the solution and unifying the microstructure of the concrete matrix. This leads to an increase in durability and crack resistance, without significant loss of technological properties, which indicates large values of the stress intensity coefficient Kr for samples containing limestone. Increased crack resistance of limestone fine-grained concrete is also achieved due to better packing of particles inside the concrete matrix, and the decrease in porosity is caused by the transfer of microcracks from air and capillary areas to plastic microzones.

References

1. V.V. Belov, I.V. Obraztsov, T.B. Novichenkova, *Capillary structure formation of raw composites based on mineral binders,* Nanotechnology in construction: scientific online journal, v. **4(8)**, 23-36 (2010)
2. U. El Shamy, C. Denissen, *Microscale energy dissipation mechanisms in cyclically-loaded granular soils*, Geotechnical and Geological Engineering, **30(2)**, 343-361 (2012)
3. G. De Schutter, *Influence of limestone filler as a mineral additive in self-compacting concrete*, Belgium Magnel laboratory of concrete research (36 Conference on Our World in concrete & Structures, Singapore, 14-16/08, 2011)
4. G. Pijaudier-Cabot, K. Haidar, J-F. Dubé, *Non-local damage model with evolving internal length*, International Journal for Numerical and Analytical Methods in Geomechanics, **28(7-8)**, 633-652 (2004)
5. Z.P. Bažant, G. Zi, *A microplane constitutive model for porous isotropic rocks*, International Journal for Numerical and Analytical Methods in Geomechanics, **27(1)**, pp. 25-47 (2013)
6. D. Scerrato, I. Giorgio, A. Della Corte, A. Madeo, A. Limam, *A micro-structural model for dissipation of phenomena in the concrete,* Wiley, **39 (18)**, pp. 2037-2052 (2015)
7. Y.S. Alam, P. Kotronis, A. Loukili, *Crack propagation and size effect in concrete using a non-local damage model*, Engineering Fracture Mechanics, **109**, 246-261 (2013)

Investigation of the effect of implantation with iron ions on the carbon content in the implanted layer of structural and instrumental materials

Viktor Ovchinnikov[1*], *Elena* Lukyanenko [1], and *Svetlana* Yakutina[1]

[1]Moscow Polytechnic University, 38 B, Semenovskaya st., Moscow, 107023, Russian Federation

Abstract. As part of the ongoing research, iron ions were implanted with targets from armco iron, steels 10, 45, U8 and 30KhGSN2A with a dose of 5 1017 ion / cm2. Implantation was performed under the conditions of an "oil-free vacuum" obtained through the use of modern turbomolecular pumps. It is shown that for tool steels with a carbon content of 0.8–1.2 mass. % the concentration of free carbon in the implanted layer does not exceed 5 at. %, and for carbon steels upon implantation with a two-component ion beam of copper and lead, the concentration of free carbon in the surface layer does not exceed 5 at. %.

1. Introduction

The development of industry and the increasing complexity of engineering objects place increasing demands on functional materials, in particular, carbon and alloy steels [1-5]. One of the promising methods for improving the physicomechanical characteristics of steels is surface treatment with concentrated energy flows, in particular, plasma treatment, laser treatment, and ion implantation. The improvement of the physicomechanical properties of ion-doped materials is associated with the formation of intermetallic compounds, nanoscale interstitial phases, and the like in the surface layers [6–11].

The problem of increasing the physicomechanical properties of functional materials can be solved by double implantation in the combination "an ion of one metal - an ion of another metal" [12-15]. As a result of this, it is urgent to develop such technologies for the modification of steels that, on the one hand, would be controllable, and on the other hand, would make it possible to obtain a composition and structure in the surface layer that vary over a wide range.

Therefore, the study of the structure, elemental composition, physico-mechanical properties, phase and morphological changes occurring in the surface layers of steels after ion implantation is an urgent and important task.

* Corresponding author: uchevatkina@yandex.ru

2. Experimental detail

As part of the ongoing research, iron ions were implanted with targets from Armco iron, steels 10, 45, U8 and 30KhGSN2A with a dose of $5 \cdot 10^{17}$ ion / cm^2. Implantation was performed under the conditions of an "oil-free vacuum" obtained through the use of modern turbomolecular pumps.

The results show that the content of free carbon in the surface layer after implantation is quite well correlated with the carbon content in the target (Fig. 1).

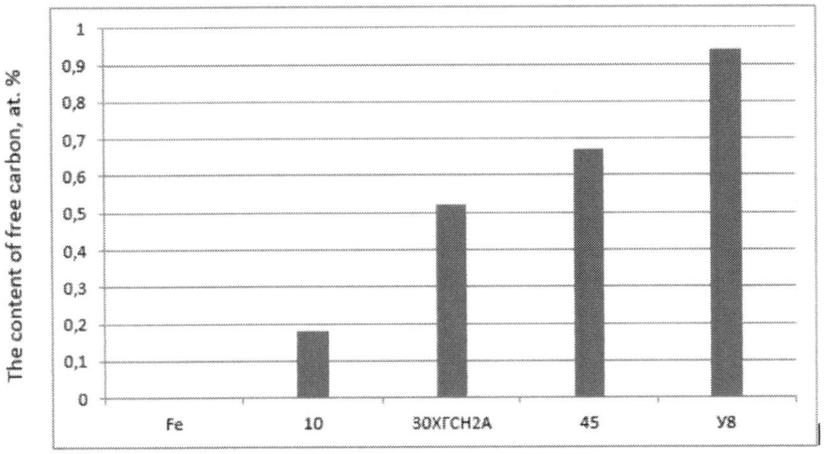

Fig. 1. The influence of the chemical composition of the target on the amount of free carbon in the surface layer after implantation with iron ions

3. Results and discussion

An analysis of the obtained data allows us to conclude that ion implantation does not introduce additional carbon into the target (as evidenced by the results on armco-iron). However, in the presence of carbon in the target, implantation with iron ions is accompanied by the formation of free carbon in the surface layer.

The amount of free carbon in the surface layer depends on the chemical composition of the target. With increasing carbon content in the target, the amount of free carbon increases.

When processing with iron ion beams of steels of the studied grades, the structural-phase state of their surface layer changes. If the irradiation regime is maintained so that the samples do not heat above 150°C, then the initial structure is crushed in the surface layer.

On electron microscopic images of foils prepared from a surface layer treated with a beam of iron ions (Fig. 2), it is clearly seen that both the length and width of the martensitic rails in packets decreases by an average of 2–3 times in the first case. In the microdiffraction patterns, the reflections from the martensitic phase, instead of the point ones, become close to ring ones.

X-ray diffraction analysis (Table 1) also indicates a decrease in the sizes of coherent scattering blocks. The totality of the data obtained indicates the nanostructuring of the martensitic phase in the ion-treated surface layer.

The refinement of the structure is also accompanied by an increase in the dislocation density in martensite crystals up to $(2.5–3.4) \cdot 10^{11}$ cm^{-2}. Also, the process is accompanied by phase transformations. Presumably - cementite partially decomposes, instead carbide is formed with the same morphology as cementite and free carbon.

When determining the tribomechanical characteristics of samples from these steels after their treatment with iron-based ion beams, an increase in fatigue life and endurance limit, ultimate tensile and bending strength, microhardness and wear resistance, as well as relative elongation was established. It should be emphasized that, in contrast to the traditional types of steel treatment [16-21], ionic nanostructuring simultaneously increases both the strength and plastic properties of steels. The effect of treatment with beams of ionic complexes is much higher than that of treatment with single-element ionic metal beams.

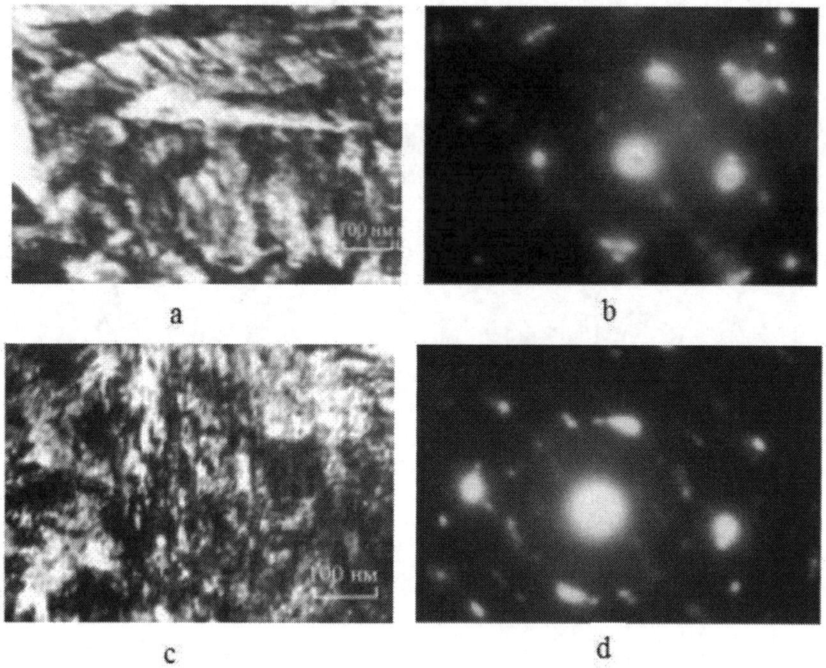

Fig. 2. The microstructure of the surface layer of steel 45 in the initial state (a, b) and treated with beams of iron ions (c, d) at an accelerating voltage of 30 kV and an irradiation dose of $5 \cdot 10^{17}$ ion / cm^2.

The increase in wear resistance, hardness and strength properties of the surface layer of carbon and low alloy steels during ionic nanostructuring is associated with the action of four main mechanisms - grain-boundary, dislocation, dispersion, and solid-solution hardening. The ratio of their contributions is determined by the type of ion beam, the processing mode, and the grade of steel.

Despite the same accelerating voltage and radiation dose, the elemental composition of the ion beam affects not only the projective range, but also the actual number of ions embedded in the surface layer.

The magnitude of the effect of increasing the tribomechanical properties depending on the type of implanted ions is consistent with the total area under the concentration profiles, which is proportional to the total number of ions embedded in the surface layer. The increase in ductility and fatigue life is due to an increase in the shear instability of the nanostructured surface layer, which relaxes subsurface stress concentrators due to accommodation deformation. As a result, the development of microcracks is inhibited and the destruction of the sample is slowed down.

Based on a study of the nature of the action of ion beams on the mechanical properties of carbon and low alloy structural steels at temperatures below 150 ° C, the effect of ion-beam nanostructuring of the surface layer was experimentally established, which, in contrast to traditional hardening methods, leads to a simultaneous increase in microhardness and strength, on the one hand, wear resistance, on the other hand, ductility and fatigue life. It reaches its greatest value when treating the surface with high-energy beams of ionic complexes based on metals with a large atomic radius than that of iron, which, due to their low mass, sputter the surface layer weakly, but penetrate to a greater depth, causing strong distortions of the crystal lattice of the material and creating a high stress gradient.

The effect is associated with the grinding of martensite to a nanoscale level due to radiation-induced recrystallization of the surface layer of steel. At the same time, a significant increase in the density of dislocations, a change in the particles of the initial phases, and the formation of fine particles of new phases that stabilize a highly defective nanostructure occur in the surface layer. The magnitude of the effect and the optimal radiation dose at which it reaches a maximum are determined by the composition of the ion beam and the type of material being processed.

To experimentally evaluate the dependence of the concentration of free carbon in the implanted layer when using a multicomponent ion beam for implantation, steels with different carbon contents were used as targets. The carbon content and grade of steels used as targets are presented in table. 1.

Table 1. The carbon content of steels used as targets.

The carbon content in steel, wt. %						
Steel grades						
10	20	45	60	У8	У10	У12
0.07-0.14	0.17-0.24	0.42-0.59	0.55-0,63	0.76-0.83	0.96-1.03	1.16-1.23

Ion implantation was carried out by a two-component ion beam containing copper and lead ions, using a 64% Cu + 36% Pb monotectic alloy implant cathode. The implantation dose was $5 \cdot 10^{17}$ ion / cm^2. The results of determining the concentration of free carbon in the implanted layer are presented in Figure 3.

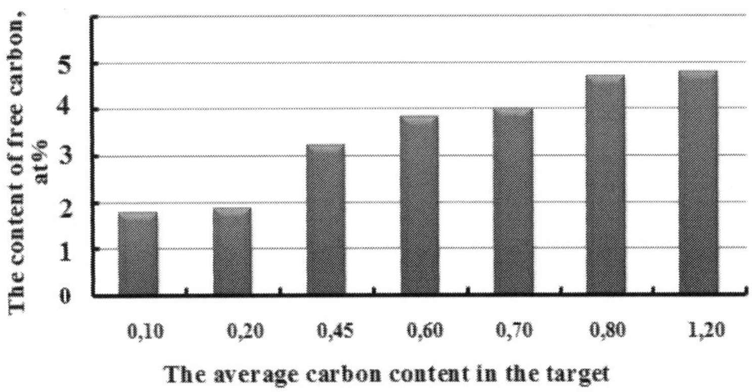

Fig. 3. Free carbon content in implanted steel layer with different carbon content.

An analysis of the results presented in Figure 1.3 shows that after implantation in steels with a carbon content of 0.10–0.20, the content of free carbon in the implanted layer does not exceed 2 at. %

In steels with a carbon content of 0.4-0.7 mass. % free carbon content in the implanted layer is 3.2–4 at. %

For tool steels with a carbon content of 0.8–1.2 mass. % the concentration of free carbon in the implanted layer does not exceed 5 at. %.

4. Conclusions

An analysis of the results presented in Figure 1.3 shows that after implantation in steels with a carbon content of 0.10–0.20, the content of free carbon in the implanted layer does not exceed 2 at. %

In steels with a carbon content of 0.4-0.7 mass. % free carbon content in the implanted layer is 3.2–4 at. %

For tool steels with a carbon content of 0.8–1.2 mass. % the concentration of free carbon in the implanted layer does not exceed 5 at. %

This work was carried out as part of the state task of the Ministry of Education and Science of Russia "The influence of magnetic fields and ion implantation on the structure, chemical composition and properties of titanium, aluminum alloys and elementary semiconductors"

References

1. S.L. Kakani, Materials Science. New Age International, 656 (2006)
2. Y.Y. Chen, H.J. Tzeng, L.I. Wei, L.H. Wang, J.C. Oung, H.C. Shih, Corrosion Science, **47(4),** 1001-1021 (2005)
3. J. Keisler, O.K. Chopra, W.J. Shack, Div. of Engineering Technology; Argonne National Lab., IL **NUREG / CR-6335; ANL-95/15** / (1995)
4. W.M. Garrison, JOM, **42(5)**, 20-24 (1990)
5. M.Z. Ibrahim, A.A. Sarkhan, F. Yusuf, M. Hamdi. Review article. Journal of Alloys and Compounds, **714**, 636-667 (2017)
6. D.I. Proskurovsky, V.P. Rotshtein, G.E. Technology of Surfaces and Coatings, **96 (1)**, 117-122 (1997)
7. S. Konovalov, H. Chen, V. Sarychev, S. Nevsky, V. Gromov, M. Trtika, Metals, **7 (1)**, 4. (2017)
8. S.N. Brother, S.V. Sokolov. Physical surface engineering, **10(2)**, 138-161 (2012)
9. D.N. Korotaev, E.V. Ivanova, Promising Materials, **2**, 98-102 (2011)
10. A.V. Kolomeychenko, Orel: Publishing house Orel GAU, **2013**, 230 (2011)
11. V.P. Perevertov, I.K. Andronchev, M.M. Abulkasimov, Proceedings of the International Symposium "Reliability and Quality", **1**, (2015)
12. R.M. Khisamutdinov, V.V. Zvezdin, I.Kh. Israfilov, R.R. Saubanov, A.A. Spirin, R.R. Rakhimov, In the physical journal: conference series. IOP Publishing, **669 (1)**, 012024 (2016)
13. G. Krauss, Heat Treatment Journal, **9 (2)**, 81-89 (1992)
14. M.K. Lei, Z.L. Zhang, Journal of Vacuum Science and Technology A: Vacuum, Surfaces and Films, **15 (2)**, 421-427 (1997)
15. P. Sioshansi, Materials Science and Engineering, **90**, 373-383 (1987)
16. D. Tobola, V. Brostov, K. Chekhovsky, P. Rusek, Wear, **382**, 29-39 (2017)
17. H.A. Youssef, John Wiley and Sons, 275 (2015)
18 S.N. Polyansky, S.V. Butakov, V.A. Alexandrov, I.S. Olkov, Agrarian Bulletin of the Urals, **12 (142)** (2015)
19. A.M. Guryev, L.G. Voroshnin, Yu.P. Kharaev, B.D. Lygdenov, E.V. Chernykh, Fundamental problems of modern materials science, **2 (3)**, 37-45 (2005)
20. A.P. Veselovsky, S.V. Kubarsep, L.A. Ushomirskaya, Metalworking, **(4)**, 29-31 (2001)
21. A.M. Guriev, B.D. Lygdenov, O.A. Vlasova, Basic research, **(8)**, 10 (2008)

Removal of residual stresses in aircraft structural elements by pulsating subsonic gas flows treatment

Gennadiy Kovalenko[1], and *Denis* Ivanov[1,*]

[1]Saint Petersburg state University of civil aviation, Saint Petersburg, Russia

Abstract. The paper presents a promising method of removing undesirable tensile residual stresses in aircraft metal structural elements; the basis of the method is the action of pulsating subsonic gas flows on the surface of these elements. The results of the research showed high effectiveness of controlling residual stresses in aircraft metal structural elements by using pulsating subsonic air flows.

Traditionally, the residual stresses, in particular dangerous tensile ones in the surface layers, which appear during manufacturing or corrective maintenance, are reduced by heating. This method has a range of considerable shortcomings. Along with the problem of granules coagulation, and if there are carbide or intermetallic inclusions, it is necessary to mention that the residual stresses are removed to the minimum value at the given temperature during 1.5-2 hours, thus considerably rising the cost of manufacturing or parts' maintenance as well as their duration. The methods of surface plastic deformation are not suitable for residual stresses removal in the parts of aircraft engines, because bead blasting and centrifugal-ball (rolling cold hardening) inadmissibly worsen the surface smoothness; while rolling by a steel ball (roller) and a diamond smoothing by arbor are hardly implemented for the parts of a complicated shape, as well as lead to the change of linear dimensions.

The application of non-deformative methods for controlling residual stresses in aircraft structural elements seems to be having prospects. These methods include the action of pulsating subsonic air flows (gas-impulsive treatment) in whole, and do not have the mentioned above shortcomings.

The stresses, which exist and balance in the solid after determining the causes of their appearance, are accepted to be called residual stresses. Residual stresses are always internal because they balance themselves inside the solid or hard aggregate. The generation of residual stresses is always connected with heterogeneous plastic deformations in adjacent dimensions of the material or the aggregate, caused by the external deformative action, by temperature gradient along the section of the product during quick cooling, or by the heterogeneity and irregularity of phase transformations, connected with the volume change.

The goal of mechanical methods of determining residual stresses is to define on the basis of the known set of movements or deformation values what has caused their stresses.

* Corresponding author: ivanov.denis.71@mail.ru

The research of residual stresses was carried by Davidenkov's method, applied to the ring samples.

In the ring samples the values of tangential residual stresses decrease from the surface to the centre, and after going through the neutral layer, they change their sign into the opposite one. During the distribution of residual stresses on the thickness of the ring, which is close to the linear one, in the layer, remote from the surface for distance a

$$\sigma_{\theta(a)} = \pm 2\frac{E\delta_c}{(1-\mu^2)D_{av}^2}(h/2-a), \qquad (1)$$

where E isYoung's modulus;

μ is Poisson's ratio;

D_{av} is an average diameter of the ring before cutting;

h is the thickness of the ring opposite the cut;

δ_c is the change of the ring's diameter after cutting.

The thickness of the thinnest part, where the value of the residual stresses was determined, was accepted as the ring thickness. The diameter of the ring was also determined by the thinnest part.

The upper sign in the formula (1) is used during the removal of the external layers, the bottom sign is used while removing internal layers.

On the surface of the ring

$$\sigma_\theta = \pm\frac{E\delta_c h}{(1-\mu^2)D_{av}^2}. \qquad (2)$$

In the latter formula the sign «+» corresponds to the external surface of the ring, while «-» corresponds to the internal one.

We researched the influence of gas-impulsive treatment [1-11] without preliminary heating on the residual stresses in thin-walled ring piped structural elements of aircraft engines, aircraft and airdrome equipment.

The research was done on the equipment which is a gas-jet generator of Galton's whistle type with a cylindrical axisymmetric resonator on the bottom of which a gas pipe is fixed in alignment with the resonator in order to supply compressed air. The opposite end of the gas pipe and a flat round cap make up a slot ring nozzle, the width of which is regulated by changeable annular collars. The captive nut fixes the position of the changeable cutting ring, having a wedge-like profile with the edge directed to the slot nozzle side. Such a generator structure allowed varying the removal of the wedge-like cutting from the nozzle cut, the nozzle slot width, and the reciprocal position of the cutting relative to nozzle edges. The radial jet, being accumulated on the cutting ring, creates the conditions for stimulation of non-stationary resonator filling and emptying, which in its turn causes an oscillatory movement of the radial jet. The resonator was located vertically, and with the help of the threaded joint a cylindrical still-water chamber was fit on its top. It was done to provide an even gas flow velocity distribution on the whole cross-section. The products, which had to undergo a pulsating gas flow treatment, were placed above the still-water chamber.

The duration of the treatment was from 10 to 20 min with the oscillation frequency of flow parameters 1-2 kHz.

The research of the residual stresses was done by Davidenkov's method applied to the ring samples.

The results of gas-impulsive treatment of the products with the surface stress state, which are made of structural metal materials, are presented in the histogram (Fig. 1).

The tensile residual stresses are marked by the sign «+», while the compressive stresses are marked by the sign «-».

The given results demonstrate that the gas-impulsive treatment influences not only the value, but also the sign of the residual stresses on the products' surface, at the same time the value of the stresses decreases.

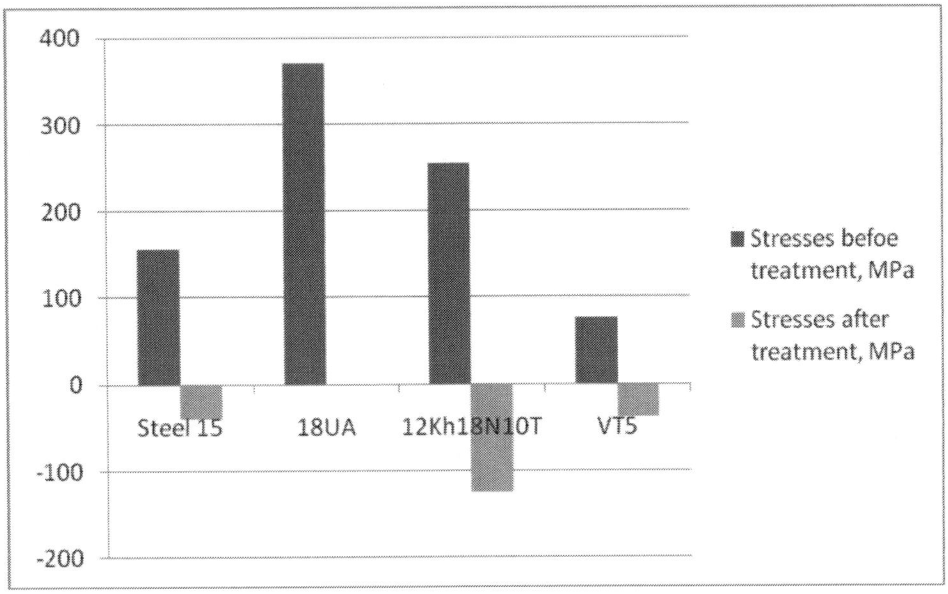

Fig. 1. The level of the tangential residual stresses on the surface of metal ring products before and after the gas-impulsive treatment.

In addition we carried out a more detailed research of the gas-impulsive treatment duration influence on the value of the tangential residual stresses in the surface layers of metal products.

The object of the research was a jointless cold-rolled pipe with the outer diameter of 38 mm and the wall thickness of 1.1 mm, produced from steel 12Kh18N9T, which is used in propulsion engineering.

The width of the cut rings is 10 mm. The tangential tensile stresses on the ring surface were +183 MPa.

The blow-off during 5 minutes by the gas-pulsating flow at ripple frequency of 1130 Hz and sound pressure of 122 dB provided the decrease of surface tangential residual stresses by 70% (to +55 MPa), and during 10 minutes by 75 % (to +46 MPa) which is a good result if we take into consideration the combination of high toughness of material with the ultimate strength of 1100-1300 MPa (in cold-worked state).

Experimentally we got the following dependence of the tangential residual stresses on the blow-off duration: 2.5 min – +293 MPa; 5 min – + 55 MPa; 7.5 min – + 110 MPa; 10 min – + 46 MPa, 12.5 min – +82 MPa;15 min – + 73 Mpa; 17.5 min – + 55 Mpa (Table 1).

The periodic increase of the tangential residual stresses level values may be connected with the decrease of axial and radial residual stresses, having the opposite sign, influence as a result of their relaxation.

Thus, it was found out that there is an optimum duration of the gas-impulsive treatment from the point of view of the level and the sign of the residual stresses on the product surface, and depending on the material, geometric parameters of the sample, and amplitude-frequency of the flow itself.

Table 1. The values of the tangential residual stresses in the surface layers of the ring samples made from steel 12Kh8N9T depending on the gas-impulsive treatment duration.

Treatment duration, min.	2.5	5	7.5	10	12.5	15	17.5
Residual stresses, MPa	293	55	110	46	82	73	55

The further continuation of the blow-off makes the process less manufacturable and may lead to the decrease of gas-impulsive treatment positive effect.

As numerous experimental researches showed, the gas-impulsive treatment of metal products, including aircraft structural elements and airdrome equipment, is an effective means of controlling the value and the sign of the residual stresses, having appeared in the process of manufacturing the product, its exploitation and corrective maintenance.

The main factor, determining the value and the sign of the residual stresses, is the blow-off duration.

Taking into consideration that at the initial moment of the blow-off the residual stresses in the surface layers of the product are positive (tensile), i. e. being the most dangerous during exploitation, we may get their zero value during the treatment duration T/4, where T is the period of about 20 minutes.

Continuing the gas-impulsive treatment we may get compressive residual stresses on the product surface, which are desirable in many cases, as they increase the steadiness to corrosion and fatigue breakdown. The negative (compressive) residual stresses grow in modulo till the moment, corresponding to T/2, the modulus of their values is not more than 40% from the initial value of the residual stresses σ_{res0} .

During further treatment of the product the residual stresses, without changing the sign, are decreasing in modulo till zero value (time moment 3/4 T) and changing the sign into the positive one, reach the maximum value at the time moment T, the value of the residual stresses at T moment is significantly less than the initial one.

The consequence of the blow-off continuation will be the reiteration of the above described process with significantly less values of the residual stresses in comparison to the first period; to the time moment 3/4T the level of the residual stresses does not practically change and remain being close to zero.

Thus, the process of residual stresses change under gas-impulsive treatment is similar to the process of oscillation damping and can be described by the corresponding equation:

$$y = A \cdot e^{-\alpha x} \sin(\omega x + \varphi_0) \tag{3}$$

where A is the initial oscillation amplitude (positive value), α is the damping coefficient, $A \cdot e^{-\alpha x}$ is the instantaneous amplitude, $(\omega x + \varphi_0)$ is the oscillation phase, $\omega = 2\pi/T$ is the cyclic frequency, φ_0 is the initial oscillation phase.

The experimental date processing allowed to get the following empirical formula:

$$\bar{\sigma}_{residual} = \sigma_{residual}/\sigma_{residual\ 0} = e^{-0,1\tau} \cdot \cos(\omega\tau) \tag{4}$$

Where τ is the time of the blow-off.

The dependence $\sigma_{res}/\sigma_{res0}$ on the blow-off duration is presented in Fig. 2.

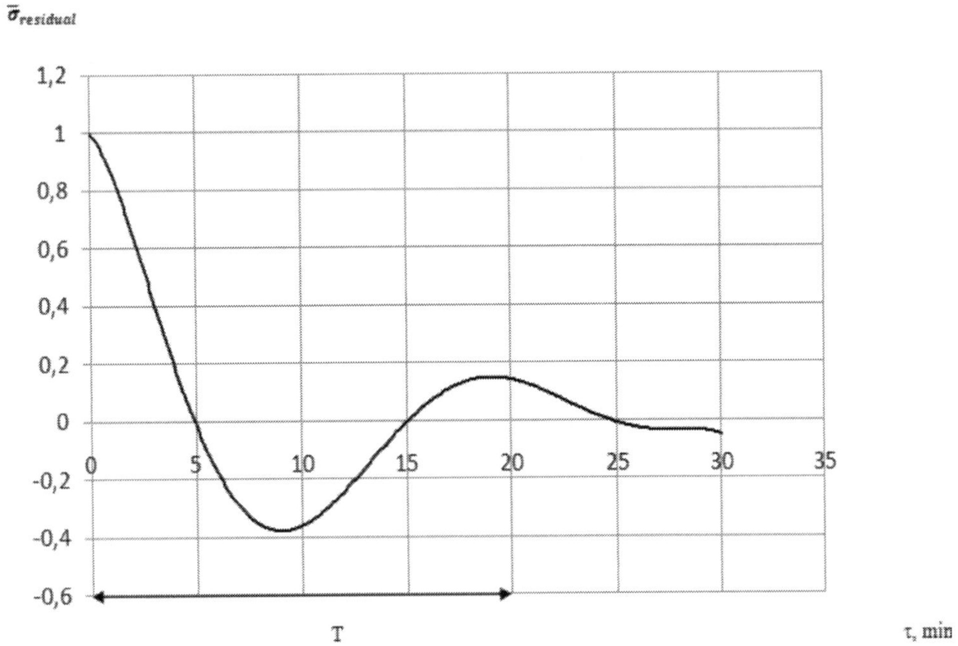

Fig. 2. The dependence of the residual stresses values relation to their initial value on the duration of pulsating subsonic gas flow treatment.

The rings of the rolling bearings underwent the gas-impulsive treatment; despite the displacement of structural steels in the aircraft structures by titanic alloys and other materials, the bearings remain steel. The rings of the rolling bearings during exploitation undergo cyclic alternate loads; that is why to provide a prolonged bearing exploitation it is necessary to exclude undesirable tensile stresses on its rings surface. This can be achieved by the gas-impulsive treatment. The bearings' rings undergo hardening by martensite with the further low tempering for preservation of high solidity and durability. The strained residual stresses often appear on the surface of the bearings' rings which lower their durability. Similar stresses may appear as a result of rings' surface polishing process in order to give the necessary smoothness to their surface.

The tangential tensile stresses on the surface of the raw ring were +27 MPa; and as a result of the blow-off during 5 minutes they showed the value +13.5 MPa; after the gas-impulsive treatment during 10 minutes the residual stresses on the surface became compressive with the value of -27 MPa; and after 15 minutes of treatment the stresses had a zero value (see the graph in Fig. 3).

This manifested the dependence character of residual stresses level on the product surface on the gas-impulsive treatment duration, which is already known from the previous researches.

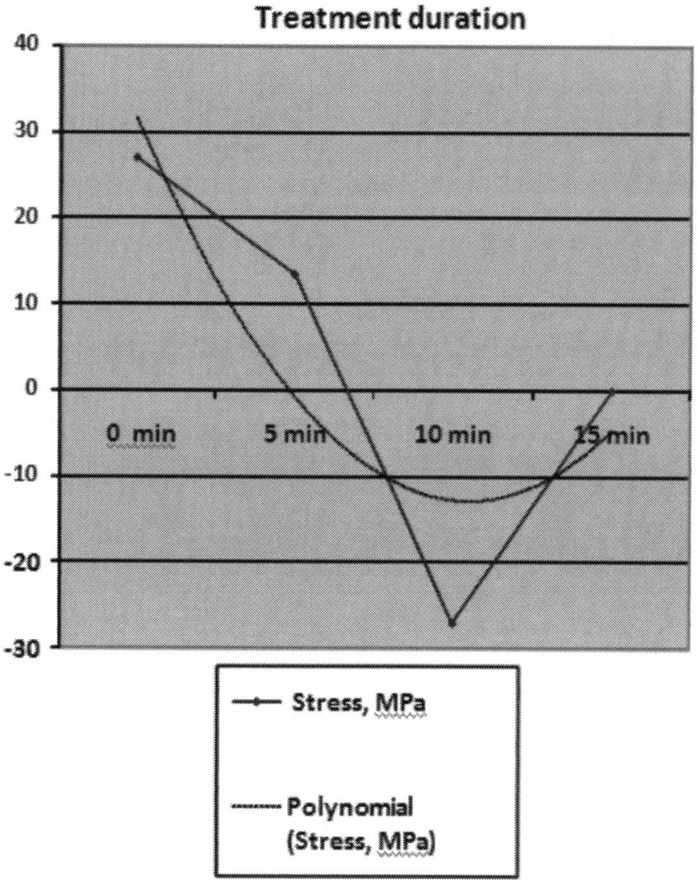

Fig. 3. The level of tangential residual stresses on the surface of the rolling bearing rings made from steel ShKh15 depending on the gad-impulsive treatment duration.

Further the ring samples of aluminum alloy AMg6 (the outer diameter is 33 mm, the ring width is 7 mm, the wall thickness is 1.2 mm) underwent subsonic pulsating air flow treatment. The tangential residual tensile stresses were 122 MPa on the surface and were nullified after pulsating gas flow treatment.

The results of the research have shown that the use of pulsating subsonic air flows has prospects in order to control residual stresses in aircraft metal structural elements.

References

1. Y. A. Kuznetsov, A. V. Kolomeichenko, V. V. Goncharenko, I. N. Kravchenko, Mater. Sci. Forum. **968,** 153-160 (2019)
2. A. V. Kolomeichenko, S. N. Sharifullin, Yu. A. Kuznetsov, V. N. Logachev, N. V. Titov, *J. Phys. Conf. Ser.* **1058,** 012074 (2018)
3. Q. P. Tran, Y. C. Kuo, J. K. Sun, J. L. He, T. S. Chin, Surf. Coat. Technol. **303,** 61-67 (2016)
4. A. A. Filippov, G. V. Pachurin, V. I. Naumov, N. A. Kuzmin, Metallurgist **59,** 810-815 (2016)
5. Y. N. Smyrnov, V. A. Skliar, V. A. Belevitin, R. A. Shmyglya, O. Y. Smyrnov, Steel Transl. **46,** 325-328 (2016)

6. A. N. Smirnov, S. V. Kuberskii, E. N. Smirnov, A. P. Verzilov, E. N. Maksaev, Steel Transl. **47** (7), 476-482 (2017)
7. E. N. Smirnov, V. A. Sklyar, A. N. Smirnov, V. A. Belevitin, S. P. Erońko, R. E. Pivovarov, Steel Transl. **48** (6), 381-387 (2018)
8. G. V. Kovalenko, D. A. Ivanov, *Advances in Engineering Research: Proceedings of the International Conference on Aviamechanical Engineering and Transport (AviaENT 2019)* **188,** 179-183 (Atlantis Press, 2019)
9. D.A. Ivanov, O.N. Zasukhin, Russian Metallurgy (Metally) **13**, 1522-1525 (2019)
10. D.A. Ivanov, O.N. Zasukhin, *Method of hardening heat treatment of carbon tool steels*, Patent 2557841 C2 Russian Federation, Bul. 21, (51) IPC C21D 1/04 (declared. 13.09.20132, publ. 27.07.2015)
11. D.A. Ivanov, O.N. Zasukhin, *Method of increasing the resistance of metal-cutting tools made of high-speed steel* (Patent 2580767 C2 Russian Federation, Bul. 10 (51) IPC C21D 9/22, decl. 24.12.2013, publ. 10.04.2016)

MATEC Web of Conferences **315**, 03001 (2020) https://doi.org/10.1051/matecconf/202031503001
ICMSSTE 2020

Design of composite materials for clothes

Sergey Koksharov[1], *Nadezhda* Kornilova[2*], and *Yulia* Shammut[2]

[1] G.A. Krestov Institute of Solution Chemistry of the Russian Academy of Sciences, 153045, Ivanovo, Akademicheskaya st, 1, Russia
[2] Ivanovo State Polytechnic University, 153000, Ivanovo, Sheremetevsky ave., 21, Russia

Abstract. Transition to digital models of garments is the first step in digitization of the light industry. Adequate digital models can be obtained by using precise parameters of stress-strain properties of a fabric and other materials. The study has revealed regression dependences of the stiffness index of the polymer – fibre composite on the stiffness of the suit fabric and the combination of fusible interlining parameters. The obtained equations can be used to design composite materials of the required stiffness, depending on the form of clothing being created

1 Introduction

The first step in the digital transformation of apparel industry is to replace physical samples with virtual garments or digital twins – an accurate digital version of the garment that designers can manipulate exactly as they would a physical version. These digital garment twins should be produced in a fast and scalable way, removing the need to wait for a physical garment to make design decisions. From design to fitting and adjustments, all the way through to sales in an online shop, using a digital garment reduces the time to market from almost a year to merely days [1].

Virtual garment design and simulation involves a combination of a large range of techniques, involving mechanical simulation, collision detection, and others. Article [2] provides an overview of the evolution of these techniques made in the last decade to bring virtual garments to the reach of computer applications not only aimed at graphics, but also at CAD techniques for the garment industry. If cloth simulations could be improved to the point that they could generate realistic cloth motion in real-time, they would find uses in many aspects of fashion design and manufacturing [3 - 5].

When applying physical simulation, CAD producers have to look for a compromise between simulation time and resource-intensity and, on the other hand, between the accuracy of material simulation and closeness to the garment fit [4]. Flexibility of the structure, compound weave, anisotropy, drapability and absence of static position make simulation of textile materials much more complicated than the simulation traditionally used in machine building, architecture and other fields of industry.

A separate problem is taking into account properties of bonded zones of garment parts whose stress-strain properties may be different from those of the neighboring (non-bonded) zones by dozens of times [6].

* Corresponding author: nkorn@mail.ru

As shown in [7, 8], these sections are a composite, which consists of several layers of textile materials connected by a thermo-fusible adhesive. To obtain it in the sewing industry, prepreg technologies of forming garments using hot-melt adhesive interlining materials are widely used [9]. The properties of the resulting composite are affected by the composition of the hot-melt adhesive polymer, the method of applying it to the textile liner, and the parameters of processes for bonding the layers and subsequent wet-heat treatment of the finished garment [10 – 12].

The required level of stiffness of the formed composite is determined depending on the shape of the garment and the properties of the shell material. The stiffer the shape of the garment being created and the softer the material of the top, the greater the load the components made of composite material must withstand.

As Fig. 1 shows, one front part is divided into several zones with different levels of required stiffness. By studying some samples of men's jackets of different models, we have established that stiffness parameters of the fabrics in each of the zones depend on the degree of plasticity and silhouette, which is demonstrated in Table 1.

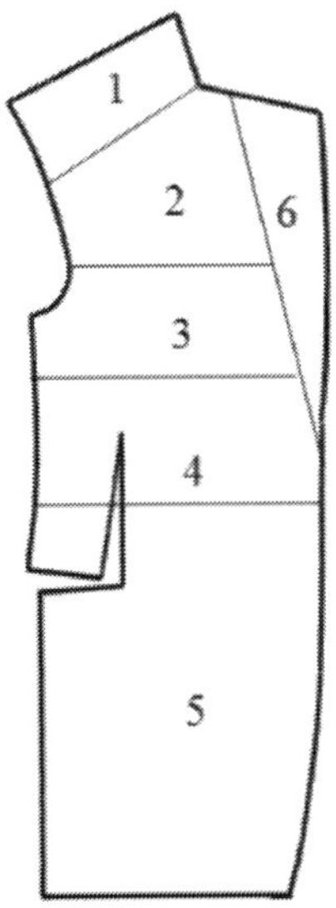

1 – shoulder;
2 – fore top printing;
3 – breast foreside;
4 – breast lower part;
5 – bottom;
6 – lapel.

Fig. 1. Different zones of the front of jacket according to a flexural rigidity

The stability of garment shape in different zones of a certain garment part is achieved by using several layers of interlinings – as a rule, at least three different articles with different stiffness and elasticity degrees – fixed to each other and to the base fabric in different ways. Currently, the most widely-used type of fixing is fusing, which employs fusible interlinings (*FI*). They differ in the structure of the textile carrier (knitted, non-woven or, more rarely, woven fabrics), nature of the adhesive polymer and method of its application [13, 14]. Presence of a thermoplastic adhesive does not only allow obtaining a strong connection of fabrics but also makes the panel stiff by forming a polymer matrix between the layers, with the matrix properties depending on the polymer type [15].

Table 1. Flexural rigidity (*EI*, 10^{-3} Ncm^2) of different zones of the front of jacket according to a three-dimensional form.

Zones	Flexural rigidity, *EI*, mN·cm²								
	Large three-dimensional form			Moderate three-dimensional form			Small three-dimensional form		
	soft-plastic	soft fixed	frame-fixed	soft-plastic	soft fixed	frame-fixed	soft-plastic	soft fixed	frame-fixed
shoulder	16.5-23.3	24.7-31.5	32.9-39.8	10.9-51.6	17.8-23.3	24.7-30.2	10.9-15.0	16.5-20.5	21.9-26.0
fore top printing	15.0-20.5	21.9-26.1	27.4-32.9	8.2-13.7	15.0-20.6	21.9-27.4	8.2-13.8	15.0-19.2	20.5-24.7
breast foreside	12.3-17.8	19.2-23.3	24.7-27.4	8.2-12.4	13.7-19.2	20.5-24.7	5.5-9.6	10.9-15.0	17.8-21.9
breast lower part	6.8-10.9	12.3-17.9	19.2-23.3	6.8-10.9	12.3-17.8	12.3-20.6	5.5-9.6	10.9 – 15.0	
bottom	4.1 – 9.6								
lapel	2.7 – 10.9								

Normally, the composite material stiffness is varied by changing the fibrous composition of the *FI*, its surface density, type and amount of adhesive polymer. However, the optimal *FI* types are chosen empirically as there is no literature data about the regularities describing the interconnection between the *FI* properties and the stiffness of the obtaining composite. This research aims to solve this problem.

2 Materials and methods

The objects of study were two types of shell materials for suits (*SM1*, *SM2*) and 4 types of fusible interlinings (*FI 1…4*).

SM1 – fibrous composition (%): viscose 55, wool 35, polyester 10; surface density MS = 240±3 g/m²;

SM2 – fibrous composition (%): viscose 50, polyester 50; MS = 180±2 g/m².

The properties of fusible interlinings are given in Table 2.

Table 2. Characteristics of fusible interlinings (*FI*)

Unit designation	Manufac-turer	Vendor code	Fiber material, %	Surface density, Ms, g/m2	Mass fraction of weft yarns, G_{WY}, %	Number of glue dots, N_{GP}, units / cm²
FI1	Germany	R161G57/090	polyester 27, viscose 73	58	70.5	70
FI2	China	DIVP 80	polyester 30, viscose 70	65	54.4	145
FI3	China	DIVP 65	polyester 30, viscose 70	80	60.8	115
FI4	Russia	216/4	polyester 60, cotton 40	75	60.8	400

Figure 2 demonstrates a typical structural feature of all the studied *FI*. It consists in the presence of cellulose-containing weft yarn and predominantly lengthwise orientation of the warp yarn.

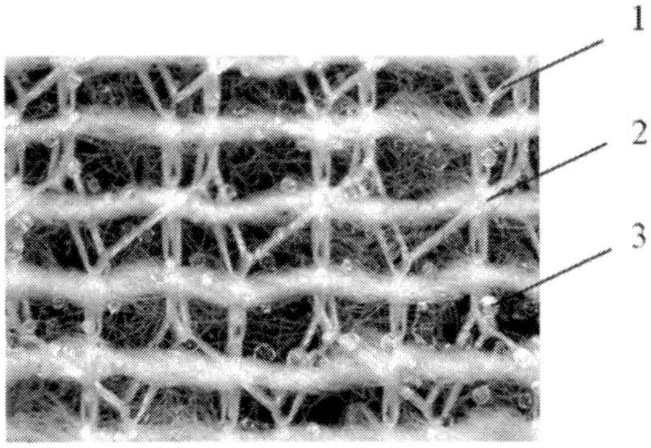

Fig. 2. View of fusible interlinings under the microscope: 1 – ground yarn; 2 – weft yarn; 3 – thermoplastic adhesive fragments (glue points).

The composites studied were obtained on a continuous fusing machine Japsew SR-600 (China) at the temperature of 80°C and subsequent steaming at 140 °C till the moisture content reached 20…30%.

The indicator stiffness of textile liner and composites was calculated in accordance with GOST RF 10550-93 using contactless console methodology using the device PT-2 (Fig. 3).

Fig. 3. View of the device PT-2.

The strips of material size of 160 x 30 mm were used. They were placed horizontally on the top side of the machine and pressed by the load size of 20 mm in the middle. After that the sides of anchor were dropped down, the ends of material hanged loose due to gravitational force. After a minute, the amount of overhang of the strips ends was measured. The value of stiffness EI (mN·cm^2) was calculated using following formula (1):

$$EI = \frac{42.046\,m}{0.5755\,f^3 - 2.411\,f^2 + 8.502\,f}, \tag{1}$$

where m is the mass of material strip in grams.

3 Results and discussion

The measurement results are shown in Fig. 4 for the composites with the corresponding type of *FI* in the structure in comparison with the original indicator level for the suit fabric.

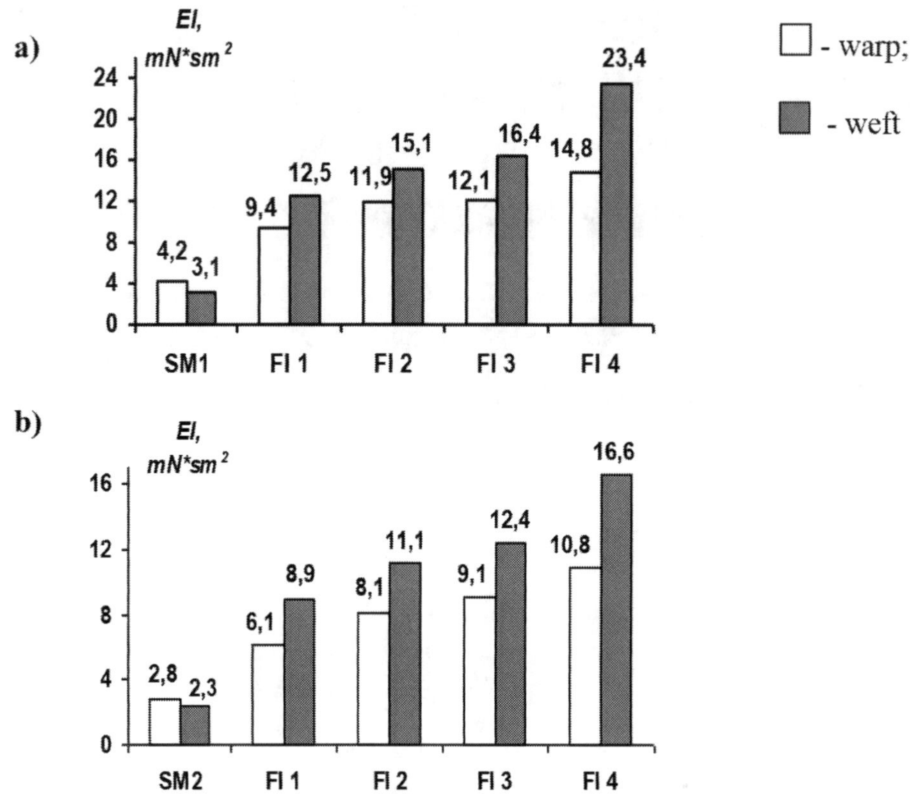

Fig. 4. The indicator stiffness of suit fabrics SM1 (a) и SM2 (б) and the composites with the corresponding type of *FI*

The analysis of the obtained values of composites stiffness EI^{fp} in combination with the *FI* characteristics given in Table 2 demonstrates an increase in the EI^{fp} as the M_S and N_{GP} parameters grow but does not reveal any correlations with any of these individual properties or any their combination. However, if the content of the weft yarn in the textile carrier (G_{WY}, %) is taken into account in the regression analysis, it is possible to reach a satisfactory degree of approximation in the description of the experimental data set. The dependence of the panel warp stiffness (EI_{wa}^{fp}) or the composite weft stiffness (EI_{we}^{fp}) on the stiffness of the shell material and structural characteristics of FI looks as follows:

$$EI_{wa}^{fp} = EI_{wa}^{sm}(0.027\,M_S + 0.017\,(100 - G_{WY}) + 0.0025\,N_{GP}); R^2 = 0.9999 \quad (2)$$

$$EI_{we}^{fp} = EI_{we}^{sm}(0.049\,M_S + 0.0086\,G_{WY} + 0.0079\,N_{GP}); R^2 = 1.0 \quad (3)$$

where EI_{wa}^{sm} и EI_{we}^{sm} are the warp and weft stiffness values of the shell material.

The physical meaning of the mathematical model allows us to say that composite stiffness is jointly affected by the stiffness of the original cloth and the interlining. All the equation members have positive values, this indicates that the composite stiffness grows with an increase in any of the input parameters. The weft stiffness in this case becomes higher as the content of the weft threads in the interlining grows, whereas the warp stiffness directly depends on the content of *FI* warp threads.

While analyzing the properties of multi-layered composites, we found a factor of great importance demonstrating the forecasting ability of the obtained mathematical models. For the considered case of applying *FI* to a knitted cloth of the same type, the resulting composite stiffness value correlates with the value of the additive increment in each of the input values. The M_S end N_{GP} values for each of the layers are summed and for G_{WY} an averaged value is used. For example, for two-layered composites, equations (2) and (3) are transformed in the following way:

$$EI_{wa}^{fp} = EI_{wa}^{sm}(0.027(M_{S1}+M_{S2})+0.017(100-0.5G_{WY1}-0.5G_{WY2})+0.0025(N_{GP1}+N_{GP2}) \quad (4)$$

$$EI_{we}^{fp} = EI_{we}^{sm}(0.049(M_{S1}+M_{S2})+0.0043(G_{WY1}+G_{WY2})+0.0079(N_{GP1}+N_{GP2}) \quad (5)$$

A comparison of the calculated values for all the possible combinations of interlining materials in multi-layered composites that we have studied with the experiment results is given in Table 3. The differences between the corresponding values do not exceed 5 %, which indicates that the model can quite satisfactorily predict the stress-strain properties of such composites. The developed approach allows us to effectively select a set of interlining materials for regulating the stiffness of certain zones of a garment in case of changing the model or the shell material. To do that, it is enough to have a set of experimentally determined stiffness values of the shell material with the interlining materials available.

Table 3. The results of modeling the stiffness of multilayer packages

Package contents	$EI_{РАСЧ}$, mN·sm^2		$EI_{ЭКСП}$, mN·sm^2	
	warp	weft	warp	weft
SM1+FI1+FI2	18.7	25.7	17.9	24.8
SM1+FI1+FI3	19.9	27.3	19.1	26.9
SM1+FI1+FI4	22.3	33.6	22.4	33.9
SM1+FI2+FI3	22	30	19.2	28.8
SM1+FI2+FI4	24.4	36.3	23.2	34.9
SM1+FI3+FI4	25.6	37.9	23.9	36.4
SM1+FI1+FI2+FI3	29	40.7	29.9	38.7
SM1+FI1+FI2+FI4	31.4	46.9	29.9	45.7
SM1+FI2+FI3+FI4	34.6	48.5	35.7	49.6
SM2+FI1+FI2	12.4	18.9	13.1	19.7
SM2+FI1+FI3	13.1	20.1	13.4	20.3
SM2+FI1+FI4	14.7	24.7	15.1	25.0
SM2+FI2+FI3	14.6	22.1	15.3	22.1
SM2+FI2+FI4	16.2	26.7	17.5	27.1
SM2+FI3+FI4	16.9	27.9	17.4	27.3
SM2+FI1+FI2+FI3	19.2	29.9	20.1	29.8
SM2+FI1+FI2+FI4	20.8	34.5	22.6	34.8
SM2+FI2+FI3+FI4	22.9	35.7	23.9	37.4

The obtained data about the elasticity modulus of the original fabric and bonded parts of a garment can be used in the programs for simulating the fit of different types of garments such as coats and suits, shirts and corsets.

4 Conclusions

We have determined the ranges of the required stiffness of the men's jacket panel for different zones of the front part depending on the garment silhouette.

We have obtained regression dependences of the stiffness index of the bonded panels of suiting fabric on the stiffness of the shell material and a combination of parameters of the fusible interlining material (surface density, content of weft threads and density of glue points), which allowed us to select interlining materials depending on the required stiffness of a certain garment part.

References

1. L. Mageean. Which PLM (2018) https://www.whichplm.com/3d-key-complete-digital-apparel-design-process/
2. P. Volino, F. Cordier, N. Magnenat-Thalmann. Computer-Aided Design **37 (6)**, 593-608 (2005)
3. K-J.Choi, H-S.Ko. Computer-Aided Design **37 (6)**, 585-592 (2005)
4. P. Apeagyei. J. of Fashion marketing and Management **11 (3)**, 349 - 365 (2007)
5. Y.-J. Liu, D.-L. Zhang and M. Ming-Fai Yuen. Computers in Industry **61**, 576–593 (2010)
6. N. Kornilova, S. Koksharov, A. Arbuzova etc. Indian J. of Fibre & Textile Res. **42**, 150-159 (2017)
7. O. Erdem, I. Paul. Woodheap Publishing Series in Textiles 377-420 (2018)
8. Q. Zhang, C-W. Kan. Polymers **10 (11)**, 1230 (2018)
9. S.A. Koksharov, N.L. Kornilova, J.A. Shammut et al. Key Engineering Materials **816**, 219-227 (2019)
10. E.A. Elnashar, P.D. Dubrovski. Autex Res. J. **8 (2)**, 41-43 (2008)
11. S. Jevsnik, J. Gersak. Fibers & Textiles in Eastern Europe **12**, **1 (45)**, 47-52 (2004)
12. K. Dapkūnienė, E. Strazdienė. Materials Science (Medžiagotyra). **12 (1)**, 73 – 78 (2006)
13. L.I. Zhen-Ying. Wool Textile J. **8**, 35-39 (2011)
14. S.S. Saleh, E.M. Kamal. J. Basic. Appl. Sci. Res. **2 (10)**, 9970-9977 (2012)
15. S.A. Koksharov, N.L. Kornilova, S.V. Fedosov. Russian J. of General Chemistry **87 (6)**, 1428-1438 (2017)

The use of non-metal materials in the filler units of cooling towers with circulating water supply system

Andrey V. Dmitriev[1], *Ilnur* N. Madyshev[2,*], and *Aliya* I. Khafizova[2]

[1]Kazan State Power Engineering University, 420066, Kazan, Krasnoselskaya str., 51, Russia
[2]Kazan National Research Technological University, 420015, Kazan, Karl Marx str., 68, Russia

Abstract. The design of cooling tower's filler unit, consisting of inclined-corrugated contact elements, and made of polymer material, is developed and described in this research paper. The experimental studies were conducted to determine the efficiency of cooling the circulating water in the proposed cooling tower. The authors obtained the results of evaluating the heat-mass transfer efficiency coefficient of cooling tower, when changing the mass flow rates of liquid and gas phases. The research results showed that the highest efficiency of cooling the circulating water is observed at relatively low irrigation densities equal to 15-25 $m^3/(m^2 \cdot h)$, while the heat-mass transfer efficiency of cooling tower achieves 72%.The use of non-metal materials as the main elements of cooling tower allows to abandon the use of chemical reagents against bacteria and microorganisms in favor of circulating water ozonation method.

1 Introduction

The circulating water supply at the enterprises of many industries, such as chemical, energy, petrochemical and others is increased every year, allowing to reduce the use of water from natural sources, thereby preventing the environmental pollution and contamination.

The circulating water supply at the industrial enterprises often acts as a cooler or heat carrier, while cooling the main process equipment. The further cooling of water takes place in cooling towers. The most common are the fan-type evaporative cooling towers, where liquid is cooled by its contact with air, as well as due to the partial evaporation of water. The evaporated part of liquid is replenished with the feed water from a natural source. The surface water sources contain huge concentrations of suspended substances. Liquid is pre-treated and settled in reservoirs, purified in settling tanks, but it is not uncommon for water to enter the cooling tower directly from a natural source. Thus, a huge amount of various deposits accumulate on the surface of equipment, which have a negative impact on the performance of cooling tower, starting with the disruption of liquid flow structure and finishing with the destruction of cooling equipment. Another source of contamination of

* Corresponding author: ilnyr_91@mail.ru

circulating water supply systems is the flow of atmospheric air, containing dust. The dust particles are captured by liquid drops and settle on the surface of cooler. Also, the environment inside the cooling tower, including oxygen in water, is favorable for the development and reproduction of bacteria and microorganisms on the surface of cooler [1, 2].

A huge number of mechanisms are developed to fight against contamination, such as chemical treatment of water with reagents, mechanical disinfection, reagent-free, etc. The chemical treatment of circulating water supply system is widely used, but this method requires huge economic costs for the purchase of chemical reagents, in addition, the reagents themselves have an adverse impact on the environment. In this regard, new ways to prevent contamination are being searched every day. The well-known methods of water treatment are magnetic and ultrasonic. However, these methods do not allow us to prevent the formation of biological and mechanical deposits completely in the circulating water supply systems [3].

There is also another highly effective method for controlling the organic contamination – the method of water ozonation. The ozonation allows to remove hard-to-reach contaminations, disinfects and softens water, washes away previously formed deposits. Ozone effectively oxidizes metal compounds, so this method cannot be applied for metal coolers [4].

The fillers are the main structural elements of cooling tower that determine its cooling capacity. The filler is a contact device on the surface and in the volume of which the processes of heat, mass and momentum transfer take place [5-18].

In order to create contact devices, various types of materials are used, such as wood, metal, and polymers [19, 20]. For a long time, wooden checker filling units were very popular. The main advantage of them is that the surface has a high wettability, and the distribution of water on the surface of unit is even with the liquid film formation. However, there is an important drawback – they are not durable (short service life). Wood, due to its nature, regardless of treatment methods, under prolonged exposure to water, quickly rottens and destroys. Wooden checker filling units are also subject to high chemical and biological effects. Very often at different enterprises you can find contact devices, made of metal. In comparison with wooden contact devices, the metal ones are more durable, while the same as wood ones have good wettability. However, the manufacturing of contact device structures from metal is very expensive and material-intensive. Moreover, carbon steels of ordinary quality are not resistant to corrosion.

The checker filling units, made of polymer materials are widely used today [21]. These units are not affected by many chemical compounds and characterized by cheapness and durability. However, the polymer materials, compared to wooden and metal ones for the checker filling units, do not have good wettability, therefore, in order to achieve high cooling efficiency, it is necessary to intensify the heat-mass transfer processes by applying various roughnesses, porosities, perforations to the surface of polymers, as well as other methods of intensification.

The organization of technological process in the circulating water supply system implies taking into account the above requirements, as well as the use of checker filling unit design, which allows to increase the efficiency of cooling process, to reduce energy costs with minimal hydraulic resistance of contact devices.

2 Description of the device and its operation

The authors of this research paper developed a fan-type evaporative cooling tower, with the use of inclined-corrugated contact device. The main elements of cooling tower are the fan that provides air flow, the filler unit for distributing the liquid flow, the catchment basin and the water pump for liquid pumping [22].

The main process (technological) elements of cooling tower are made of polymer material, which allows to abandon the use of chemical reagents against bacteria and microorganisms in favor of water ozonation method.

The filler unit is a block of inclined-corrugated plates with round-shaped holes, executed along the entire working surface (Fig. 1). The plates have rounded profile corrugations, in the hollows of which the liquid film flows down. The corrugations on the surface of heat-mass transfer create significant turbulence of moving liquid at a relatively low flow rate. The material for the elements of cooling tower's checker filling unit, in particular, the inclined-corrugated plates, is polyethylene of low-pressure. The main reason for choosing the polymer materials for the cooling tower's checker filling unit is durability, cost-effectiveness, and ease of manufacture. In addition, it can be noted that polymers have good corrosion resistance, thermal and electrical insulation properties, high strength, and are resistant to temperature impact. Very often, chemical reagents are used to purify the circulating water from bacteria and microorganisms, which have almost no impact on the composite materials, due to their resistance to chemicals. The developed design of contact devices, made of polymer materials, provides convenience and compactness for long-distance transportation [23].

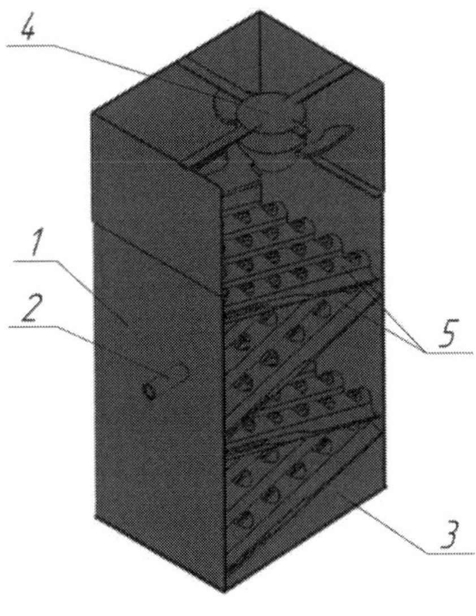

Fig. 1. The fan-type evaporative cooling tower with self-distribution of liquid: *1* – shell\casing; *2* – liquid supply manifold; *3* – basin for liquid collection; *4* – fan; *5* – filler, consisting of corrugated plates.

The circulating water in cooling tower is cooled as follows: hot used water enters the inclined-corrugated checker filling unit through the distribution device. The design of cooling tower allows you to avoid using a huge number of switchgears that is, undoubtedly, an advantage. Liquid enters onto the surface of corrugated plate and contacts with the upward air flow, then breaks and flows down the recesses of plate to the next stage and thus reaches the catch basin. A part of liquid falls through the holes within the plate. Atmospheric air enters the cooling tower, moves in the volume of checker filling unit under the action of fan, passes through the holes within the inclined-corrugated plates, while involving a part of liquid with it, which is further sprayed throughout the entire volume of contact device. After that it is thrown on the walls of filler unit, while breaking the surface of moving liquid film and, thereby, creates a highly developed contact surface of gas and liquid phases. Water is cooled by a direct contact of phases in the form of film, splashes, drops, liquid bubbling, water flow turbulence on the surface of plates, while there is an active evaporation of a part of liquid.

Due to the given design solution, water in the volume of filler unit is distributed independently and evenly, filling almost the entire volume of contact device, while the manufacturing of device from polymer material is the most appropriate.

3 Description of the study and its results

In order to determine the efficiency of developed fan-type evaporative cooling tower with the inclined-corrugated contact elements, an experimental unit for conducting physical experiments was developed. This unit allows to conduct the heat-mass transfer processes upon condition of countercurrent movement of two phases, and to study the hydrodynamic processes in the contact device. In the course of experiments, the hydraulic resistance of inclined-corrugated contact elements, its heat efficiency coefficient and the efficiency of heat-mass transfer processes were determined.

The experimental unit consists of two-stage block of inclined-corrugated plates with a total height of 340 mm. The corrugated plates have a rounded profile with a radius of 7.5 mm. The round-shaped holes with a diameter of 6 mm are executed on the side surfaces and within the upper part of corrugations. The cross-section dimensions of apparatus, under the study, are of 100x100 mm.

The experiments were conducted in a two-phase air-water system. Water entered the upper central part of the first contact stage at various mass flow rates, which was measured by means of rotameter. At the inlet and outlet of apparatus, the liquid temperature was measured by a two-channel meter-regulator OBEH 2TPM1, while the temperature of hot water varied within the range of 37.1–43.8°C. The fictitious cooling air flow rate was measured by means of testo 405i thermal anemometer with a relative error of not more than 5%. The measured value range of this portable device is from 0 to 30 m/s. The fictitious gas rate for the full cross-section of contact device during the physical experiment varied within the range of 1.4–3.5 m/s. The irrigation density varied within the range of 12–37 $m^3/(m^2 \cdot h)$. In the course of experiment, the relative air humidity, determined by means of testo 605i thermohygrometer, varied within the range of 49.7–57.2%. The pressure drop along the height of contact device was measured by means of testo 510i differential pressure gauge.

In the course of experiment, it was found that the range of operating air rates for the stable operation of contact device without liquid entrainment is quite wide and varies from 1.7 m/s to 3 m/s, the maximum value of which corresponds to the beginning of "suspension" regime. The liquid bubbling is revealed during visual observation that has a significant impact on the efficiency of water cooling in the proposed contact device. The value of hydraulic resistance of device does not exceed 2.5, and is quite low among existing analogues [24-28].

For a more complete understanding of the proposed design efficiency, the heat efficiency coefficient and water evaporation efficiency coefficient were determined [29].

Analyzing the graphs shown in Fig. 2, the following conclusions can be made: the heat efficiency coefficient of cooling tower with inclined-corrugated contact elements reaches 32%, which corresponds to the technical data for the modern filler units with high-efficiency, widely used at the enterprises of different industries. It is obvious that with an increase in the irrigation density, the value of heat efficiency coefficient decreases, and with an increase in fictitious air flow rate, vice versa, increases. This is primarily due to the fact that with an increase in water consumption, liquid fills the whole volume of inclined-corrugated contact device, thereby preventing the movement of air flow. With an increase in the fictitious air flow rate, the phenomenon of liquid bubbling is observed; a large number of splashes and drops are formed that fly out from the liquid surface, which leads to an increase in the cooling efficiency of circulating water.

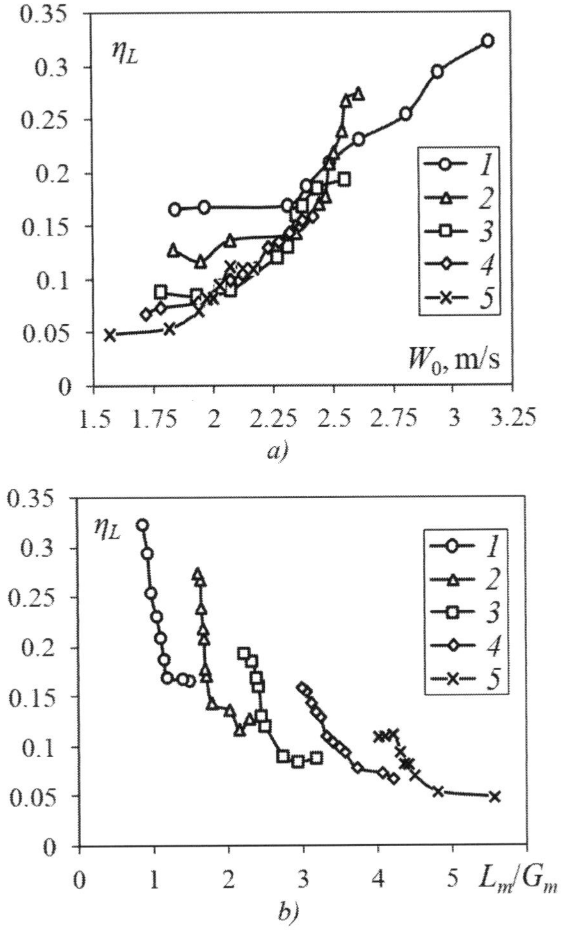

Fig. 2. The dependency of change in the heat efficiency coefficient on the fictitious air rate (*a*) and the ratio of mass flow rates of liquid and gas phases (*b*) at different irrigation densities *q*, $m^3/(m^2 \cdot h)$: *1* – 12; *2* – 18; *3* – 24; *4* – 31; *5* – 37.

When determining the value of efficiency of heat-mass transfer processes during water evaporation in the inclined-corrugated contact elements, it was found that the highest efficiency is equal to 72% (Fig. 3). It is observed at relatively low irrigation densities, equal to 15–25 $m^3/(m^2 \cdot h)$. With further increase in the irrigation density, the heat-mass transfer efficiency of cooling tower is usually reduced. For example, an increase in the irrigation density from 18 $m^3/(m^2 \cdot h)$ to 37 $m^3/(m^2 \cdot h)$ reduces the average heat-mass transfer efficiency by 1.45%.

4 Conclusion

The efficiency of developed inclined-corrugated contact devices, made of polymer materials, was experimentally confirmed in practice. In the course of experiments, the following conclusions were made: the hydraulic resistance of proposed device is relatively low; the efficiency of heat-mass transfer processes at an irrigation density equal to 12–31 $m^3/(m^2 \cdot h)$ is at the level of existing analogues, while there is a wide range of operating gas rates from 1.7 to 3 m/s without significant entrainment of liquid drops. The disinfection of liquid in the developed design of cooling tower is carried out by ozonation.

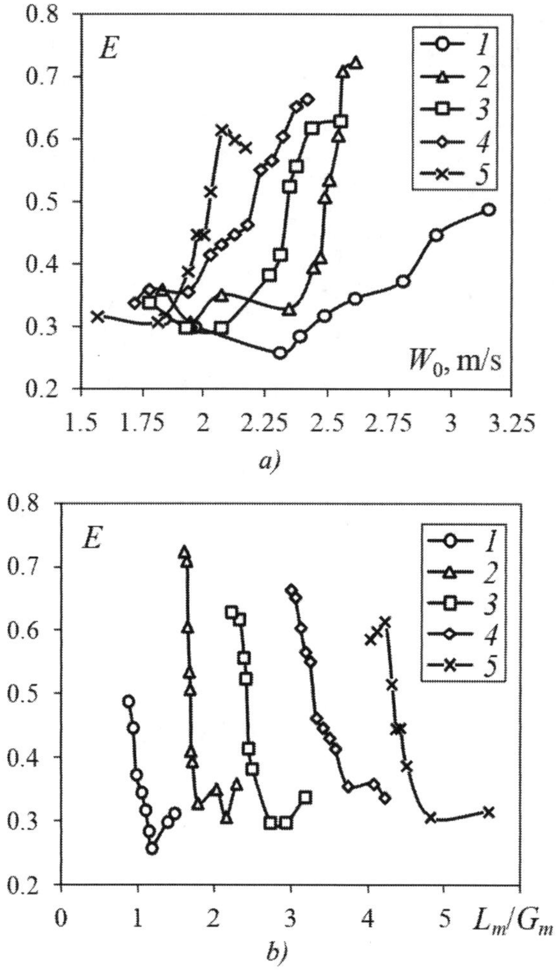

Fig. 3. The dependency of change in the heat-mass transfer coefficient on the fictitious air rate (*a*) and the ratio of mass flow rates of liquid and gas phases (*b*) at different irrigation densities *q*, $m^3/(m^2 \cdot h)$: *1* – 12; *2* – 18; *3* – 24; *4* – 31; *5* – 37.

The research was conducted with funding from the RF President`s grant project No. MK-417.2019.8.

References

1. E.V. Boev, S.P. Ivanov, V.G. Afanasenko, E.A. Nikolaev, Chem. and Pet. Eng. **45**, 454 (2009)
2. V.S. Ponomarenko etc., *Cooling towers of industrial and energy enterprises: a reference manual* (Energoatomizdat, Moscow, 1998).
3. N.A. Merentsov, A.V. Persidskiy, V.N. Lebedev, N.A. Prokhorenko, A.B. Golovanchikov, Advances in Intelligent Systems and Computing **983**, 547 (2019)
4. A.R. Iskandarova, N.D. Chichirova, Heat Supply News **165**, 05 (2014)

5. A.E. Ivanov, M.G. Berengarten, M.I. Klyushenkova, Chem. and Pet. Eng. **45**, 526 (2009)
6. N.A. Merentsov, V.N. Lebedev, A.V. Persidskiy, V.A. Balashov, IOP Conf. Ser.: Earth and Environmental Science **288**, 012106 (2019)
7. A.B. Golovanchikov, N.A. Merentsov, M.V. Topilin, A.V. Persidskiy, IOP Conf. Ser.: Earth and Environmental Science **288**, 012089 (2019)
8. A.B. Golovanchikov, N.A. Merentsov, V.A. Balashov, Chem. and Pet. Eng. **48**, 595 (2013)
9. M.I. Shilyaev, E.M. Khromova, A.R. Bogomolov, MATEC Web of Conf. **115**, 08006 (2017)
10. G.V. Kuznetsov, D.V. Feoktistov, E.G. Orlova, S. Y. Misyura, V.S. Morozov, A.G. Islamova, Int. J. Heat Mass Transfer **126**, 161 (2018)
11. N. Merentsov, A. Persidskiy, V. Lebedev, M. Topilin, A. Golovanchikov, Advances in Intelligent Systems and Computing **983**, 560 (2019)
12. A.V. Persidskiy, N.A. Merentsov, V.N. Lebedev, A.B. Golovanchikov, IOP Conf. Ser.: Earth and Environmental Science **288**, 012110 (2019)
13. A.V. Dmitriev, O.S. Dmitrieva, I.N. Madyshev, A.N. Nikolaev, Chem. and Pet. Eng. **53**, 501 (2017)
14. A.V. Dmitriev, O.S. Dmitrieva, I.N. Madyshev, Chem. and Pet. Eng. **53**, 430 (2017)
15. I.N. Madyshev, O.S. Dmitrieva, A.V. Dmitriev, A.N. Nikolaev, Chem. and Pet. Eng. **52**, 299 (2016)
16. N.A. Merentsov, V.A. Balashov, D.Y. Bunin, V.N. Lebedev, A.V. Persidskiy, M.V. Topilin, MATEC Web of Conf. **243**, 5 (2018)
17. N.A. Merentsov, A.V. Persidskiy, M.V. Topilin, V.N. Lebedev, V.A. Balashov, A.B. Golovanchikov, Journal of Phys.: Conf. Ser. **1278**, 012024 (2019)
18. N.A. Merentsov, V.N. Lebedev, A.B. Golovanchikov, V.A. Balashov, E.E. Nefed'eva, IOP Conf. Ser.: Earth and Environmental Science **115**, 012017 (2018)
19. A.B. Golovanchikov, V.A. Balashov, N.A. Merentsov, Chem. and Pet. Eng. **53**, 10 (2017)
20. O.S. Dmitrieva, I.N. Madyshev, A.V. Dmitriev, J. Eng. Phys. Thermophy. **90**, 651 (2017)
21. V.G. Afanasenko, F.Sh. Khafizov, N.F. Khafizov, S.P. Ivanov, E.V. Boev, Chem. and Pet. Eng. **43**, 653 (2007)
22. I.N. Madyshev, A.I. Khafizova, O.S. Dmitrieva, E3S Web of Conf. **126**, 00031 (2019)
23. A.V. Moshkarin, V.A. Kalatuzov, Bulletin of Ivanovo State Energy University **2**, 1 (2006)
24. K.V. Chizh, A.S. Pushnov, M.G. Berengarten, Chem. and Pet. Eng. **50**, 244 (2014)
25. A.S. Sokolov, A.S. Pushnov, M.V. Shapovalov, Chem. and Pet. Eng. **53**, 1 (2017)
26. A.B. Golovanchikov, N.A. Merentsov, Advances in Intelligent Systems and Computing **983**, 635 (2019)
27. Sokol B.A., Chermyshov A.K., Baranov D.A., *Mass-transfer column packed-type devices* (Infokhim, Moscow, 2009).
28. Kagan A.M., Laptev A.G., Pushnov A.S., Farakhov M.I., *Contact packings in industrial heat-and-mass transfer apparatuses* (Otechestvo, Kazan, 2013).
29. A.G. Laptev, V.A. Danilov, I.V. Vishnyakova, Thermal Eng. **51**, 661 (2004)

Collecting finely-dispersed particles from the gas flow in a centrifugal separator with coaxially arranged pipes

Vadim E. Zinurov[1], *Oksana* S. Dmitrieva[2,*], and *Oksana* S. Popkova[1]

[1]Kazan State Power Engineering University, 420066 Krasnoselskaya str. 51, Kazan, Russia
[2]Kazan National Research Technological University, 420015 Karl Marx str. 68, Kazan, Russia

Abstract. The article deals with the problem of increasing the efficiency of dedusting the gas flow from the finely dispersed particles smaller than 10 μm. In order to solve this problem, a design of centrifugal separator with coaxially arranged pipes is proposed. The described principle of operation includes the large values of centrifugal forces, which take place inside the device when the flow is swirled, and these forces throw the finely dispersed particles to the walls of device. This scientific paper shows a numerical simulation of gas flow dedusting process by means of ANSYS Fluent software package. The efficiency of dedusting the gas from the finely dispersed particles of up to 10 μm in the device is on average within the range of 53.8–76.7%. The exponential function, describing the changes in the pressure loss from the input gas rate, is obtained. In the course of studies, it was found that the pressure loss in the device is not more than 800 Pa at the input gas rate from 3 to 19 m/s.

1 Introduction

Improving the quality of dedusting the gas flows from the finely dispersed dust particles is an urgent task for the production enterprises, which have sources of dust emissions [1]. For example, the main sources of dust emissions in the metallurgical industry are the activities related to the unloading of furnace burden into the bunker, unloading into the mobile weighbridge for wagons; a large number of particles of unburned fuel and ash dispersion, in the electricity-producing industry, is contained in the fume gases; dust, in the food industry, is emitted during the production of various food products; dusty gas flows are formed in the construction industry during air transportation of different materials etc. [2-7].

The stricter standards for dedusting the gas flows are caused by the increased requirements for maximum permissible emissions into the environment and internal sanitary norms. Also, many enterprises revised the ways of handling with collected dust. It can be re-used in the technological processes after preliminary treatment, for example, dry collected dust from the fume gases is re-fed to the combustion chamber furnaces at power and metallurgical facilities. In some cases, there is an economic profitability to use the

* Corresponding author: ja_deva@mail.ru

collected dust as a material for the production of other products. For example, the collected ash and slag are used for concrete mortars. Therefore, effective collection of dust particles from gases is of a great demand from environmental and economic point of view [8].

As of today, a large number of various devices with high efficiency of dedusting the gas flows were developed, for example, various modifications of inertial dust collectors, such as dust-collecting chambers, cyclones, etc. [9-18]. Easy operation, high reliability and ability to operate at high temperatures and pressures are important advantages of these devices. However, one of the main drawbacks is the low efficiency of dedusting the gas flows from the particles smaller than 10 μm [19]. In order to eliminate this drawback, the technological (process) line for the gas flow purification is equipped with additional devices, for example, a wet purification device or a bag filter shall be installed after an inertial dust collector. However, high capital and operating costs, increased maintenance requirements for these devices, etc. are the limiting factors for the use of them. Therefore, the development of new technological designs of devices for dedusting the gas flows from the particles smaller than 10 μm with high efficiency and low pressure loss inside them is an urgent and relevant task [20].

2 Device description

The authors developed a design of centrifugal separator with coaxially arranged pipes (Fig. 1).

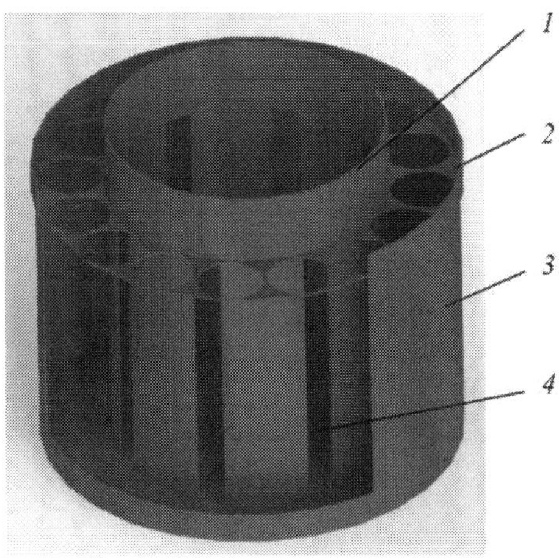

Fig. 1. Three-dimensional model of centrifugal separator with coaxially arranged pipes: *1* – inlet pipe, *2* – ring with holes, *3* – casing of centrifugal separator, *4* – slots.

This device is designed for dedusting the gas flows from the particles smaller than 10 μm. It is proposed to use a centrifugal separator as the second purification stage for dedusting the gases from the finely dispersed particles after inertial dust collectors, such as, cyclones. The developed design consists of cylindrical body *3*, inside which there is an inlet pipe *1*, intended for feeding the dusty flow into the device and its uniform distribution in an axisymmetric direction through the rectangular slots *4* and rings with holes *2* for the output of purified gas flow. The principle of device operation is the following: the dusty gas flow enters the centrifugal separator through the inlet pipe *1*, after that the flow is randomly divided into equal volumes and changes its direction towards the closest slots *4*. Further, each gas jet, moving through the slot, is divided into two flows, which are swirled, due to the design of separator, namely the geometric arrangement of slots *4* and the exit holes, executed in the form of rings with holes *2*. When swirls are formed, the centrifugal forces arise, throwing the particles from the structured flow to the "dead" zones, located between the pipes. When getting there, the particles settle to the bottom of centrifugal separator or adhere to the surfaces of inlet pipe *1* or the inner wall of housing *3*. It should be noted that for the actual design of device, instead of separator bottom, the dust bag shall be used. After that, the purified gas flow leaves the centrifugal separator through the exit holes, which are used as rings with holes *2*.

The purpose of this paper is a numerical simulation of gas flow dedusting process in a centrifugal separator with coaxially arranged pipes (Fig. 1). The numerical simulations were conducted by means of ANSYS Fluent software package. In order to conduct the studies, a three-dimensional model of centrifugal separator was created, shown in Fig. 1. In the process of creation of three-dimensional model, the following geometrical dimensions were taken: the height of device – 125 mm, the diameter of inlet pipe – 100 mm, the diameter of ring – 24 mm, the width and height of slot – 10 mm and 97 mm, respectively, the diameter of device – 150 mm. The number of rings with holes and slots was 16 and 8, respectively. In the course of studies, two variants of gas flow dedusting process were considered: adhesion of particles to the surfaces when they are in direct contact, and reflection of particles from the walls. Consideration and analysis of these two variants are extremely important. On the one hand, the finely dispersed particles adhere to almost all surfaces due to electrostatic and intermolecular forces, Brownian motion, and other forces. On the other hand, the finely dispersed particles can be re-drawn into the flow structure from the "dead" zones due to high values of centrifugal forces in the device.

In the course of studies the gas flow rate was set at the inlet of device, which varied from 3 to 19 m/s. At the outlet of device, the ambient pressure was set as equal to 101325 Pa. The number of particles n in the gas flow was set as equal to 1000, and the density of finely dispersed particles 1075 kg/m^3. In order to determine the dependencies, the diameter of particles a was changed from 1 to 10 μm.

The operation efficiency of centrifugal separator was determined by the following formula [21]:

$$E = 1 - n_k/n, \tag{1}$$

where n_k – the number of particles that remain in the gas flow after dedusting process in the centrifugal separator.

3 Results and discussion

The results of studies are shown graphically in Figs. 2–4. The efficiency of dedusting the gas flow from the particles of up to 10 μm, upon a given condition of their adhesion to the surfaces of centrifugal separator, while in a direct contact of dust particles with any surface inside the device, was on average equal to 76.7% (Fig. 2).

Upon the condition of reflection of finely dispersed dust particles from the walls of separator, the efficiency of dedusting the gas flow was on average equal to 53.8% (Fig. 3). In the course of studies, it was found that the pressure loss in the centrifugal separator is not more than 800 Pa at the input gas flow rates from 3 to 19 m/s (Fig. 4). The high efficiency of dedusting the gas flow from the finely dispersed particles of up to 10 μm is explained by the design simplicity, which allows to create high values of centrifugal forces in the gas flow swirls between coaxially arranged pipes.

When the gas flow moves to the upper part of device, i.e. to the exit from it, each swirl, during its rotation, additionally accelerates the other two symmetrically relative to itself that increases the efficiency of dedusting the gas. When the gas flow is swirled, each swirl has a minimum number of points of contact with the surfaces of centrifugal separator, resulting in a minimum boundary layer, which is the reason for the low value of pressure loss in the device. The difference between the efficiency values of dedusting the gas flow upon the conditions of adhesion and reflection was about 23%. As a rule, when the finely dispersed particles fly out of swirling gas flow, during its rotation and direct contact of particles with the device surfaces, they adhere to them due to their size and the large influence of intermolecular, electrostatic and other forces. However, due to the high values of centrifugal force, it is possible to overcome these forces and carry away the finely dispersed particles from the device along with the flow. Therefore, it is necessary to consider both cases when simulating the processes of dedusting the gas flows from the finely dispersed particles, which allows us to conclude that the efficiency of dedusting the gas in the device is on average within the range of 53.8–76.7%.

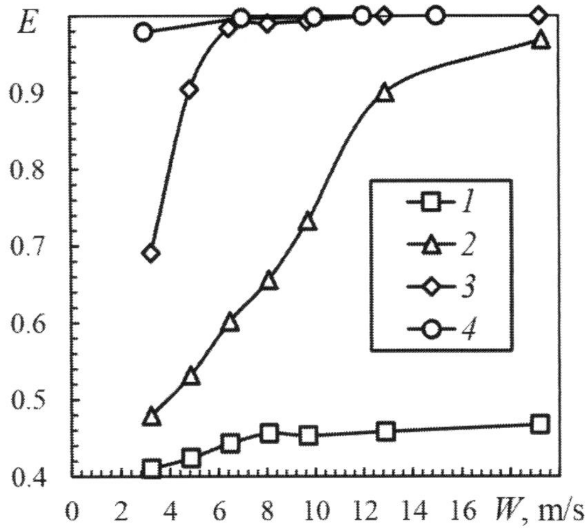

Fig. 2. The dependency of change in the efficiency of collecting the solid particles from dusty flows on the input gas rate at different sizes of particles, μm: *1* – 1, *2* – 4, *3* – 7, *4* – 10.

The efficiency of dedusting the gas flow at its input rate within the range of 3–19 m/s from the finely dispersed particles of 1, 4, 7 and 10 μm was on average equal to 44.4, 69.6, 93.7 and 99.6%, respectively, upon the specified condition of adhesion to the walls. It is worth noting that the efficiency of dedusting the gas flow from the particles larger than 7 μm was on average equal to 99.9% at the input gas rates equal to more than 6.5 m/s. The high degree of dedusting the gas from the particles of 4 μm was achieved at a gas rate of more than 9.7 m/s, corresponding to the efficiency of at least 70%. For the particles of 1 μm, the efficiency of dedusting the gas flow did not exceed 50% at its rate from 3 to 19 m/s. It is worth noting that with an increase in the gas rate of more than 10 m/s, the efficiency of its dedusting by means of separator from the particles of 1 μm practically did not change, moving asymptotically to a value equal to 46.1% (Fig. 2).

The numerical simulation was conducted under the condition of particles adhesion to the walls. The efficiency of dedusting the gas flow at its input rate within the range of 3–19 m/s from the finely dispersed particles of 1, 4, 7 and 10 μm was on average equal to 15.2, 34.4, 75.4 and 90.1%, respectively, under the specified condition of reflection of particles from the walls. Similar to the simulated gas flow dedusting process, shown in Fig. 1, the highest efficiency, corresponding to the values of at least 90%, was related to the particles larger than 7 μm at the gas rate within the range of 6.5–19 m/s. It should be noted that for the particles of 4 μm, the gas dedusting efficiency equal to at least 50 60 and 70% was achieved at its rate of at least 12.9, 16.1 and 19.2 m/s. The efficiency value of dedusting the gas from the particles of 1 μm, similar to the numerically simulated case, shown in Fig. 1, moved asymptotically to 15.5% at the gas rate of more than 10 m/s (Fig. 3).

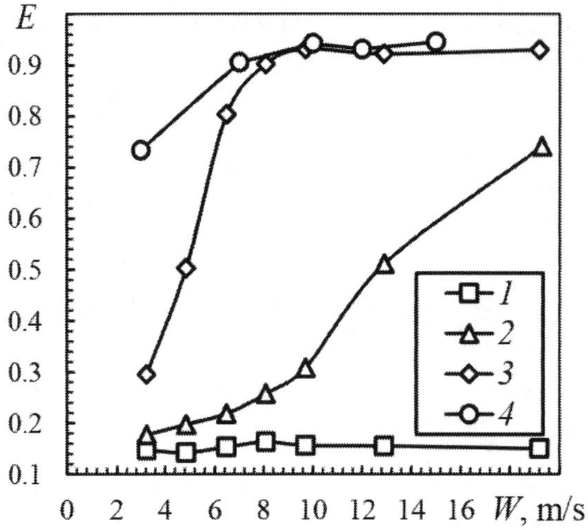

Fig. 3. The dependency of change in the efficiency of collecting the solid particles from dusty flows on the input gas rate at different sizes of particles, μm: *1* – 1, *2* – 4, *3* – 7, *4* – 10.

The numerical simulation was conducted upon the condition of particles reflection from the walls. As noted earlier, due to the ordered structure of movement, almost absence of boundary layer and the minimum number of local resistances in the centrifugal separator, the pressure loss in it was not more than 800 Pa at the input rate of gas flow from 3 to 19 m/s.

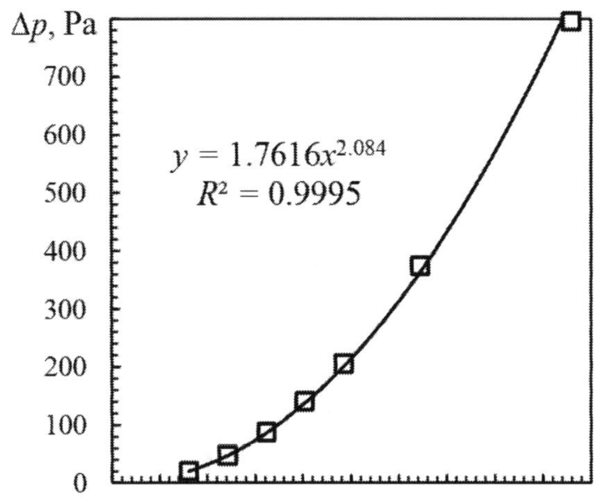

Fig. 4. The dependency of change in the pressure loss in a centrifugal separator with coaxially arranged pipes on the gas input rate.

The exponential function was determined to describe the changes in pressure loss inside the device due to the input gas rate with a high reliability value of its approximation equal to 99.9%. It should be noted that a significant increase in the pressure loss in the device was achieved at the input gas rate from 10 to 19.2 m/s, the values of which varied from 205 to 795 Pa (Fig. 4).

4 Conclusion

The conducted studies showed that the use of centrifugal separator with coaxially arranged pipes for dedusting the gases from the finely dispersed particles as the second stage of technological (process) line after inertial dust collectors, in particular, the cyclones, is advisable. When using the single centrifugal separator, the efficiency of dedusting the gas from the finely dispersed particles is of at least 50% at its input rate from 3 to 19 m/s. The use of single centrifugal separator in a process treatment line is the most rational solution mainly for dedusting the gases from the finely dispersed particles larger than 4 μm, since at moderate values of input gas rate of up to 10 m/s, the dedusting efficiency is of 40–70% with a pressure loss of not more than 205 Pa in the device. When the gas flow rate is from 10 to 19.2 m/s, the efficiency of dedusting the gas from the finely dispersed particles larger than 4 μm is of 66–87% with a pressure loss in the device of not more than 800 Pa. When dedusting the gas flow from the particles of 1 μm, the efficiency is on average of 15–44% at its input rate from 3 to 19 m/s. In order to intensify the dedusting of gases from the finely dispersed particles, several developed devices can be installed, connected to each other, as the second stage of process line.

The advantages of developed centrifugal separator design with coaxially arranged pipes are the high efficiency of dedusting the gas, equal to at least 50%, from the finely dispersed particles of less than 10 μm at low pressure loss in the device, not exceeding 800 Pa; the wide range of operating rates from 3 to 19 m/s, the low metal consumption, the ease of manufacture, the minimum capital and operating costs.

The research was conducted with funding from the RF President`s grant project No. MK-616.2020.8.

References

1. M. Fang, N. Yi, W. Di, T. Wang, Q. Wang, Int. J. Greenhouse Gas Control, **93**, 102904 (2020)
2. J. Chen, Z. Zhong, Y. Xia, X. Ke, W. Xing, Chin. J. Chem. Eng., **27**, pp. 1391-1402 (2019)
3. R. Singh, A. Shukla, Renewable Sustainable Energy Rev., 29, pp. 854-864 (2014)
4. F. Sh. Khafizov, V. G. Afanasenko, I. F. Khafizov, A. Sh. Khaibrakhmanov, E. V. Boev, Chem. Pet. Eng., **44**, pp. 425-428 (2008)
5. N. A. Merentsov, V. A. Balashov, S. A. Bokhan, E. Nefed'eva, D. A. Tezikov, V. V. Groshev, IOP Conf. Ser.: Earth Environ. Sci., **224**, 012041 (2019)
6. M. G. Ziganshin, A. M. Ziganshin, A. V. Dmitriev, Chem. Petrol. Eng., **45**, pp. 230-235 (2009)
7. A. B. Golovanchikov, V. A. Balashov, N. A. Merentsov, Chem. Petrol. Eng., **53**, pp. 10-13 (2017)
8. S. A. Miller, P. R. Cunningham, J. T. Harvey, Resour., Conserv. Recycl., **146**, pp. 416-430 (2019)
9. A. V. Shvab, N. S. Evseev, Theor. Found. Chem. Eng., **49**, pp. 191-199 (2015)
10. J. Wu, Y. Zhang, H. Wang, Sep. Purif. Technol., **132**, pp. 541-551 (2014)
11. V. S. Shvydky, Y. G. Yaroshenko, N. A. Spirin, V. V. Lavrov, IOP Conf. Ser.: Mater. Sci. Eng., **150**, 012021 (2016)
12. X. Li, T. Wei, D. Wang, H. Hu, L. Kong, W. Xiang, Chem. Eng. Sci., **151**, pp. 79-92 (2016)

13. B. K. Lee, K. R. Jung, S. H. Park, J. Aerosol Sci., **39**, pp. 1079-1088 (2008)
14. U. Heck, Filtr. Sep., **53**, pp. 14–16 (2016)
15. M. Kr. Prabhansu, P. C. Karmakar, P. Kr. Chatterjee, J. Environ. Chem. Eng., **3**, pp. 689-702 (2015)
16. K. Reijnen, J. V. Brakel, Powder Technol., **40**, pp. 81-111 (1984)
17. S. Koshkarev, A. Evtushenko, S. Pushenko, Procedia Eng., **165**, pp. 1057-1069 (2016)
18. V. N. Azarov, S. A. Koshkarev, Magazine of Civil Engineering., **54**, pp. 18-32 (2015)
19. A. V. Dmitriev, O. S. Dmitrieva, S. V. Dang, V. L. Nguen, Chem. Petrol. Eng., **55**, pp. 329-335 (2019)
20. A. V. Dmitriev, V. E. Zinurov, O. S. Dmitrieva, E3S Web Conf., **126**, 00007 (2019)
21. O. S. Dmitrieva, V. L. Nguen, N. D. Yakimov, E. G. Sheshukov, IOP Conf. Ser.: Earth and Environmental Science, **337**, pp. 012057 (2019)

The evaluation of possibilities of modern level casting bindings on the technological lignin basis practical appliance

Yu.A. Svinoroev[1], *V.V.* Dyadichev[2*], *K.A.* Batyshev[3], *K.G.* Semenov[3], *A.V.* Dyadichev[2], *S.Ye.* Chornobay[2], and *S.G.* Menyuk[2]

[1] Kamensky Technological Institute (branch), Platov SRSPU (NPI), Kamensk-Shakhtinsky, Russia
[2] V.I. Vernadsky Crimean Federal University, Simferopol, Russia
[3] Bauman Moscow State Technical University, Moscow, Russia

Abstract. Nowadays the ways of overcoming a problem of resource deficit include rationalization of the existing resource potential. In this respect it is expedient to use the secondary products of vegetable raw material processing due to expansion of lignosulfonate materials application.

1 A problem statement

The experience of industrial enterprises work shows that technologies based on the resource-saving conceptions [1] will allow obtaining cheaper, environmentally friendly, and consequently competitive castings of high quality [1, 2]. Let us consider the sphere of binding materials use as the main component of the technological process of casting manufacture [3]. At the industrial plants, where they are applied as the technology component, i.e. casting manufacture, metallurgy, productions of wood chipboards and wood fiberboards, plywood, in construction industry and automobile industry, it is reasonable to adopt the usage of bindings produced on the basis of inorganic materials [4, 5] or vegetable raw materials [6, 7].

The absence of such advances in this area leads to the emergence of a range of negative problems: the rise in cost of the final product [8] due to its prime cost increase; the decrease of its competitiveness because of that reason [1, 6, 8, 9]; worsening of the ecological situation in the location of such enterprises [6, 10]; lowering of the industrial safety [6, 10, 11]; the growth of occupational diseases [11, 12] connected with the use of expensive and environmentally dangerous binding materials on the basis of various synthetic resins [13] and oil bindings [6, 7]. The use of technological lignins, a large-tonnage waste of the pulp-and-paper industry, due to their processing into binding materials [7, 14], which differ from inorganic materials by their high manufacturability at the final stages of casting

* Corresponding author: author@email.org

manufacture (pliability of casting rods and mould elements, having light knockout and regeneration [6, 7]), is a promising trend in the development of new binding materials.

1.1 The objective of the work

Evaluating the developed ways of improving binding ability of lignosulfonate (LST) bindings [7], it is reasonable to consider the practical aspects of their application in the casting manufacturing processes as the tool for improving the resource efficiency of the casting technology.

2 Materials and the results of the research

The most significant ways of improving LST binding ability, from the point of view of their successful usage in the casting manufacturing processes, is their modification by nonionic SAS and mechano-activation.

The obtained results [6, 7] give the opportunity to broaden the application of LST due to its advantages with the consequent positive effects; however the practical appliance requires doing the adaptation work which will include:

- the evaluation of casting technology, where expensive and dangerous synthetic resins and oil bindings are used, for the purpose of their substitution for lignosulfonate materials processed according to the suggested algorithms;
- the development of the mixture compositions for these technologies and their practical testing instead of the existing ones for the purpose of providing the required quality level;
- the optimization of casting processes' parameters, where the binding material on the LST basis is "built into";
- doing the experimental-industrial tests under the foundry conditions, the correction of the manufacturing procedures' content (setting the working areas and assembly of disintegrating machineries, determining and "provision" of the modifier input spots during their usage, etc.);
- the development of the technical requirements for the application of mixture compositions with LST NSAS instead of mixtures with oil and resin materials.

The most important stage is the first one, listed above, as it sets the rest actions and the obtained results. The content of this stage work will include:

- the comparison of strength characteristics of the bindings (LST and the applied one);
- the economical grounds for the reasonability of application according to all the aspects of the manufacturing process (manufacturability, economy, qualities, ecological compatibility, the change of working conditions, etc.).

Estimating the level of the obtained results and the set of the existing technology of binding materials appliance, as well as considering the specificity of lignosulfonate materials, which consists in the fact that this binding is on the water basis, requiring thermal influence for hardening, it is reasonable to analyze the whole "coordinate system: the LST abilities – the technology of molding-sand preparation" (Table 1).

A consolidated evaluation of the possibilities of LST with improved strength properties practical appliance in the casting manufacturing processes is presented in Table 1. It is necessary to mention that nowadays the results of the work are implemented in the cast-iron molded casting manufacturing process; the method of LST modification by nonionic surface active substances (NSAS) is implied. The application of the developed core sand mixture of such type can be used and for small steel casting, the laboratory experiments of the indicators of its strength correspond to the required regulations, however the industrial tests are required.

The improvement of the composition has to be concentrated on the issues of providing the required thermal stability for guaranteeing the quality of the steel casting.

It is more reasonable to start implementing technologies where LST are traditionally used (technologies of manufacturing rods and moulds hardened by convective drying) for cast-iron casting manufacture. In this case the improvement tendency will be a partial or complete removal of oil bindings of SB or USK type from the technology. It is enough to use the method of LST modification by nonionic SAS.

Let us consider in more details the manufacturing process of getting rods using heat for their hardening. The technology of producing rods in the heating set and by convective drying applies different toxic synthetic resins, oil bindings and their combinations. The disadvantage of this manufacturing process is the necessity of using expensive and very often deficit binding materials which leads to the rise of the cost for core sand mixture. Besides, during the manufacturing rods process a considerable amount of toxic compounds, creating unfavorable sanitary and hygienic working conditions, are emitted into the working space and the environment [6, 15]. That is why the task of improving the technical and economic as well as sanitary and hygienic characteristics of this technology becomes urgent.

The set problem can be solved in different ways:

- by combining synthetic resins with low-toxic bindings;
- by selection of special catalysts of hot hardening;
- by creating absolutely new low-toxic binding on the basis of modified LST with the properties of resins and oil materials [7].

The first and the second way allow reducing the content of toxic resins in the mixtures a bit, but they do not solve the problem of considerable improvement of sanitary and hygienic working conditions. That is why the creation of low-toxic bindings on the basis of LST with the properties of high-performance bindings is an effective way for solving a complex of problems connected with economy – LST are cheaper, with ecological compatibility – LST are cleaner, and with sanitary and hygienic working conditions in the foundries. This offers a prospect of emergence of new generation core sand mixtures, meeting the ecological requirements of the modern production [10, 16].

The methods of LST mechano-activation are reasonable to be applied for the improvement of molding sands quality indicators. Mechano-activation leads to the increase of binding ability of LST from 0.4 to 0.9 MPa, i.e. twice; this gives an opportunity to use this effect for improving the molding sands compositions, both for the production of cast-iron and steel casting (Table 2). The experimental compositions of facing molding sands for dry-molding, oriented to the production of small cast-iron molded casting, are developed. LST were processed in the recommended modes of high-energy influences on the disintegrating set UDA 100. The given compositions of the molding sands are used both for the production of grey cast-iron casting, and for the manufacture of steel castings from carbon steel.

The task of the searching experiments was to find out the possibility of applying methods of LST mechano-activation for the optimization of the molding sand composition, and at the same time reducing the content of the "pure" sand due to the increase of reclaim applying. The possibilities of reducing the usage of molding clay and the hardening cycle change were also studied.

The issues of reducing the duration of hardening were also considered; for drying half-molds on the basic (used in the manufacture) mixture composition this period lasts 4 hours at the temperature of 200 – 220 ^{0}C.

Thus, the application of the LST mechano-activation methods leads to the rationalization of the molding sand composition: the reduction of the amount of "pure" molding sand is possible due to the increase of the reclaim part and due to the decrease of molding clay usage twice.

Table 1. The evaluation of the possibilities of LST with improved strength properties practical appliance in the casting manufacturing processes

The technology of the molding-sand preparation and hardening of the casting rods and moulds	**Cast-iron casting**	**Steel casting**	**Notes**
Technologies of hardening by convective drying	1. The composition of the core sand mixture is developed and implemented in the production of small cast-iron molded casting (modification method) 2. The experimental composition of the molding sand with a lower content of SB binding and an accelerated hardening cycle is developed (mechano-activation method)	1. The composition of the core sand mixture is developed and tested for small steel casting manufacture (modification method). 2. The experimental composition of the molding sand with a lower content of SB binding and an accelerated hardening cycle is developed (mechano-activation method)	Ecological effect, the reduction of harmful emissions into the environment, decrease of casting prime cost due to the reduction of SB binding usage
Technologies of hardening in the heating set	The composition of the core sand mixture is developed for the production of bell's rods (weighed not more than 10 kg) in manufacturing cast-iron pressure-water pipes and sewer pipes (modification method)	The researches were not carried out	Ecological effect, the reduction of harmful emissions into the environment, decrease of casting prime cost due to the reduction of carbamide-formaldehyde resins (CF-O, CF-MT) usage
Technologies of hardening by hot gas blowing	The composition of the core sand mixture is developed for the production of bell's rods (of any weight) in manufacturing cast-iron pressure-water pipes and sewer pipes (modification method)	The researches were not carried out	The reduction of casting prime cost and the improvement of the ecological state due to the usage of modified LST without resin binding application. Technological difficulties
CHM technologies	The searching experiments for molding sands compositions were carried out (modification method)	The searching experiments for molding sands compositions were carried out (modification method)	The difficulty of implementation is the necessity of using effective hardeners
LSM Technologies	The researches were not carried out	The researches were not carried out	

Table 2. A comparative evaluation of facing molding sands for dry molding on the basis of LST processed by mechano-activation methods

Composition components:	**Compositions and quality indicators**			
	Experimental compositions			**Basic mixture composition, corresponding quality indicators**
	1	**2**	**3**	
Silica sand of brand 1K02B, Verkhnedneprovsky, %	30	35	40	42
Burnt soil after regeneration, %	53	48	43	38
Molding clay, %	3	3	3	6 – 8
LST, %	4	4	4	4 – 8
Sawdust, %	10	10	10	10
Quality indicator:				
Humidity, %	6.9	6.8	6.9	6.0 – 7.0
Gas permeability,units	100	100	100	101
Strength in the damp state, kg-force/cm2	0.25	0.27	0.31	0.27 – 0.33
Strength in the dry state, MPa	0.42	0.44	0.48	0.33 – 0.45
The duration of drying, min (at the temperature of 200°C)	30	30	30	45

The duration of drying can be reduced by 20 – 25%, however for more accurate judgments additional industrial experiments are required.

The presented results allow claiming a potential possibility of applying the method of LST mechano-activation for optimization of molding sand compositions.

An urgent issue is the input of the modifier under dosing it into the LST composition.

The modification variants can be implemented in three ways:

- variant 1- into the LST storage tank;
- variant 2 – into the accumulation tank in the foundry;
- variant 3 – into the metering device of the chaser mills.

Variant 1 is reasonable to be applied when LST in its full volume has to be used in a modified form. Its disadvantage is the technological inconveniences: during the mixing process it is necessary to adhere to the temperature conditions.

Variant 2 is the most acceptable as in this area it is possible optimally to create the necessary conditions for implementing the modification technology. It is conditioned by the reasonable for technological process implementation capacities (tanks from 400 liters to 2000 liters) and conditions for process implementing – mixing and heating.

Variant 3 is reasonable while implementing the LST modification process by NSAS under conditions of single (repair) production when the manufacturing processes are done on a non-regular basis.

The search for possibilities of implementing the obtained results for preparing CHM and LSM is promising. The searching experiments on creating CHM on the basis of the modificated LST by NSAS demonstrated the perspective of this tendency; however, the difficulties in this case are connected with the search of effective hardeners: non-toxic and allowing to compete in the hardening speed with CHM on the basis of resin bindings.

From the point of view of regeneration, the practical experience of the core sand mixture composition implementation for the production of cast-iron facing casting indicates that the expansion of using bindings with LST increases the regenerative properties of the worked-out mixtures by 12 – 14%.

The presented examples allow viewing from a new angle the possibilities of lignin containing materials as the potential raw materials for creating new casting bindings in principle [17].

3 Conclusions

As a result of the carried out research we found out:

- creating low-toxic bindings on the basis of LST with the properties of high-performance bindings is an effective instrument for solving a complex of problems connected with the increase of resource efficiency of casting processes.
- applying the method of LST mechano-activation leads to the rationalization of the molding sand composition: the decrease of the amount of using "fresh" core sand mixture is possible due to the increase of reclaim part and reduction of molding clay usage by two times. The duration of drying can be reduced by 20 – 25%, however, for more accurate judgments concerning this issue, additional experimental tests are required.

The study was carried out with the financial support of the Ministry of Education and Science of the Russian Federation within the framework of the scientific project FZEG-2020-0030.

References

1. A.I. Batyshev, K.A. Batyshev, Foundry engineering **10,** 23-27 (2017)
2. M. Holtzer, M. Górny, R. Dańko, *Microstructure and properties of ductile iron and compacted graphite iron castings* p. 348 (London, Springer (Eds.), 2015)
3. S.S. Zhukovsky *Cold-hardening bindings and mixtures for casting rods and moulds: a reference book* p. 256 (Moscow: Machine building, 2010)
4. M. Chvojka, L. Mlynarcik, P., Schmiedt *Progr. zlievaren. technol. a metalurgii. – Konf., Zlata Idka* p. 84 (May 29-31, 2009)
5. I.E. Illarionov, Yu.P. Vasin, *Moulding materials and mixtures* Part 1, p. 223 (Monography, Cheboksary: The publ. house of Chuvash University, 1992)
6. O.V. Rodionov, Yu.S. Svinoroev, et al. *Diagnostics of developing the enterprise potential: monography* p. 292 (Lugansk: "Knowledge", 2012)
7. Yu. Svinoroev, V. Kostrub, O. Klimova, Teka commission of motorization and power industry in agriculture Lublin university of technology Volodymyr Dal East-Ukrainian national university of Lugansk **XB**, 227-231 (Lubin, 2010)
8. I.A. Dibrov, The foundry worker of Russia **10,** 7-14 (Chelyabinsk, 2017)
9. *Forecast: U.S. Casting Sales to Reach $33 Billion, A Modern Casting Staff Report*, Modern Casting 22-27 (2018)
10. *The Directive of the European Parliament and Council of the 24-th November 2010 concerning industrial emissions* (2010)
11. H. Löfstedt, H. Westberg, A.I. Seldén, I.L. Bryngelsson, M. Svartengren, Journal of Occupational and Environmental Medicine, **53**, 1425-1429 (2011)
12. H. Löfstedt, H. Westberg, A.I. Seldén, S. Rudblad, I.L. Bryngelsson, Y. Ngo, M. Svartengren, *Journal of Occupational and Environmental Medicine*, **53,** 43-48 (2011)

13. *The information resource «Hüttenes-Albertus», Germany*, access mode: http://www.huettenes-albertus.ru/
14. Yasuyuki Matsushita , J Wood Sci **61**, 230–250 (2015)
15. H. Löfstedt, H. Westberg, A.I. Seldén, C. Lundholm, M. Svartengren, American Journal of Industrial Medicine, **52**, 455-463 (2009)
16. A.N .Bolding, A.I. Yakovlev, S.D. Teplyakov *Engineering ecology of the foundry: a tutorial* p. 352 (Ed. by A. N. Boldin, Mscow.: Machine building, 2010)
17. Yu.A. Svinoroev, R. Ber, Yu.I. Gutko, The foundry, **12**, 23-26 (2016)

Fracture features of dielectric nanoceramic with graphene by high-voltage discharge

Aleksandr Kuznetsov[1], *Georgii* Savenkov*[1], *and Vladimir* Stolyarov[2]

[1]JSC "Armalit Mechanical Engineering Plant," 198097 St. Petersburg, Russia
[2]Mechanical Engineering Research Institute of RAS, 101990, Moscow, Russia

Abstract. Results of dynamic strength investigation for Al_2O_3 nanoceramic with different content of five-layer flake graphene under action of nanosecond duration high-voltage pulse are given. It has been found that the graphene content increasing results in embrittlement of the nanocomposite during of nanosecond loading.

1 Introduction

Interest in nanostructured materials in fundamental-applied science and in industry is increasing. In many industries, cermet composites are widely used with greater strength but brittleness than metal materials. The solution to the problem of increasing the viscosity of composites can be the use of carbon nanoparticles of various types, including graphene (G) [1]. Composites with graphene have a number of unique electrophysical, mechanical and thermal properties [2-6]. Of particular interest is a composite based on aluminum nanopowder and small additives of multilayer graphene obtained by the SPS method [5].

The SPS method allows one to obtain a ceramic composite with a density close to the theoretical value and grain size close to the particle size of the initial powder. It is assumed that graphene flakes located at the grain boundaries of the ceramic matrix can cause a significant decrease in the fragility of the Al_2O_3 composite due to an increase in the adhesive properties of submicron elements [5]. Some works confirm this in bending tests under static compression [7]. However, there is practically no information on the features of the destruction of such compositions under extreme conditions of dynamic loading in the nanosecond region.

The aim of the work is to study the strength and fracture nature of the Al_2O_3 / G nanocomposite with an increase in the graphene content under conditions of a high-voltage discharge.

2 Experimental materials and methods

A composite based on dielectric Al_2O_3 nanoceramic (particle size 45 nm) with the content of conductive additives of five-layer flake graphene 0.5, 1.5, and 2.0 wt.% Was chosen as the studied material. The composite was obtained by mixing powders and subsequent sintering by the SPS method [7].

*Corresponding author: sav-georgij@yandex.ru

The samples were disks with a diameter of 15 mm and a thickness of 2.5 mm. The microhardness of the nanocomposite was measured on an FM-800 Future-Tech under a load of 300 g for 10 s. The flexural strength to failure was determined on a Quasar 50 machine using methods for brittle materials [8].

To study the dynamic strength of a nanocomposite to destruction by a high-voltage discharge, we used the solid dielectrics destruction method under conditions of surface breakdown by nano- and submicrosecond durations [9], which is based on the discovery [10]. The general block-diagram of the experimental setup is shown in Fig. 1.

As a source of voltage pulses, a short-pulse generator GKVI-300 was used. The generator allows you to receive rectangular pulses with voltage 100-300 kV, duration 30-300 ns, current amplitude 2 kA and energy 60 kJ. The distance between the electrodes corresponded to the thickness of the samples.

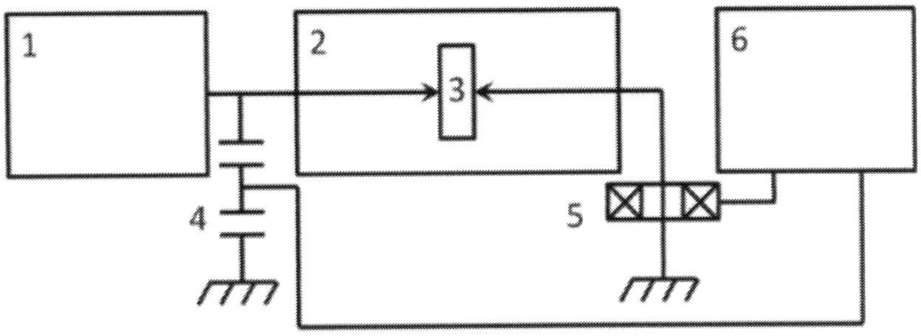

Fig. 1. Block-scheme of the experimental installation: 1 – generator, 2 – experimental camera with electrodes, 3 – sample, 4 – voltage meter, 5 – current meter, 6 – oscillograph.

In order to prevent breakdown on the sample surface, the destruction was carried out in a chamber filled with distilled water. In each experiment, a voltage pulse applied between the electrodes was monitored using a capacitive voltage divider, and the current passing through the sample was measured using a Rogowski belt. Signals from both devices were fed to a Tektronix TDS 2012 digital 2-channel oscilloscope.

The essence of electric-pulse destruction of disk samples is the formation of conducting plasma channels in the material, their expansion after they overlap the interelectrode gap, the generation of waves of predominantly tensile stresses and brittle cracking of the disk material [11]. Under single-pulse loading above the fracture threshold, the sample splits into two or more fragments. The closer to the fracture threshold, the fewer fragments and at the fracture threshold the number of fragments becomes equal to two.

The mechanical energy that directly goes to the destruction of the sample is part of the total electrical energy stored in the pulse generator, is determined by the relation [12]:

$$E_m = k_{em} E_{el} \tag{1}$$

where $k_{em} = 0.06 - 0.08$ is the coefficient of electrical energy conversion into mechanical energy. Electric energy is determined by the recorded waveforms of current and voltage.

3 Results and discussion

Microhardness and bending strength of composite ceramics are presented in table 1. It is seen that the microhardness of aluminum ceramics increases significantly with the use of

Al_2O_3 nanopowder and the introduction of graphene. As for the bending strength, it decreases sharply when switching from a large-sized powder to a nanosized powder.

However, the introduction of graphene up to 2% can partially compensate for the decrease in bending strength.

Table 1. Microhardness and bending strength of pure and composite Al_2O_3 ceramics.

Powder dispersity	Graphene, wt.%	Microhardness, GPa	Bending strength, MPa
polycrystalline	0	15.5 [13]	550 [10]
subcrystalline	0	22.1 [7]	200[7]
nanocrystalline	2	27.4 [7]	260 [7]

In Fig. 2 shows typical oscillograms of current and voltage during high-voltage breakdown of nanocomposites with different graphene contents.

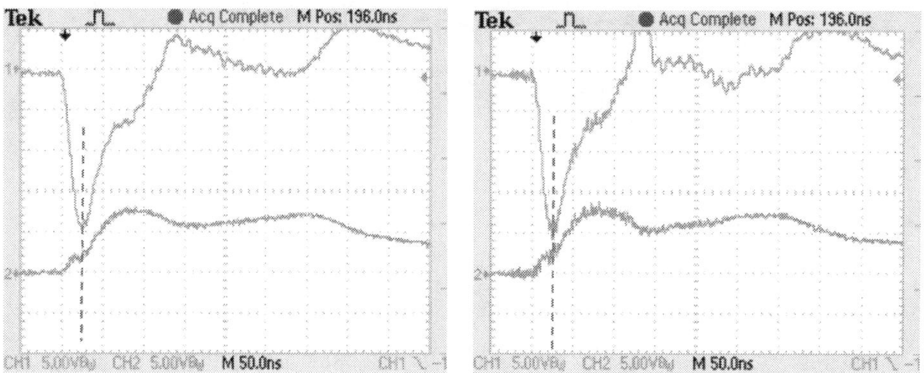

a b

Fig. 2. Typical oscillograms of voltage (1) and current (2) pulses: a – Al2O3 + 0,5% G; b – Al2O3 + 2,0% G. The dotted line marks the moment of electrical breakdown.

As the experiments showed, the breakdown voltage and the current at which it occurred turned out to be almost the same in all experiments, regardless of the graphene content. Since the thickness of the samples is quite small, their fragmentation took place far beyond the threshold of destruction. Therefore, the conditional criterion for the dynamic strength of the samples, which indicated the effect of graphene additives, was the number of fragments into which the sample was destroyed [14].

The constant value of the electrical breakdown is obviously due exclusively to the high power of the GKVI-300 voltage generator and the small thickness of the samples.

The number of fragments into which the samples were destroyed, depending on graphene content, significantly varied: at a content of 0.5% graphene, the number of fragments was 7– 9; at 1.5% – 11 – 14; at 2% – 17 – 20 (Fig.3). Only those fragments whose characteristic size was not less than 0.5 mm were counted. In this case, the mass deficit of the samples after their destruction was determined. In all cases, the mass deficit was not more than 1.5%.

Fig.3. Fragments of the destroyed samples: from left to right - 0.5%, 1.5%, 2.0% G.

Thus, with an increase in the graphene content, the number of fragments into which the samples are destroyed upon exposure to a high-voltage pulse increases sharply. The result obtained is consistent with the data of [15], in which the addition of copper oxide nanosized particles to the studied materials led to general embrittlement of the samples upon exposure to the products of an electric explosion of a conductor that was in the sample from an energy-storage phase material (paraffin or wax).

The results of the study showed that an increase in graphene content in the range of 0.5-2.0 wt. % contributes to an increase in the number of fragments upon destruction. A similar fact is typical for ceramics experiencing brittle fracture. Comparison of properties in Table 1 with the results of fragmentation during destruction also confirms this conclusion. Thus, the introduction of graphene into the nanocomposite sharply increases the microhardness, which generally indicates embrittlement of the material. In a nanocomposite, it is due to the inhibition of the formation of the main transcrystalline crack and its replacement by intergranular fracture.

The work has been supported by Russian Science Foundation [grant № 16-19-10213].

References

1. K. S. Novoselov, A.K. Geim, S. V. Morozov, D. Jiang, M. I. Katsnelson, I. V. Grigorieva, S. V. Dubonos, A. A. Firsov, Nature, **438,** 197 (2005)
2. H. Xia, X. Zhang, Z. Shi, Z. Zhao et al., Y. Li J. Wang G. Qiao, Mater. Sci. Eng., **639,** 29 (2015)
3. H. J. Kim, S.M. Lee, Y.S. Oh, Y.H. Yang, Y. S. Lim, D.H. Yoon, Ch. Lee, J.Y. Kim, R. S. Ruoff Sci. Rep., 2014, **4,** 5176 (2014)
4. A. Centeno, V. G. Rocha, B. Alonso, A. Fernández, C. F. Gutierrez-Gonzalez, R. Torrecillas, A. Zurutuza, J. Eur. Ceram. Soc., 33, 201(2013)
5. A. Borrell, R. Torrecillas, V. G. Rocha, A. Fernández, V. Bonache, M. D. Salvador, Wear. **274,** 94 (2012)
6. S. Stankovich, D. A. Dikin, G. H. B. Dommett, K. M. Kohlhaas, E. J. Jimney, E. A. Stach, R. D. Piner, S. T. Nguyen, R. S. Ruoff, Nature, 442, 282 (2006)
7. E. A. Klyatskina, A. Borrell, E. G. Grigoriev, A. G. Zholnin, M. D. Salvador, V. V. Stolyarov, J. Ceram. Sci. Technol., **9,** 215 (2018)
8. A. T. Tumanov, M.: Mashinostroenie, **320** (1974) in Russian
9. A. F. Usov, Vestnik Kol'skogo nauchnogo tsentra RAN, **4**, 166 (2012) in Russian
10. A. A. Vorobyov, G. A. Vorobyov, A.T. Chepikovet, Russian Discovery Certificate, A-122 (1998)
11. V. V. Lopatin, M. D. Noskov, G.Z. Usmanov, A.A. Cheglakov, Phys. Mesomechanics, **13** 89 (2010)
12. A. A. Lukin, V. A. Morozov, Yu.V. Sudyenkov, Bull. St. Petersburg State Uni. Ser. **2**, 133 (2008)
A. G. Zholnin, I. V. Kovaleva, M. S. Yurlova, A. M. Ilyina, E. G. Grigoriev, E. A. Olevsky, Phys. i Chim. Obr. Mater., **2**, 73 (2015) in Russian
13. M. Michálek, M. Michálková, et al, Ceram. Inter. **44**, 3255 (2018)
14. G. G. Savenkov, V. A. Morozov, A. A. Lukin, Tech. Phys. Lett., **42,** 1110 (2016)

Study of the inhibitory ability of inorganic di- and polyphosphate compositions

Aktoty Niyazbekova[1], *Timur Shakirov*[1*], and *Mayra Almagambetova*[1]

[1]Zhangir khan West Kazakhstan agrarian-technology University, Higher School of Petroleum, Gas and Chemical Engineering, 090000 Uralsk, Kazakhstan

Abstract. The article considers the anticorrosive properties of inorganic di-and polyphosphate compositions in relation to St-3 steel, depending on the pH of the medium, the nature and concentration of phosphate, and the nature of the modifier ion. The research was conducted using GOST-based methods: gravimetry, potentiometry, photocolorimetry, infrared spectroscopy and scanning electron microscopy. On the basis of experimental data, quantitative indicators of corrosion process were determined: the rate of corrosion process, the degree of protection, the depth index, the coefficient of inhibition and the assessment of the stability of the formed film on the ball scale of corrosion resistance against steel. The analysis of experimental data allows to establish the influence of the above factors on the corrosion processes in the systems under study. The experimental data are supplemented by thermodynamic calculations of the corrosion process parameters, the results of which correlate well with the kinetic data of the process under study. In the course of the research work, the analysis of corrosion deposits was also carried out. The regularities established during the work contribute to the creation of effective di-and polyphosphate inhibitors with the highest degrees of protection.

1 Introduction

The problem of corrosion is one of the most important in industry, transport and agriculture. Experts estimate the total annual cost of corrosion control at 3-5 % of the gross domestic product [1, 2]. According to research by the international Association of corrosion engineers (NACE), the cost of US corrosion control in 2011, for example, amounted to almost $ 1 trillion. The numbers are not comforting, and encourage large companies around the world to engage in thorough anti-corrosion protection.

Most of the costs are actually the cost of anti-corrosion protection, losses in case of equipment failure due to corrosion, as well as the cost of replacing damaged components and, of course, all sorts of accidents. One of the most common ways to reduce the level of corrosion losses in the operation of industrial equipment is the use of corrosion inhibitors [3, 4]. To date, considerable practical experience has been accumulated in their use. However, the difference in the corrosive aggressiveness of working environments and

* Corresponding author: shakirov_1985@mail.ru

changes in the operating conditions of equipment and structures at different stages of development put forward new requirements for the selection of inhibitors and improvement of the inhibitor protection technology [5].

Despite the wide range of reagents available, there is a constant search for new inhibitors and inhibiting compositions that can provide a comprehensive protective effect. In this regard, the relevance of this work is to find and implement complex inhibitors to ensure efficient and reliable operation of industrial equipment.

2 Research methodology

The purpose of this work is to study effective corrosion inhibitors of complex action.

The following were studied as inhibitors: sodium dihydrodiphosphate $Na_2H_2P_2O_7$, sodium diphosphate $Na_4P_2O_7$, sodium cyclotriphosphate $Na_3P_3O_9$, sodium cyclohexaphosphate $Na_6P_6O_{18}$. Trivalent ions of iron, aluminum, and lanthanum were chosen as modifier ions.

The corrosion test procedure was generally accepted [6, 7, 8]. Rectangular steel plates 30×20×3 mm in size were used. The duration of the experiments is 24-480 hours. The corrosion rate was estimated by the weight loss of the samples after 24, 48, 72, 96, 120, 240 hours.

Quantitative indicators of corrosion processes were calculated according to the formulas, the measurement uncertainty was estimated according to the algorithm, using the Student's coefficient with a confidence probability of 0.95 [9, 10].

In the course of the work, potentiometric determinations of the pH of corrosive media were carried out using a combined glass electrode and ionomer, the conductometric method, and photocolorimetric determination of the iron (III) content with potassium thiocyanate [11, 12]. And also analyzes of corrosion deposits were carried out on infra-red specroscopy and scanning electron microscopy.

3 Results and its discussion

36 multicomponent systems were constructed in order to conduct research on the inhibitory ability of phosphate compounds.

In diphosphate model systems, the passage of reactions is indicated by changes in the concentration of modifier ions, iron (III) ions, and phosphate ions. This relationship is shown in figure 1 for a system with a lanthanum ion, where the inhibitory effect is most effective.

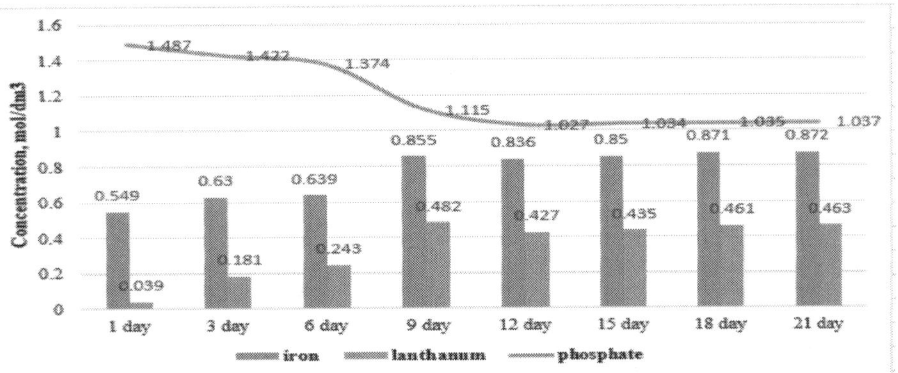

Fig. 1. The dependence of changes in the concentration of the ion of the modifier of lanthanum, iron, phosphate on time for sodium dihydrodiphosphate.

As shown in the figure, the plate dissolves and the iron passes into the solution under the influence of the medium and the system components. There is an increase in the concentration of iron in the solution in this system on the ninth day. The concentration of phosphate ions at the same time decreases. This indicates that there is a process of interaction between phosphate ions and iron and the formation of corrosive deposits. On the twelfth day, the concentration of iron, lanthanum and phosphorus ions decreases in this system. Apparently, a compound is formed in the system, which inhibits the process of dissolution of iron. Subsequently, the system again shows a slight increase in the concentration of all components of the system. Hence, it can be assumed that the formed compound does not have sufficient strength and degree of protection. The effectiveness of the inhibitory action is only 8.43%. This is the highest degree of protection observed among 18 diphosphate systems.

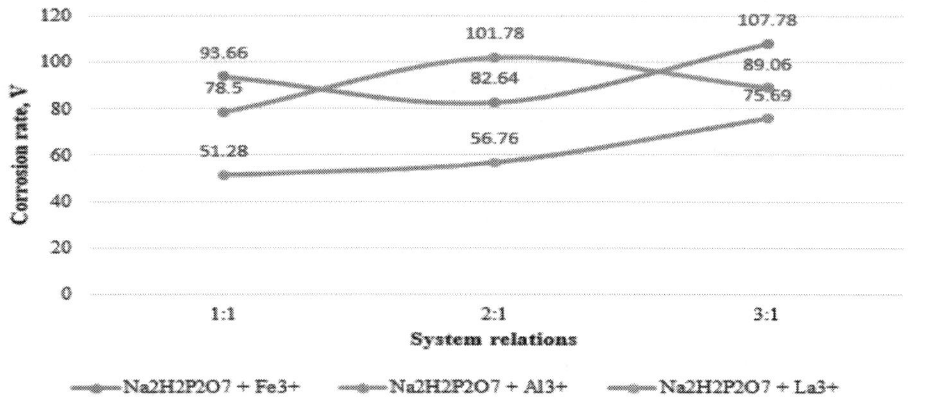

Fig. 2. Dependence of the corrosion rate on the nature of the modifier ion at a different ratio for sodium dihydrodiphosphate.

The change in the corrosion rate and the course of the dependence curve in the presence of modifier ions at different ratios for sodium dihydrodiphosphate is shown in figure 2. As can be seen from the figure, in the presence of iron ions, when the concentration of phosphate increases, there is a decrease in the corrosion rate at a ratio of 2:1, and at a ratio of 3:1, it increases back. For a system with an aluminum modifier ion, the inverse relationship is observed. In the system with lanthanum ion there is a gradual increase in the speed of the corrosion process with increasing concentration.

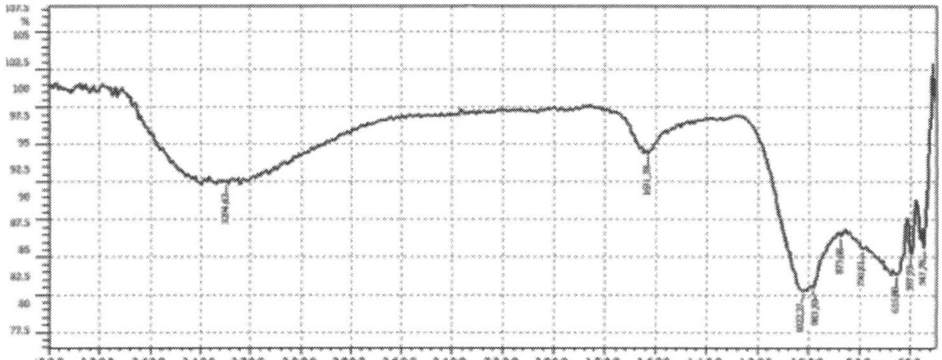

Fig. 3. Spectrogram of the corrosion deposition of the composition of sodium dihydrodiphosphate with lanthanum in a ratio of 1:1.

Figure 3 shows a spectrogram of the corrosion deposition of the sodium dihydrodiphosphate system with an ion modifier in a ratio of 1:1. According to the literature [13, 14], sodium dihydrodiphosphate is characterized by bands of 1270 cm^{-1} corresponding to asymmetric ν (P = O), and a band of 1092–990 cm^{-1} to symmetric ν (P = O) vibrations. The bands at 884 cm^{-1} correspond to asymmetric ν (P-O-P), and the bands at 792-595 cm^{-1} correspond to symmetric ν (P-O-P) vibrations. The figure shows that in the IR spectrum of the corrosion deposition peaks are formed in the range of 1030–980 cm^{-1}, as well as 1620 cm^{-1}, confirming that another unstable compound is formed.

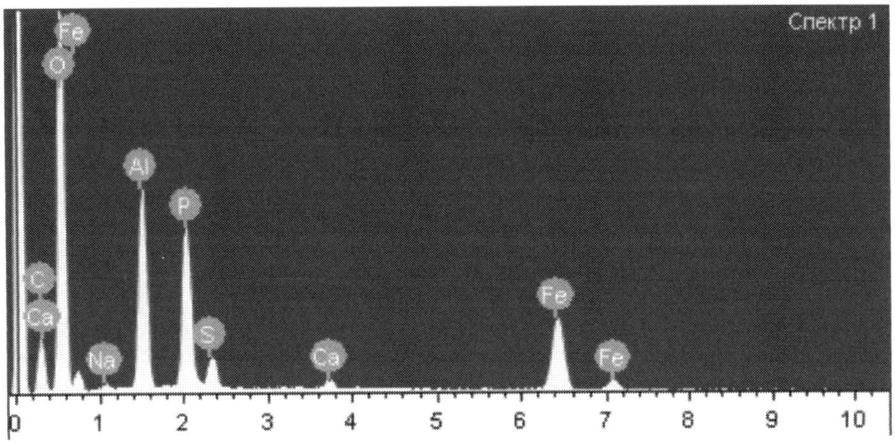

Fig. 4. Spectral scale of elements for corrosion deposition of the composition of sodium dihydrodiphosphate with lanthanum in a ratio of 1:1.

Analysis of corrosion deposits using a scanning electron microscope (Figure 4) showed that the corrosion deposits of this system contain oxygen - 46.58%, Fe - 13.36%, P - 7.92%. A small amount contains Na - 0.44%.

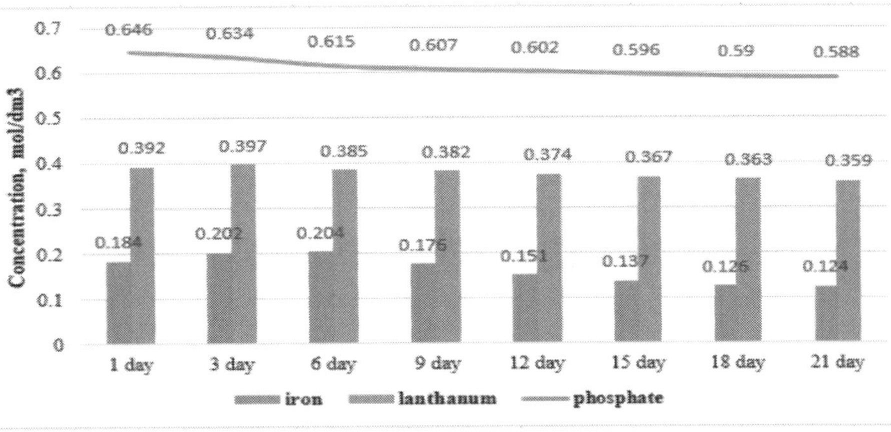

Fig. 5. The dependence of the change in the concentration of the ion of the modifier of lanthanum, iron, phosphate on time for sodium cyclotriphosphate.

Of the eighteen cyclophosphate systems, the sodium cyclotriphosphate system with the lanthanum modifier ion in the ratio 3:1 has the greatest inhibitory effect (Figure 5). The effectiveness of the inhibitory effect is 45.86%. From the above dependence it is seen that at the beginning of the experiment, a gradual increase in the concentration of iron and a decrease in the concentration of lanthanum and phosphate ions are observed. On the ninth day and thereafter, a gradual decrease in iron concentration is observed. At the same time, the concentration of lanthanum and phosphate ions decreases. This, possibly, indicates that a complex compound is formed, which reduces the transition of iron ions from the steel plate to the solution, and therefore reduces the corrosion process.

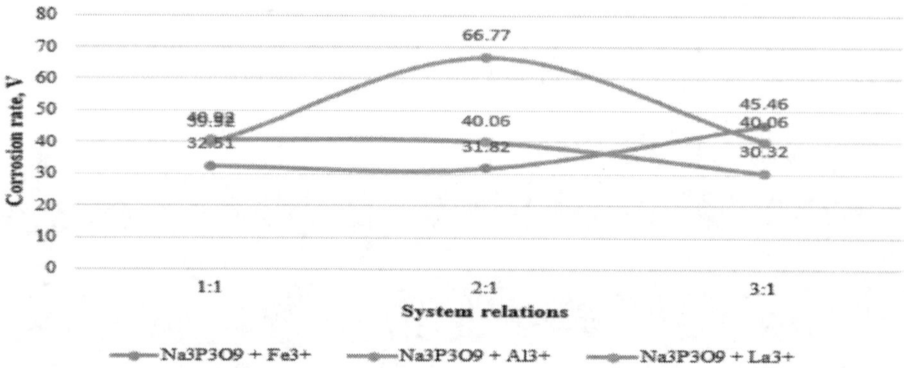

Fig. 6. Dependence of the corrosion rate on the nature of the modifier ion at a different ratio for sodium cyclotriphosphate.

As can be seen from Figure 6, as well as for diphosphates, cyclophosphates in the presence of iron ions with an increase in the concentration of phosphate are characterized by a decrease in the corrosion rate at a ratio of 2:1, and at a ratio of 3:1 it increases back. An inverse relationship is observed for a system with an aluminum modifier ion. In the system with the lanthanum ion, a gradual decrease in the rate of the corrosion process with increasing concentration is observed. As mentioned above, perhaps this is due to the formation of a complex compound, which reduces the corrosion process.

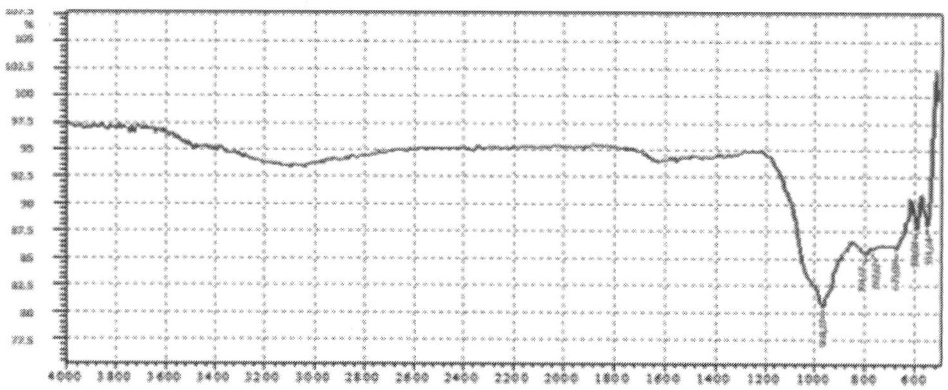

Fig. 7. Spectrogram of the corrosion deposition of the composition of sodium cyclotriphosphate with lanthanum in a ratio of 3:1.

Figure 7 shows a spectrogram of the corrosion deposition of the sodium cyclotriphosphate system with lanthanum in a ratio of 3:1. From literary sources it is known that for sodium hydrogen phosphate, the bands of 1225 cm^{-1} are characteristic for asymmetric ν (P = O), and the band for 1294-1262 cm^{-1} - for the symmetric ν (P = O) vibrations. The bands at 1168 cm^{-1} correspond to asymmetric ν (P-O-P), and the bands at 1162-1121 cm^{-1} correspond to the symmetric ν (P-O-P) vibrations. In the IR spectrum of the corrosion deposition, the peaks are not similar to the data from literature, which suggests that another compound was formed as a result, which forms a protective film.

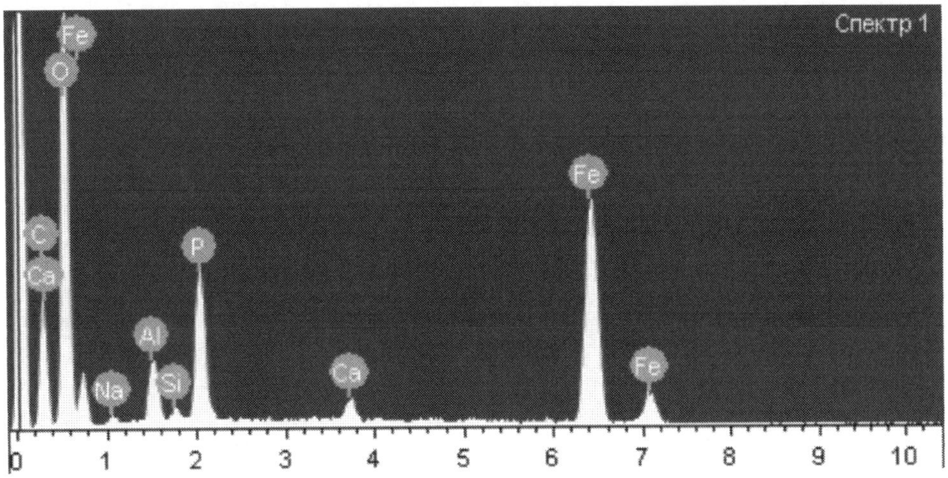

Fig. 8. Spectral scale of elements for corrosion deposition of the composition of sodium cyclotriphosphate with lanthanum in a ratio of 3:1.

Figure 8 presents an analysis of corrosion deposits on a scanning electron microscope and showed that oxygen - 36.97%, iron - 28.68%, phosphorus - 5.24% prevail in the corrosion deposits of this system. A small amount contains sodium, carbon.

Table 1. Basic thermodynamic indicators of multicomponent diphosphate and cyclophosphate systems.

№	Inhibitor + La^{3+} in the ratio 1: 1	- $\Delta G^0_{298.15}$, kJ/mol	ΔH, 10^{-5} kJ/mol	ΔS,10^{-3} kJ/mol	lgK
1	$Na_2H_2P_2O_7$	21.25*10^4	-12.71	713.36	37.34
2	$Na_4P_2O_7$	21.17*10^4	-1.51	750.56	37.28
3	$Na_3P_3O_9$	21.77*10^4	-10.95	730.64	38.24
4	$Na_6P_6O_{18}$	21.3*10^4	-4.71	714.91	37.42

Table 1 presents the thermodynamic characteristics of diphosphate and cyclophosphate systems with ion modifiers. The highest Gibbs energy for diphosphate systems is characteristic of the sodium dihydrogen diphosphate system.

Diphosphate ions are metastable in an aqueous solution and their protective effect under the experimental conditions is somewhat low [15, 16]. The mechanism of action of diphosphates is that, when the corrosion process proceeds in background media, according to literature data, they form relatively soluble complex compounds of the $Me[Me_2(P_2O_7)_2]$ type, which are stable in weakly acidic media pH = 3-5, which shifts the equilibrium of the corrosion process towards metal dissolution [17, 18].

$\Delta G^0_{298.15}$ of the cyclotriphosphate system is -21.77*10^4.This confirms that this composition has the greatest inhibitory effect among cyclophosphate compositions.

4 Conclusion

Based on the obtained experimental data, the following conclusions can be drawn:

1. An increase in the concentration of phosphate in most cases leads to a deceleration of the corrosion process and the pH is shifted to the slightly alkaline side in all the studied systems.

Diphosphate systems showed the lowest rates of corrosion rate and caused a shift in the pH to the acidic region, which is associated with the metastable state of the phosphates themselves [19, 20].

2. Quantitative indicators of the effectiveness of modified di- and polyphosphate inhibitors are determined. The degree of protection in the $Na_3P_3O_9 + La^{3+}$ system is 45.86% and the value of the corrosion resistance of the metal based on the value of the depth index of corrosion is 4 points (according to GOST 5272-90).

3. The experimental data were confirmed by thermodynamic calculations to determine the stability of the system in the presence of modified inhibitors. The highest value of Gibbs energy is also characteristic of the sodium tripolyphosphate system with an ion modifier lanthanum in a ratio of 3:1 ($\Delta G^0_{298.15} = -21.77*10^4$).

36 compositions were studied and based on the experimental data obtained, it was established that the most practically significant protective effect was shown by $Na_3P_3O_9$ systems with a concentration of 0.05 mol/dm^3 with La^{3+} and a concentration of 0.025 mol/dm^3 with Fe^{3+}.

Currently, Kazakhstan corrosion inhibitors are inferior to international standards in technological and economic parameters. The main task of the coming years is the creation in Kazakhstan of our own effective reagents and the expansion of industrial production, and their introduction into the national economy. The production of our own corrosion inhibitors will reduce the dependence on the import of expensive imported supplies, which necessarily require additional analytical quality control at the places of consumption.

The correct and effective use of the country's natural resources will contribute to sustainable economic growth and better people's lives.

The regularities established during the work expand and supplement the existing ideas about the protective effect of compositions based on inorganic phosphate compounds. The results obtained significantly contribute to the creation of effective, environmentally friendly and economically feasible steel corrosion inhibitors based on Kazakh raw materials. The data obtained make a certain contribution to the scientific field of effective inhibitory protection of metals.

References

1. E. McCafferty, Introduction to Corrosion Science. Springer, New York, NY, (2010).
2. J. Vosta and J. Eliasek, Study on Corrosion Inhibition from Aspect of Quantum Chemistry, Corros. Sci., **11**, 223 (1971)
3. N.P. ZHuk, *The course of the theory of corrosion and metal protection* (Moscow, Metallurgy, 1976)
4. G.G. Ulig, R.U. Revi, *Corrosion and the fight against it. Introduction to Corrosion Science and Technology* (Leningrad, Chemistry, 1989)
5. D.I. Bregman, *Corrosion inhibitors* (Moscow, Chemistry, 1966)
6. S.M. Reshetnikov, *Inhibitors of acid corrosion of metals* (Leningrad, Chemistry, 1986)
7. T.N. Vladimirskaya, M.L. Chepelevetsky, *Studies of complexation in the systems* Fe^{3+} - $P_3O^{5-}_{10}$ - H_2O *and* Fe^{3+} - $P_2O^{4-}_7$ - H_2O (In: Chemistry and technology of condensed phosphates, 1989)

8. A.B. Niyazbekova, N.V. Akatyev, G.K. Sulekeshova, T.A. Shakirov, *Chromatographic study of systems cyclotri-, cyclotetra- and cyclohexaphosphate with two and trivalent cations of p- and d-elements,* Technology «Materials of the vi international research and practice conference», Materials of the VI International research and practice conference Munich, Germany, pp. 186-194 (2013)
9. G.S. Kuanysheva, G.R. Makasheva Г.P., G. Kamalova, A.B. Niyazbekova, *Complexation of salts of some d-elements with diphosphate anion*, Bulletin of Karaganda State University, v. **1 (13)**, pp. 71-73 (1999).
10. D.E.C. Corbridje, *The structural of chemistry of phosphates*, Bull. soc. fr. miner. et. cristallog. v. **94**, pp. 271–299 (1971)
11. U.F. Zhdanov, *Chemistry and technology of polyphosphates* (Moscow, Chemistry, 1979)
12. A.V. Laserov, L.V. Kubasova, N.N. Chudinova, *Income and research and alkali metal phosphates,* v. **18**, pp. 127–131 (1982)
13. A. Durif, *The development of cyclophosphate crystal chemistry*, Solid State Sciences, v. **7**, pp. 760–766 (2005)
14. R. Pearson, *Hard and soft acids and grounds*, Successes of chemistry, v. **15**, pp. 1259–1271 (1974)
15. I.B. Tananaev, A.B. Lavrov, N.N. Chudinova, *The study of phosphates,* v. **33**, pp. 2041–2048 (1988)
16. N. Narayana, M. Rao, R. Gómez-García, A. Kornberg, *Inorganic Polyphosphate: Essential for Growth and Survival,* An- nual Review of Biochemistry, v. **78**, pp. 605–647 (2009)
17. M.Kh. Karapetov, S.I. Drakin, *General and inorganic chemistry* (Moscow, Chemistry, 1981)
18. Kura Genichiro, *Hydrolysis reaction of inorganic cyclophosphates at various acid strengths*, Polyhedron, v. **6**, pp. 531–533 (1987)
19. G.S. Kuanysheva, S.S. Dauletkalieva, *Experimental and quantum-chemical studies of complexes of iron (III) with diphosphate in solution,* // Mater. XXV Int. Chugaev Conf. in coordination chemistry, pp. 337-338 (2011)
20. A. Niazbekova, N. Akatyev, M. Mukasheva, A. Rakhova, *Quantum- chemical calculations of electronic structure of polyphosphate complexes of manganese, cobalt, copper and zinc*, Materials of the international research and practice conference «European Science and Technology», Wiesbaden: Germany, pp. 82-85 (2012)

Continuous production of multicomponent powder mixtures

Wasem Mansour[1], *Denis* Melekhin[2], and *Alexander* Pasko[3]*

[1]Post-graduate student from the Syrian Arab Republic, Tambov State Technical University, 106 Sovetskaya Street, 392000, Tambov, Russian Federation
[2]Architecture Faculty, Czech Technical University in Prague, 1903/4 Zikova, 166 36 Prague 6 Czech Republic
[3]Department of Technology and Methods of Nanoproducts Manufacturing, Tambov State Technical University, 106 Sovetskaya Street, 392000, Tambov, Russian Federation

Abstract. Mixing of powders is an important process in various industries, from chemical, pharmaceutical to food and construction. The main characteristic of the mixture is its homogeneity, which is usually estimated by the coefficient variation and the lower this ratio, the better the mixture. Typically, particles of the components of the mixture differ in size and / or density. These differences cause segregation, i.e. the formation of zones in which there is a high concentration of one of the components and this affects the quality of the mixture. However, through the use of special technological methods, the negative result of segregation can be transformed into a positive one and not only reduce the mixing time, but also improve the quality of the mixture. This article discusses one of the options for the production of a multicomponent mixture with a specific schedule for loading components. It is shown that by organizing the correct sequence of loading components, you can control the mixing process. Two-step technology is used to feed components into the mixer.

1 Introduction

Mixing of powders is an important process in various industries, from chemical, pharmaceutical to food and construction. The main characteristic of the mixture is its homogeneity, which is usually estimated by the variation of the coefficient. The greater the coefficient of heterogeneity, the worse the quality of the mixture. In addition, when implementing mixing processes, it is necessary to minimize the process time and energy consumption to a minimum. Often the mixing process is complicated by the presence of liquid components, which requires other technologies and mixer designs. For example, in the production of concrete, graphene structures are introduced in the form of an aqueous suspension [1]. A similar situation arises when modifying epoxy resin [2] or lubricants with graphene nanostructures [3-5]. Sometimes the mixing process directly determines the quality of the finished product.

*Correspondingauthor:ttmp2005@yandex.ru

Carbon black is an important component of the electrodes of lithium-ion batteries, as it significantly improves the electrical conductivity in the coating. It is shown in the paper [6] that the structure of soot in the electrodes of a lithium-ion battery can be precisely controlled by careful changes in parameters in the upward dry mixing process. Significant changes in important properties of electrodes, such as mechanical, structural and electrical characteristics are achieved and can be described as functions of the process parameters "Mixing time" and "mixing intensity". Thus, the proper organization of the mixing process can improve the performance of the finished product.

In addition to the large difference in the percentage of individual components in the mixture, often there is a large difference in the particle sizes of the various components. For example, in the production of a multicomponent powder mixture Fe - Cr - Co - Ni – Mn[7], the following materials were used as starting components: radio engineering carbonyl iron powder (R-10 with an average particle size d = 3.5 μm), nickel powder (NPE-1 , d = 150 μm), cobalt powder (PK-1u, d <71 μm), chromium powder (PH-1M, d <125 μm) and manganese powder (MR0, d <400 μm). In this case, the particle size differs by more than 100 times. Such a difference in size causes active segregation of these components in the mixer, i.e. particles with a maximum size are concentrated in one zone of the mixer, and particles with a minimum size in another zone of the mixer and this greatly reduces the quality of the finished mixture. To solve the problem of separation of the prepared mixture, the mixing process was carried out in an AGO-2 mechanical activator with water cooling, using steel balls with a diameter of 9 mm and acceleration in air of 90 g. The best results were obtained with a mixing (alloying) time of 90 minutes.

Majority of the active ingredients and adjuvant used for the manufacture of pharmaceutical dosage forms constitutes in the powder form, the market for solid dosage forms involving powder processing is large and is growing day by day [8]. Handling and processing of powders, in particular feeding and mixing is central to pharmaceutical operations. But it poses numerous problems due to their unpredictable and irregular behavior. The article deals with sequential approaches to be enforced by the pharmaceutical industries to draw out new technical responses to change with respect to mixing.

Scientists and engineers are constantly creating new mixer designs. For example, based on an analysis of methods for mixing bulk materials and mixer designs, the method of continuous mixing of bulk materials and the design of a spiral mixer are justified [9]. The main idea is that a multilayer component stream is formed in the desired ratio, followed by dividing the stream into small portions and mixing the components into portions. In our opinion, the authors of this work did not pay enough attention to the dosing accuracy of the component, but if there is no exact ratio of the components, it is impossible to prepare a good mixture. By the way, when modeling the mixing process, the dosing errors of the components are also not taken into account.

In recent decades, for the production of new structural and functional materials, not only micro, but also nanoscale components have been used. For example, in the manufacture of catalysts for the production of gasoline, a mixture of iron and platinum particles with sizes less than 100 nanometers is used.

In this article, we analyze the possibility of preparing multicomponent mixtures with a specific schedule for loading components using two-step feeding technology.

2 Multi-component mixing technology

Actually, the components of the mixture have a difference in size and/or density, which causes particle segregation. The result of segregation is the formation of several zones, in each of which one of the components prevails. Small differences in either size or density lead to flow-induced segregation, a complex phenomenon without parallel in fluids [10].It is believed that segregation negatively affects the mixing process, however, in some cases; the effect of segregation can be used to intensify the mixing process and improve the quality of the finished mixture. An analysis of articles on mixing bulk materials [10-16] shows that, regardless of the type of mixer, the process of mixing particles has a deterministic-stochastic nature. At the same time, depending on the type of mixer, either the deterministic component or the stochastic component prevails.Obviously, the deterministic mixing process is easier to control than the stochastic mixing process. A striking representative of mixers in which deterministic particle motion can be observed, and therefore deterministic mixing, is a drum mixer, i.e. horizontal rotating cylinder, inside which a mixture of different particles moves. It is best when the bulk material moves in a rolling / cascading mode. In this case, the particles move in a closed loop. Part of the particles that are closer to the drum rise up, and the other part of the particles rolls down. Numerous experimental studies show that if the particles of the two components differ in particle size or density, then regardless of how to initially load these particles into the drum, as a result of the rotation of the drum, small or heavy particles form a region in the centre of the circulation circuit, i.e. particle segregation will occur.

In order to convert the negative effect of the segregation effect into a positive one, a specific procedure for loading components into a batch mixer is proposed [17-19]. The idea is very simple and understandable, it is necessary to load the particles of a specific component into a zone that is diametrically opposite to that zone to which this component will be located as a result of a long rotation of the drum. Let us consider the schedule for loading components into a continuous drum mixer. Suppose that we need to prepare a mixture of three components A, B and C, and the particles of these components have different diameters; LJJ.Figure 1 shows a diagram of the loading of components into a drum mixer and a number of cross-sections of the mixer, which show the characteristic distribution of the components.

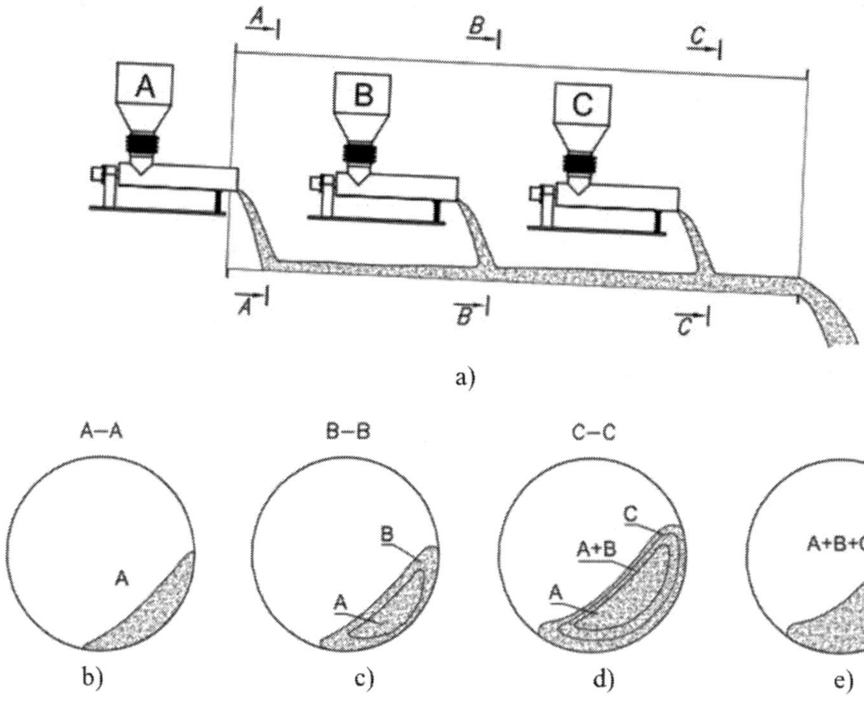

Fig. 1. Scheme of loading components into a drum mixer.

The loading of components into the mixer is as follows. First of all, component A is loaded, since it has the largest particles. The rotation speed of the drum is selected so that there is a circulation mode of movement of the bulk material, as shown in figure 1b. Then there is a continuous loading of component B, as shown in figure 1c. Component B is distributed evenly along the external boundary of the circulation circuit and the process of mixing components A and B begins. After that, component C is continuously loaded, which is initially evenly distributed at the external boundary of the circulation circuit. As noted above, since the particle size of component B is smaller than the particle size of component A, these particles will move toward the center of the circulation circuit. By the time component C is loaded, partial mixing of components A and B will occur, as shown in Figure 1e. Since the particles of component C are the smallest size, they will begin to move toward the center of circulation. At the outlet of the mixer, a mixture is formed in which components A, B and C are evenly distributed. The components are loaded similarly if they differ in density. First you need to load the lightest component, and then a heavier one and lastly you need to load the heaviest component. This is a general strategy for loading components, but the location of the dispensers of components A, B and C along the axis of the drum depends on the ratio of the diameters and/or densities of these components, as well as the percentage of components in the mixture.

The most difficult problem is solved when the particles of the components differ not only in particle sizes [20, 21], but in the densities of the materials of these particles, and large particles are also the heaviest. Unfortunately, in the scientific literature, we did not find reliable dependences for determining the tendency of particles to segregate while simultaneously differing in the diameters of these particles and in the densities. So far, the only reliable option is to conduct research on a laboratory drum in a batch mode.

3 Experiment, Results, and Discussion

We conducted experimental studies on a laboratory drum with a diameter of 400 mm and a length of 50 mm, since it was previously shown that the continuous mixing process in a drum mixer can always be adequately implemented in a batch mixer. Indeed, the movement of bulk material along the axis of a drum that operates in continuous motion is the movement of material at different points in time in a drum that operates in a batch mode. The main condition is to determine the degree of filling the material with material, since this degree decreases from the beginning of the drum to the discharge edge. This problem was solved analytically and experimentally verified [22, 23].

This setup consisted of a drum with a transparent front end wall so that it was possible to observe the movement of the material and the distribution of components in the volume of the mixture. The components were fed into the drum by a device that implemented a two-stage dosing technology. The essence of this technology is that at the first stage, the batch dispenser forms separate batches of material with a given weight and at regular intervals feeds these batches into a cylindrical tray, which performs circular oscillations. At the second stage, under the influence of vibration, individual portions are converted into a continuous stream and enter a rotating drum. This design allows you to accurately simulate the operation of a continuous mixer.

First of all, we checked the initial hypothesis: regardless of the component loading, small or heavy particles are collected in the centre of the circulation circuit. Figure 2a shows the loading of millet in the outer zone of the circulation circuit, which was formed by particles of polyethylene in the cross section of a rotating drum. Figure 2b shows the distribution of components after 3 minutes of rotation of the drum. Despite the fact that the zone in which the millet particles were concentrated is not pronounced, it is seen that most of the millet particles are located near the centre of circulation.

a) b)

Fig. 2. Distribution of the key component in a rotating drum: a – immediately after loading; b – 3 minutes after loading.

In real experiments, we used a transparent end wall with a large number of holes. These openings were covered with lids, but they were opened for sampling. Through these holes, samples of the mixture were taken and the number of individual components in each sample was determined. The quality of the mixture was evaluated by the coefficient of variation, which for mixtures is usually called the coefficient of heterogeneity:

$$k_c = \frac{100}{\bar{c}} \sqrt{\frac{1}{n-1} \sum (c_i - \bar{c})^2} \quad (1)$$

where k_c is the heterogeneity coefficient,%; $\bar{c}$ is the average concentration of the key component in the sample; c_i concentration in i sample; n is the number of analyzed samples.

As you know, the lower the coefficient of variationis, the better the mixture is, i.e. the more uniformly all components are distributed throughout the volume of the mixture.

As components, we used materials with different particle sizes and different colors, which facilitated the determination of the concentration of individual components in the samples. For example, we used: component A – polyethylene granules with a particle diameter of 1.5 mm to 3 mm (black); component B – millet with a particle diameter of 1 mm (yellow); table salt with a particle diameter of 0.5mm (white). In addition, we used river sand of various sizes. The concentration of any fraction was easily determined using a sieve analysis.

As an example, figure 3 shows the variation of the coefficients of variation for mixtures A + B and A + C. All components are sand. The particle size of component A is from 0.15 to 0.20 mm, component B is from 0.08 to 0.09 mm, and component C is from 0.04 to 0.045 mm.

As you can see the minimum value of the coefficient of variation for the mixture A + C occurs faster than for the mixture A + B. This is because the difference in particle sizes of components A and C is two times larger, the difference in particle sizes of components A and B. It should be noted that after reaching the minimum value of the coefficient of variation, it begins to increase and the quality of the mixture becomes worse. In the final result, there will be almost complete separation of the components in the sub-zones, as shown in figure 2b.The change in the coefficient of variation for the mixture B + C is not shown, because the dependence is fully consistent with the mixture A + B.

The figure 4 shows the change in the coefficient of variation for the mixture A + B + C.

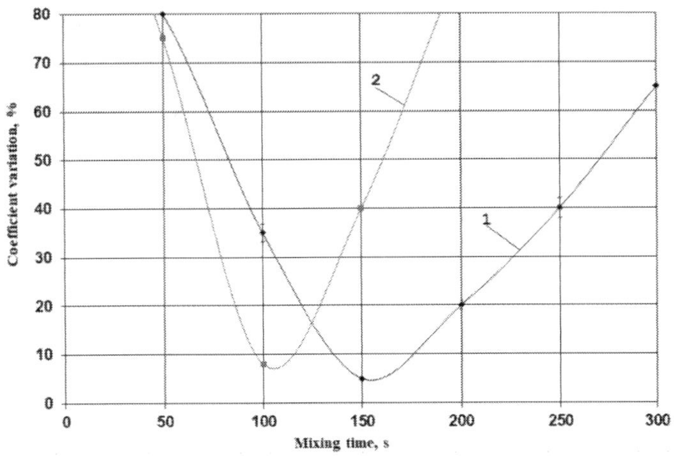

Fig. 3. Change in the coefficient of variation over time: 1 - a mixture of components A and B; 2 - a mixture of components A and C.

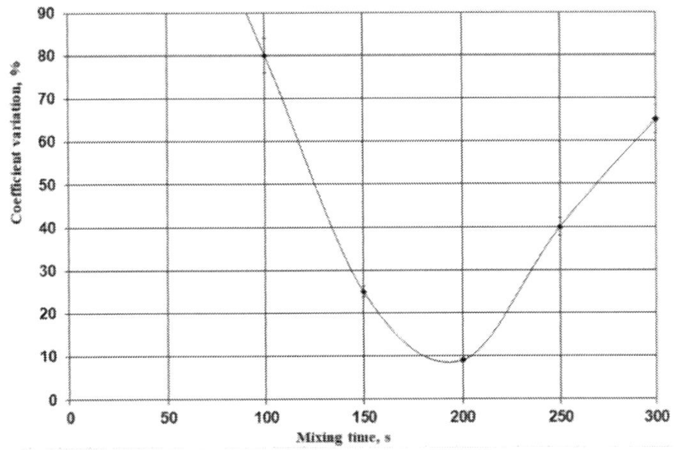

Fig. 4. Change in the coefficient of variation over time of a mixture of components A, B and C.

By nature, the curve shown in figure 4 is similar to the curves shown in figure 3. The curves differ in the minimum values of the coefficient of variation and the time in which these minimum values are reached. From a practical point of view, it should be noted that these two values depend only on the ratio of the sizes of the components and their percentage in the mixture, but also on the intensity of the supply of these components to the mixer.

The graphs show that the coefficient of variation decreases first (the components are evenly distributed over the entire volume of the mixture) and then increases (small particles gather near the center of circulation. The larger the difference between the particle sizes is, the faster the process goes. You can see that when mixing components A and B (particles B are approximately 2 times smaller than particles A), the minimum coefficient of variation is achieved in more than 150 s, and when components A and C are mixed (particles C are approximately 4 times smaller than particles A), the minimum coefficient of variation is attained in less than 100c. A similar conclusion can be drawn regarding the segregation rate. The mixture of components A and B almost completely segregate approximately 400c, and a mixture of components A and C is segregated in approximately 250s.

It should be especially noted that in the production of multicomponent mixtures it is very important that the components are accurately feeding. Currently, the most commonly used feeders, which implement the "Weight Loss" technology. Despite the claims made by the manufacturers of these feeders that they have very high metering accuracy, this is not so. The reason is that the weighing of the material is carried out in motion and dynamic effects on the weight sensor reduces the accuracy of feeding.

Figure 4 shows the change in the coefficient of variation in the mixture of components A, B and C. In this case, the concentration of component B was 10%, and component C 5%. We evaluated the quality of the mixture by the coefficient of variation of component C. The components were loaded according to the scheme shown in Figure 1a.

For continuous feeding of components, as noted above, we used the technology of two-step feeding [24-27]. This technology has been studied in sufficient detail, but for the organization of the continuous production of multicomponent mixtures, additional theoretical and experimental studies are needed. Despite the fact that there are dependencies for determining the parameters of material motion [28-31] and various models of the process of mixing bulk materials [32-36], including in a rotating drum [37-40], these questions require further theoretical and experimental studies. Particular attention should be paid to the relationship of component dispensers and a drum mixer in order to make the best possible industrial use of the technology proposed in this paper.

4 Conclusion

The negative effect of segregation can and should be used to reduce the mixing time of components that differ in particle size and / or density, as well as to improve the quality of the finished mixture. The main condition is the correct loading of the components in the mixer. Particular attention should be paid to the accuracy of continuous dispensing of components and the placement of dispensers along the length of the mixer. In addition, to calculate the parameters of industrial plants, it is necessary to develop mathematical models of joint processes for the continuous supply of components and their mixing in a continuous mode.

References

1. V. Pershin, K.A. Al-Shiblawi, A.M.R. Al-Mashhadani, A. Pasko, D. Melekhin, IOP Conf. Ser.-Mat.Sci.**693(1)**, 012046 (2019)
2. K.A. Al-Shiblawi, V.F. Pershin, V.P. Jarcev, T.V. Pasko, AIP Conf. Proc.**2041**, 020015 (2018)
3. V. Pershin, G. Zhumagalieva, A. Tkachev, A. Pasko, N. Memetov, AIP Conf. Proc.**2101**, 020011 (2019)
4. V. Pershin, K. Ovchinnikov, Z. Al-Hilo, N. Memetov, A. Tkachev, E. Galunin, AIP Conf. Proc.**2041**, 020016 (2018)
5. V.F. Pershin, K.A. Ovchinnikov, A.A. Alsilo, R.A. Stolyarov, N.R. Memetov, Nanotechnol. Russ.**13(5-6)**, 344-348 (2018)
6. H.Bockholt, W.Haselrieder, A.Kwade, Powder Technol.**297**,266-274 (2016)
7. N.A. Kochetova, A.S. Rogacheva, A.S. Shchukina, S.G. Vadchenkova. I.D. Kovaleva, Russ. J. Non-Ferr. Met.+,**60(3)**, 268-273 (2019)
8. R.Deveswaran, S.Bharath, B.V.Basavaraj, S. Abraham , S. Furtado, V.Madhavan, Research J. Pharm. and Tech.**2(2)**, 245-249 (2009)

9. I. Dudarev, R. Kirchuk, Y.Hunko, S.Panasyuk,*Modeling of Mixing Bulk Materials*,in Proceedings of the 2nd International Conference on Design, Simulation, Manufacturing: The Innovation Exchange, DSMIE-2019, 11-14 June 2019, Lutsk, Ukraine (2019)
10. J.M. Ottino, D.V. Khakhar,Annu. Rev. Fluid Mech. **32**, 55-91 (2000)
11. O.V. Dyomin, V.F.Pershin, D.O.Smolin, Chem. Petrol. Eng.+,**48(3-4)**, 213-216 (2012)
12. V.F. Pershin, Yu.T.Selivanov, Theor. Found. Chem. Eng.**37(6)**, 629-636 (2003)
13. V.F. Pershin, Theor. Found. Chem. Eng. **23(3)**, 249-255 (1990)
14. Yu.T. Selivanov, V.F.Pershin, Chem. Petrol. Eng.+,2, 7-10 (2004)
15. Yu. Selivanov, V.F.Pershin, Theor. Found. Chem. Eng. **38(1)**, 99-101 (2004)
16. Yu. Selivanov, V.F. Pershin, Theor. Found. Chem. Eng.**38(1)**, 103-105(2004)
17. V.F. Pershin, Theor. Found. Chem. Eng.**38(1-2)**, 53-57 (2002)
18. V.F. Pershin, Yu.T.Selivanov, A.V.Orlov, Chem. Petrol. Eng.+,**39(1-2)**, 60-65 (2003)
19. V.F. Pershin, Yu.T. Selivanov, Chem. Petrol. Eng.+,**40(1-2)**, 77-82 (2004)
20. V.F. Pershin, Yu.T. Selivanov, Theor. Found. Chem. Eng. **35(2)**, 206-208 (2001)
21. V.F. Pershin, Yu.T. Selivanov, Theor. Found. Chem. Eng.**35(1)**, 86-89 (2001)
22. Yu.T. Selivanov, V.F. Pershin, Chem. Petrol. Eng.+, **2**, 7-10 (2004)
23. V.F. Pershin, Yu.T. Selivanov, Theor. Found. Chem. Eng.**37(6)**, pp. 629-636 (2003)
24. T. Alsayad, V.Pershin, A.Pasko, T. Pasko, AIP Conf. Proc.**1899**,060011 (2017)
25. V.F. Pershin, A.M.Vorob'ev, V.M.Nechaev, A.A.Pas'ko, T.K.K.Alsaiyad, Chem. Petrol. Eng.+, **54(5-6)**, 386-391 (2018)
26. N.A. Alsayyad, V.Pershin, A.Pasko, T.Pasko, J.Phys.Conf. Ser.**1084(1)**, 012005 (2018)
27. T. Al-Sayyad, V.Pershin, A.Vorobiev, E.Galunin, AIP Conf. Proc.**2041**,020001 (2018)
28. V.F. Pershin, Chem. Petrol. Eng.+,**22(11)**, 606-608 (1986)
29. V.F. Pershin, Chem. Petrol. Eng.+,**24(6)**, 261-264 (1988)
30. V.F. Pershin, Theor. Found. Chem. Eng.**22(2)**, 196-201 (1988)
31. V.F. Pershin, G.A.Minaev, Theor. Found. Chem. Eng.**23(5)**, 432-435 (1990)
32. R. Liu, Z. Zhou, R. Xiao, A. Yu, Powder Technol.**361**, 315-325 (2020)
33. R.Cai, Z.Hou, Y. Zhao, Powder Technol. **352**, 193-208(2019)
34. F.Göbel, S.Golshan, H.R.Norouzi, R.Zarghami, N.Mostoufi, Powder Technol. **346**, 171-179 (2019)
35. M.Kashif, S. Muhammad, S.Siraj, Powder Technol.**344**, 617-627 (2019)
36. H. Fan, D.Guo, J. Dong, X. Cui, M. Zhang, Z. Zhang, Powder Technol. **327**, 223-231 (2018)
37. V.F. Pershin, Theor. Found. Chem. Eng.**20(4)**, 324-329 (1986)
38. V.F. Pershin, Yu.T.Selivanov, Theor. Found. Chem. Eng. **37(6)**, 629-636 (2003)
39. V.F. Pershin, Yu.T.Selivanov, A.V.Orlov, Theor. Found. Chem. Eng.**38(2)**, 209-215 (2004)
40. V.F. Pershin, Theor. Found. Chem. Eng. **23(3)**, 249-255 (1990)

Effect of sol-gel solution parameters on the formation of films on the surface of carbon fibers

Rida Gallyamova[1,*], *Rustam* Safiullin[1], *Vladimir* Dokichev[1], and *Fanil* Musin[1]

[1]Ufa Institute of Chemistry UFRS RAS, 450054, Ufa, Russia

Abstract. In this study to apply an oxide film to the surface of carbon fibers a sol-gel solution was used. Sol-gel solutions were prepared according to the described procedure from tetraethoxysilane (TEOS) and ethanol at various H2O concentrations. The molar ratio H2O/TEOS ratio was changed in the range from 2 to 20. After coating and drying, the surface morphology of carbon fibers was investigated. The relationship between the concentration of sol-gel solutions and the amount of deposited SiO2 on the surface of carbon fibers was analyzed.

1 Introduction

In recent years, the sol-gel method has been one of the most promising methods for producing oxide films. Using the sol-gel method, it is possible to obtain thin ceramic films not only on various kinds of fibers, but also on any surfaces of various materials [1,2].

It is of interest to use ceramic films as diffusion barriers in metal matrix composite materials to prevent chemical interaction at the fiber/matrix interface, such as carbon/aluminum alloy [3,4]. Due to their high chemical resistance, oxide films should improve the wettability of the carbon fiber and matrix, and prevent the formation of carbides, which adversely affects the properties of the entire composite.

The properties of the films primarily depend on the physicochemical characteristics of the sol-gel solution. To obtain films uniform in composition, the sol-gel solution must contain a certain ratio of the initial film-forming substance, solvent, and catalyst. Recent works [5–9] showed that hydrolysis and polycondensation processes are accompanied by a change in the viscosity of the sol-gel solution, which is an important criterion for the formation of uniform films on the fiber surface. In [8], the study of the sol-gel solution viscosity from its lifetime was performed and some regularities of the formation of films when they are applied to the surface of carbon fibers were determined.

At the same time, one of the main factors affecting the sol-gel process is the concentration of water in the solution. In [10, 11], it was found that sols with a low water concentration contribute to the formation of a linear structure — long chains of macromolecules, which are interconnected by Van der Waals forces, which is preferable for producing oxide fibers. At a high concentration of water in the sol, a three-dimensional

* Corresponding author: rida_gallyamova@mail.ru

structure of colloidal particles is formed, which is preferable for producing glasses. It is important to note that the ratio of the components in the sol-gel solution should, on the one hand, provide hydrolysis and condensation while preserving the resulting hydrolysis products in the form of a sol, and, on the other hand, the final hydrolysis in a thin layer of the obtained product when are applied an oxide film to the surface of carbon fibers.

In this work, we studied the effect of water concentration on the properties of a sol-gel solution and the formation of a SiO_2 film on the surface of carbon fibers.

2 Materials and method

A solution of tetraethoxysilane (TEOS) in ethanol as a source of silicon dioxide was used. The sol-gel solution was prepared according to the following procedure: a mixture of TEOS and ethanol solutions in a ratio of 1: 0.5 was intensively mixed for 30 minutes. A solution containing water and ethanol (another part) acidified with a 1M HCl solution was slowly added dropwise to the first mixture. The resulting solution was stirred for 30 minutes. In this study, various sols with a molar ratio of H_2O/TEOS (MR) in the range from 2 to 20 were used (table 1).

Table 1. The composition of the sol-gel solution at various concentrations of water.

MR	TEOS, ml	EtOH, ml	H_2O, ml	HCl, ml
2	22	23,3	3.6	0.13
4			7.2	
6			10.8	
8			14.4	
10			18.0	
20			36.1	

The kinematic viscosity of the sol-gel solution was evaluated at room temperature using a capillary viscometer. The calculation was carried out according to the following formula:

$$\upsilon = \frac{g}{9{,}807} \cdot \tau \cdot K, \qquad (1)$$

where υ is the kinematic viscosity of the solution, mm^2/s; K = 0.01060 mm^2/s^2, viscometer constant; τ is the time of expiration of the solution through the capillary, s; g - acceleration of gravity at the measurement site in m/s^2.

In order to determine the amount of SiO_2 formed at different molar ratios of H_2O/TEOS, films were deposited on the surface of carbon fibers. In this study, carbon fibers from a PAN precursor of the brand UMT49-12K-EP (Russia) were used; the average diameter of one filament is 5.5 μm. Carbon fibers were preliminarily subjected to heat treatment at a temperature of 400 °C for 20 minutes to remove polymer size from the surface [12]. Carbon fibers were immersed in the obtained sol-gel solution and held for 15 seconds. Then, the coated fibers were dried in air for 30 minutes and then subjected to heat treatment at a temperature of 500 °C (the heating rate to a given temperature was 2.5 °C/min). The amount of precipitated SiO_2 was determined by burning carbon fibers and measuring the oxide residue. The surface morphology of carbon fibers was studied using a scanning electron microscope (JEOL, Japan).

3 Results and discussion

3.1 Effect of water concentration on the viscosity of the sol-gel solutions

The effect of water concentration on the processes occurring in the sol-gel solution was evaluated by measuring the kinematic viscosity of the resulting solutions. Dependence of the kinematic viscosity at various molar ratios of H_2O/TEOS is shown in fig. 1. An increase in water concentration leads to an increase in the viscosity values of the sol-gel solution. This is primarily due to the fact that an increase in water concentration accelerates hydrolysis processes. At the same time, a slight decrease in viscosity is observed above MR = 8 (at MR = 10 – 3.6 mm²/s) and a further increase in MR does not lead to sharp changes in viscosity. At low values of water (MR = 2), partial hydrolysis of TEOS takes place, which suggests a low concentration of particles and the corresponding low viscosity values. An increase in viscosity with increasing water concentration indicates an acceleration of gelation processes. Previous kinematic viscosity studies [8] showed that the sol is in a stable state in the range of viscosity values of 3.6-3.7 mm²/s and is preferred for applying films to the surface of carbon fibers.

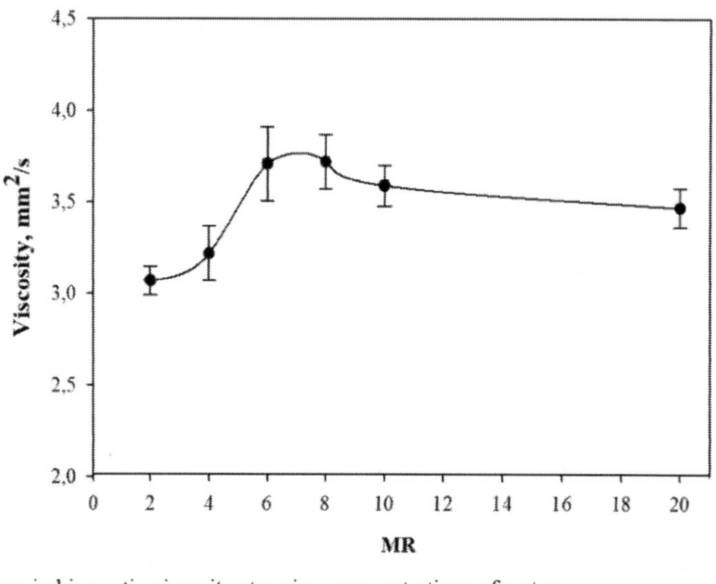

Fig. 1. Change in kinematic viscosity at various concentrations of water.

Thus, the use of low concentrations of water (MR below 6) in the sol-gel solution during film deposition leads to an excess of TEOS in the solution and, consequently, to the appearance of cracks during the drying of the film. A higher concentration of water in the solution, with MR values above 6, contributes to the formation of crack-free SiO_2 films on the surface of the carbon fiber.

3.2 Effect of water concentration on the formation of SiO_2 films

To estimate the amount of SiO_2 formed at different molar H_2O/TEOS ratios were deposited film on the surface of carbon fibers. The surface of carbon fibers with a deposited SiO_2 film is shown in Fig. 2. In the initial state, the surface of the fiber subtle longitudinal bands are observed Ffig. 2 a). After deposition of the oxide film, electron microscopy studies showed that at MR = 6 (Fig. 2 b) and above, the SiO_2 film is applied uniformly on the surface of carbon fibers (the arrows indicate the SiO_2 film boundary). This is due to the fact that high concentrations of water contribute to the complete hydrolysis in the sol-gel solution, which does not complicate the deposition of SiO_2 on the fiber surface.

Low water concentrations (at MR = 2 and MR = 4) SiO_2 film is cracked and peeled off. This is due to the fact that low concentrations of water slow down the hydrolysis processes, that is, i.e. it extends partially. In turn, partial hydrolysis indicates an excess of TEOS, which evaporates during the heat treatment and leads to cracking in the oxide film. This is the reason for the smaller amount of deposited silicon dioxide on the surface of carbon fibers (Fig. 3).

a b

Fig. 2. The surface of carbon fibers: a) the initial fiber without SiO_2 film; b) with a SiO_2 film at MR = 6 (arrows indicate the boundary of the oxide film).

Dependence of the mass of deposited SiO_2 on the surface of the fibers at various concentrations of water is shown in fig. 3. Analyzing the obtained dependence, one can see that at MR = 2 and MR = 4, the amount of silicon oxide is lower than at MR = 6. The largest amount of silicon oxide is deposited at MR = 6 (17%), which corresponds to an increase in viscosity. A further increase in water concentration leads to a decrease in the amount of SiO_2.

At low water concentrations, as described above, the hydrolysis rate decreases, which contributes to the formation of a weakly branched structure corresponding to low values of the deposited SiO_2 mass and viscosity. High concentrations of water slow down the polycondensation processes, which indicate a decrease in the amount of SiO_2 and a slight decrease in the viscosity of sol-gel solutions. Thus, at high concentrations of water (MR above 6), the formation of an oxide film on the surface of carbon fibers occurs uniformly.

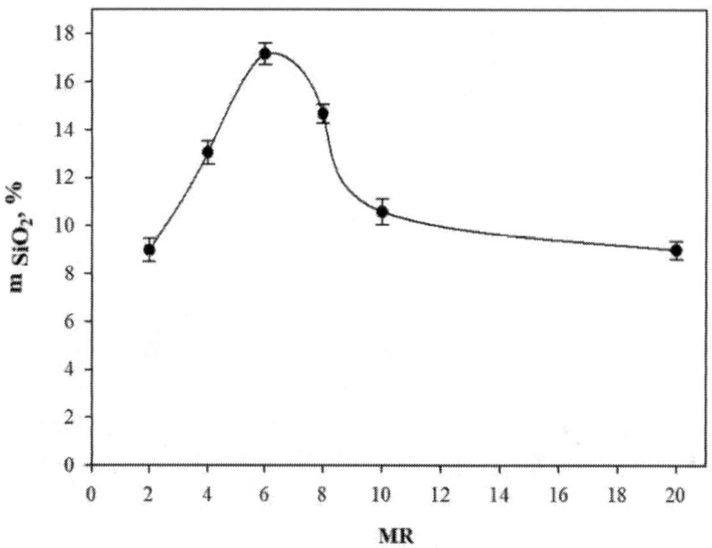

Fig. 3. The amount of deposited silicon oxide on the surface of carbon fibers, depending on the molar ratio of H_2O/TEOS.

4 Conclusion

The results of the studies showed that the concentration of water significantly affects the formation of SiO_2 films on the surface of carbon fibers:

- low water concentrations (MR < 6) lead to a slowdown in hydrolysis and an excess of TEOS, a decrease in the viscosity of the obtained sol-gel solutions and, accordingly, form cracks in the SiO_2 film;

- at high water concentrations MR ≥6, the viscosity of the sol-gel solutions does not change significantly and the formation of the SiO_2 film occurs uniformly.

Thus, the optimal molar ratio H_2O/TEOS at which SiO_2 films are formed without visible defects on the fiber surface is in the range from 6 to 20.

This work was performed on the subject № AAAA-A17-1170011910021-8 of the state assignment of the Ufa Institute of Chemistry Russian Academy of Sciences and with the financial support of the grant to young scientists of the Republic of Bashkortostan (No. 5GR/2020).

References

1. T. Wen, J. Gao, J. Shen, Z. Zhou, J. Mater. Sci. **36**, (2001)
2. R.F. Gallyamova, S.N. Galyshev, F.F. Musin, A.G. Badamshin, V.A. Dokichev, Solid state phenomena, **284** (2018)
3. T. Rajan, R. Pillai, B. Pai, J. Mater. Sci. **33**, 3491 (1998)
4. C. R. Durkin, Dissertation, Georgia Institute of Technology, (2007)
5. V.V. Kozik, L.P. Borilo, A.G. Malchik, Journal of Applied Chemistry, **69**, (1996)
6. L.N. Novoselova, Journal of Applied Chemistry, **8**, (1982)
7. I. Strawbridge, P.F. James, Journal of Non-Crystalline Solids, **86**, (1986)

8. R. Gallyamova, S. Galyshev, F. Musin, V. Dokichev, MATEC Web of Conferences, **298**, (2019)
9. E.N. Poddenezhnyy, A.A. Boyko, A.A. Alekseyenko, Vestnik GGTU, **4**, (2003)
10. S. Sakka, *Formation of Glass and Amorphous Oxide Fibers from Solution* (Elsevier Science Publishing, New York, 1984)
11. H. Tototzintle-Huitle, Journal of materials science, **37,** (2002)
12. M.I. Valuyeva, I.V. Zelenina, M.A. Khaskov, A.I. Gulyayev, Trudy VIAM, **10**, 2017

https://doi.org/10.1051/matecconf/202031505002

Investigation of the effect of ultrasonic surface treatment before ion implantation on the structure of the surface layer of aluminum alloy 1201

Viktor Ovchinnikov[1*], *Irina* Kurbatova [1], and *Nadezda* Uchevatkina[1]

[1]Moscow Polytechnic University, B. Semenovskaya st., 38 Moscow, 107023, Russian Federation

Abstract. The aim of this study was to study the properties of 1201 aluminum alloy after ultrasonic treatment and their evolution during subsequent ion implantation using the monotectic Cu-Pb alloy as the cathode material of the implant. It is shown that during ultrasonic treatment the surface layer of aluminum alloy 1201 undergoes significant changes. A nanocrystalline structure with a grain size of less than 200 microns is formed in it. Ultrasonic treatment of the surface of the target from alloy 1201 before implantation leads to a decrease in the depth of penetration of ions to 160–180 nm and the appearance of amorphization in the implanted layer.

1. Introduction

Significant progress in the development of the fundamental foundations of ion implantation in recent decades has been achieved thanks to intensive research in the field of modification of the properties of metallic materials. The study and generalization of the laws of physical processes caused by ion bombardment in a surface doped layer is an urgent task [1-3].

During ion implantation, the level of properties of the surface layer obtained as a result of irradiation depends on the mode and types of introduced ions, but also on the initial state and level of properties of the target surface before implantation. One of the areas of research is ion implantation after ultrasonic treatment.

Ultrasonic processing (RCD) is an advanced technology for finishing and hardening of metals by pressure, allowing you to replace the classical static methods of PPD according to the schemes of rolling and sliding - rolling and smoothing [4-7].

The deformation of the surface layer is performed by an acoustic head oscillating with an ultrasonic frequency, into which a carbide indenter is soldered. Due to the influence of ultrasonic vibrations, the resistance of the metal to plastic deformation is significantly reduced. At the final stage of the manufacturing process of the part, as a rule, various finishing methods are used for technological support of the quality of the surface layer -

* Corresponding author: uchevatkina@yandex.ru

abrasive finish, heat quenching, coating, surface alloying and plastic deformation. To achieve the maximum hardening effect, combined processing methods are used [8].

When solving the problem of increasing the wear resistance of parts made of titanium alloys, it is important to bear in mind that the wear resistance of the finished parts will be determined not only by the conditions and modes of implantation, but also by the type of surface treatment before implantation.

The aim of the study was to assess the effect of ion implantation using alloy implant cathodes based on immiscible components on the structure of an aluminum alloy.

2. Experimental details

As a material for the study, aluminum alloy plates with a thickness of 5 mm and dimensions 100x200 mm were used.

Ultrasonic processing was carried out using an IL10-0.63 generator (Russia), connected to an oscillating system consisting of a magnetostrictive transducer, a waveguide hub, and an acoustic head with a welded carbide indenter.

Ion implantation of alloy samples was carried out in an ion implantation unit equipped with two independent ion sources: a metal ion source and a gas ion source (the so-called double-beam installation).

The source of metal ions refers to the type of sources with a vacuum arc in metal vapors and represents a new type of source, where a vacuum arc in metal vapors is used as the plasma medium from which ions are extracted. The source operates in a pulsed mode and generates pulsed beams of metal ions.

A distinctive feature of this type of ion sources is that they form a polyenergy ion beam, i.e. the beam composition includes not only singly charged, but also multiply charged ions.

During operation, the source of metal ions provides a wide-aperture beam of circular cross section with a diameter of 200 mm at an ion beam current of 0.1 ... 1 A. The accelerating voltage is 40 kV. In this source, the arc burns with pulses with a frequency of 50 Hz and a pulse duration of ~ 300 μs. The implantation dose of the studied samples was 10^{17} ion / cm^2. A monotectic Cu – Pb alloy with a lead content of 36% was used as the cathode material.

To study the effect of processing on the microhardness of the surface of 1201 aluminum alloy, the Affri 251 VRSD universal digital hardness tester (Italy) was used. Determination of the level of residual stresses in the samples after ultrasonic treatment (RCD) was carried out on a general-purpose automatic multifunctional X-ray diffractometer DRON-7 (Russia).

Determining the nature of the microstructure and the distribution of dislocations in the treated surface layer of the samples, a JE E L JEM-2100 transmission electron microscope (Japan) was used.

Ultrasonic processing refers to dynamic PPD methods. It is a process of impulse penetration, oscillating with the ultrasonic frequency of the indenter in the surface layer of the workpiece, and causes plastic flow of the metal in the contact zone. The generator, powered by a single-phase AC network with a frequency of 50 Hz and a voltage of 220 V, generates oscillations of an electromagnetic field with a frequency of 20 kHz, which are transformed by means of an electro-acoustic transducer into mechanical vibrations of a waveguide-concentrator and, after increasing the amplitude, are transmitted to a working body - an acoustic head with a soldered carbide indenter [8].

The ultrasonic treatment mode (frequency and amplitude of vibrations, shape and radius of the indenter, speed of the main motion and feed rate, static force) determines the quality parameters of the surface layer (roughness, waviness, microhardness, residual stresses, hardening depth, dislocation density) that affect operational properties of machine parts (strength, wear resistance, contact stiffness, fatigue resistance, corrosion resistance, tightness of joints, friction coefficient, etc.) [9-12].

As the basic mode of the RCD, a mode with a static force value of 150 N and a feed rate of 180 mm / min was selected according to the results of the work.

3. Results and discussion

Samples of the aluminum – copper 1201 alloy (aluminum content 3.8 at.%) were subjected to intense ultrasonic irradiation in an aqueous medium using a VIP 1000hd ultrasound generator (Heilscher, Germany) operating at a frequency of 20 MHz. The calculated radiation power was 57 W / cm^2. The surface microrelief of the irradiated samples was studied using a LE · -1420 scanning electron microscope with an R · ntec X-ray microanalyzer. Corrosion studies were performed by electrochemical impedance spectroscopy using a Gamry FAS2 femtostat in the frequency range 5 × 104 - 10-3 Hz.

As it can be seen from the figure 1 electron-microscopic images of the surface of the samples of aluminum-copper alloy, the effect of intense ultrasound leads to a change in the morphology of the surface, which acquires a scaly-like structure. Studies by X-ray fluorescence spectroscopy showed that the result of ultrasonic exposure is surface oxidation, and when the radiation power is used, the amount of surface oxide formed is saturated after the first 10 minutes of exposure.

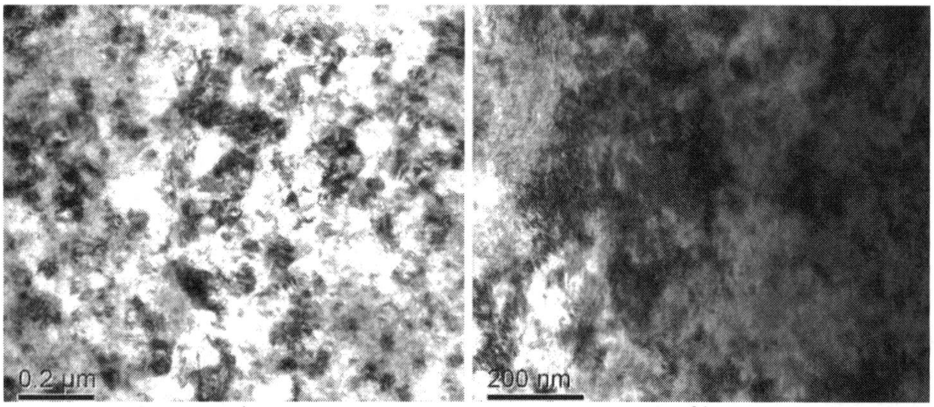

a) b)

Fig. 1. Electron microscopic images of the surface of a sample of aluminum-copper alloy after (a) and before (b) irradiation in an ultrasonic field for 10 min.

The observed oxidation of the alloy can be explained by the action of radical particles (various forms of active oxygen), as well as the destruction of the initial oxide layer by cavitation jets [13-16]. The morphology of the surface formed as a result of processing also indicates that the formation of the oxide layer includes the processes of depassivation-reoxidation.

According to x-ray phase analysis, the resulting oxidation products are bayerite Al (· H) 3; at the same time, as the dose load increases, additional boehmite lines Al · (· H) appear on X-ray diffraction patterns, this indicates the gradual development of dehydration processes under ultrasound exposure.

A detailed study of the structure of the oxidized layer was carried out using the electron microprobe method, which allows simultaneous recording of the relative contents of aluminum, oxygen, and copper at various points on the irradiated surface. From the profiles of the spatial distribution of elements along the scanning line by the electron beam, it follows that the bound oxygen is uniformly distributed over the surface of the sample subjected to ultrasonic irradiation, with the exception of areas corresponding to inclusions of the Al2Cu intermetallic compound. Near inclusions, oxidation proceeds much more efficiently, as a result of which an oxide layer of greater thickness is formed. The observed intensification of the process of aluminum oxidation can be explained by the effective reduction of the molecular oxygen process on copper-enriched inclusions under the influence of an ultrasonic field [17–20].

Light spots, correspond to areas enriched with copper, dark spots - elements of surface morphology (dimples); the line in the image corresponds to the trajectory of the electron beam when obtaining surface concentration profiles.

Sonochemical anodization is completed after the formation of a sufficiently thick and dense oxide layer encapsulating intermetallic inclusions. The latter circumstance should contribute to increasing the corrosion resistance of the alloy, since it is these inclusions that are the points of development of pitting corrosion in depassivating media. Electrochemical measurements indicate that the barrier layer created as a result of sonochemical treatment provides a sharp increase in the corrosion stability of an aluminum-copper alloy.

4. Conclusion

Thus, the ultrasonic treatment of an aluminum-copper alloy in an aqueous medium lead to the formation of a surface protective film, the process of its growth being highly selective and adaptive with negative feedback.

During ultrasonic treatment, the surface layer of aluminum alloy 1201 undergoes significant changes. A nanocrystalline structure with a grain size of less than 200 microns is formed in it. Metallographically, the nanocrystalline layer is detected as a strongly etched region, the thickness of which is determined by the parameters of ultrasonic treatment and is approximately 20–30 μm.

Implantation of aluminum alloy 1201 ions by cathode of Cu-Pb system without ultrasonic treatment and after it showed that the depth of penetration of copper and lead ions during irradiation of the target without preliminary treatment is 320-380 nm. Ultrasonic treatment of the surface of the target from alloy 1201 before implantation leads to a decrease in the depth of penetration of ions to 160–180 nm and the appearance of amorphization in the implanted layer.

This work was carried out as part of the state task of the Ministry of Education and Science of Russia "The influence of magnetic fields and ion implantation on the structure, chemical composition and properties of titanium, aluminum alloys and elementary semiconductors".

References

1. A.V. Byeli, Wear, **203,** 596-607 (1997)
2. P.J. Wilbur, Surface and Coatings Technology, **83 (1-3)**, 250-256 (1996)
3. A.N. Didenko, Yu.P. Sharkeyev, E.V. Kozlov, A.I. Ryabchikov, NTL Publishing House, 328 (2004)
4. Yu.I. Lopukhov, M.A. Lakhina, Engineering - from theory to practice, **4 (41)**, 51-59 (2015)
5. A.V. Kirichek, D.L. Soloviev, A.G. Lazutkin, Mechanical Engineering, 288 (2004)
6. M.A. Balter. Engineering, 184 (1978)
7. A.G. Tkachev, Publishing house of Tamb. state tech. University, 164 (2009)
8. O.V. Abramov, I.G. Khorbenko, S. Shvegla, Mechanical Engineering, 280 (1984)
9. V.Ye. Panin, Physical Mesomechanics, **9(2),** (2006)

10. V.E. Panin, Physical Mesomechanics, **8** (2005)
11. V.A. Klimenov, Physical Mesomechanics, **7** (2004)
12. O.V. Taratynov, V.V. Poroshin, V.V. Kharchenko, High technology in mechanical engineering, **10**, 13-19 (2012)
13. D.D. Papshev, *Finishing and hardening by surface plastic deformation,* Mechanical Engineering, 152 (1978)
14. A.G. Suslov, R.V. Gurov, E.S. Tishevskikh, Hardening technologies and coatings, **9**, 20-22 (2008)
15. V.P. Fedorov, M.N. Nagorkin, I.L. Pyrikov, Bulletin of the Bryansk State Technical University, **1**, 22-31 (2011)
16. G.G. Poznyak, V.V. Kopylov, V.A. Rogov, *Modern methods of designing nanotechnological processes in mechanical engineering,* RUDN, 152 (2008)
17. V.V. Popova, *Surface plastic deformation and physicochemical treatment,* Rubtsovsky Industrial Institute, 98 (2013)
18. A. I. Lotkov, Letters in ZhTF, **31 (21)**, 24-29 (2005)
19. A.D. Pogrebnyak, Questions of atomic science and technology, **29(6)** 571-577 (2008)
20. Yu.F. Ivanov, Fundamental problems of modern materials science. **5 (1)**, 47-52 (2008)

Wear and friction dependence on grain size

Vladimir Stolyarov[1]*

[1]Mechanical Engineering Research Institute of Russian Academy of Sciences, 101990, Moscow, Russia

Abstract. The tribological characteristics of pure metals and ceramics with graphene obtained by electrochemical, deformation and powder metallurgical methods are compared for a coarse-grained and nanocrystalline state. Reducing the grain size to the nanoscale region in the structure helps to reduce wear and friction coefficient compared to conventional analogues. The possibilities of modelling the contact of solids with the atomic structure of the surface are considered.

1 Introduction

Nanocrystalline (NC) materials, as a result of a significant decrease in grain size and a substantial increase in the volume fraction of their boundaries, exhibit unusual mechanical, physical, and electrochemical properties compared to ordinary polycrystals [1]. High strength and hardness of NC materials is the basis for promising structural applications and therefore stimulates the development and research of new metals, alloys, and compounds. There is a significant number of works devoted to studies of the mechanical behaviour of NC materials in standard tests for hardness, compression, or tension [2, 3]. However, the mechanisms of friction and wear in NC materials are poorly studied, possibly due to the difficulty of creating massive samples sufficient to study friction and wear. Most of the studies on the wear of the surface layer or coatings were carried out in metal alloys, which showed increased wear resistance of NC materials in comparison with their coarse-grained (CG) analogues [4]. At the same time, systematic studies of wear in model pure NC metals nanocomposites, in which the effect is not associated with phase transformations, are of great value. Until recently, they were few in number due to difficulties in the synthesis of macroscopic samples suitable for testing. The own and published data for pure NC metals ceramics (copper [5], nickel [6], titanium [7], Al2O3/graphene) obtained by electrodeposition, severe plastic deformation and powder metallurgy methods are presented below.

2 Materials and research methods

In pure copper, the NC state was obtained by electrodeposition of the initial CG (50-100 μm) copper with a purity of 99.99%. The average crystal size in NC copper was 20 nm. Both states were tested for wear under dry friction conditions at room temperature according to the "ball-disk" scheme, in which a WC – Co alloy ball with a diameter of 10

* Corresponding author: vlstol@mail.ru

mm slid over a copper disk of ∅7x2 mm with an amplitude load of 5 to 40 N frequencies of 5 Hz.

NC nickel was also obtained by electrodeposition; the processing method is described in detail in [6]. The crystallite size in NC nickel was estimated by the X-ray method and ranged from 8 to 28 nm; the average crystal size was 20 nm. Tribological tests were performed without lubrication according to the “steel ball-disk” scheme, on surfaces cleaned by ultrasound, at a load of 1 N, a frequency of 8 Hz, tangential displacement amplitude of 100 μm and a number of cycles of 10000. CG nickel used as a reference element was obtained cold rolling and annealing to form a structure with an average grain size of 61 μm. Microhardness was measured at a load of 1 N, so that the penetration depth was less than 10% of the thickness of the deposited layer.

For the study, commercial pure titanium was selected in the form of a hot-rolled bar of ∅50 mm containing impurities 0.07% C, 0.18% Fe, 0.10% Si, 0.12% O, 0.04% N and 0.01% H. The average grain size was 15 μm. After turning to ∅40 mm, a bar 100 mm long was subjected to warm equal channel angular pressing (ECAP), the process of which was described in detail in [7]. Part of the ECAP samples was deformed by cold rolling (CR) by 75% in order to further refine the structure. As a result, an ultrafine-grained (UFG) microstructure was obtained with a grain size of 0.3 μm (after ECAP) and NC microstructure of 0.1 μm (after ECAP + CR). A typical NC microstructure in titanium after ECAP and cold rolling is shown in Fig. 1.

a b

Fig. 1. Microstructures of NC titanium in the transverse (a) and longitudinal (b) section.

It is known [8] that the tendency to set and stick (typical for titanium) depends on the resistance of the material to adhesion under shear stresses, which can be estimated by the ratio of tangent τ_{nn} and normal stresses P_m at the contact of solids. This ratio ($\mu_a = \tau_{nn} / P_m$) is known as the adhesive component of the friction coefficient which the smaller, the smaller setting and sticking. The measurement of μ_a was carried out according to the method of [9] on a setup simulating friction conditions between a hemispherical indenter made of WCo alloy (∅2.5x25 mm) and a titanium disk (∅12x6 mm) at various contact temperatures. Titanium was studied in three states: CG (initial), UFG (ECAP) and NC (ECAP + CR). The disk rotation speed was 36 degree / s. The contact area was heated to temperatures of 150, 350, 550, and 800 °C by passing an electric current, the duration of which was less than 1 min.

A ∅15x2mm disk-shaped composite Al_2O_3/G was processed by the SPS method of spherical powder of alumina ceramics of various dispersion and five-layer scaly graphene (G) with a content of 0.5 wt.%.

3 Research Results

3.1 Copper

Figure 2 (a) shows the friction coefficient dependence on distance at a load of 5 N for CG and NC copper. Each curve is characterized by two friction modes. First, the friction coefficient rises sharply, and then reaches the saturation stage. The transition to a stable stage in NC samples takes a longer time than for CG samples. Although the situation changes the other way around with increasing load up to 10 N, the friction coefficient at the saturation stage for NC copper is always less than for CG copper.

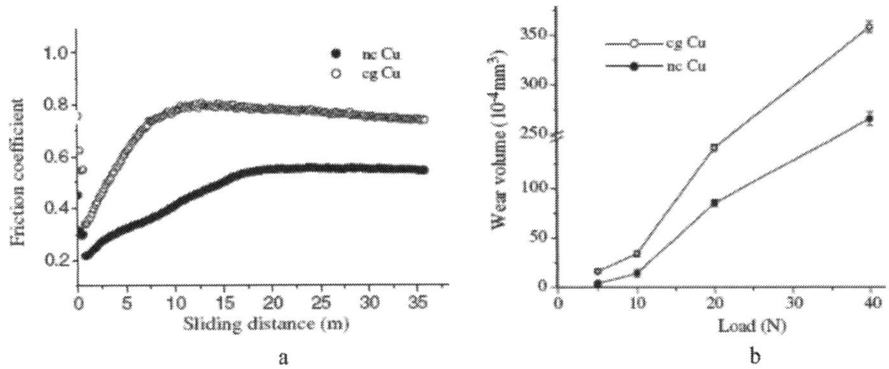

Fig. 2. Friction coefficient at a load of 5 N (a) and wear (b) for CG and NC copper [5].

The change in the value of wear and tear from the load for short circuit and NS copper is shown in Fig. 2 (b). Depreciation increases with increasing load, but the wear rate for NC copper was lower than for CG copper. The microhardness measurement showed that it is twice as high for NC copper (1050 MPa) as for CG copper (500 MPa). Thus, a lower friction coefficient and wear in NC copper than for CG copper is associated with increased microhardness, which in turn is due to the smaller grain size. It is assumed that other reasons may be features related to the refinement of the structure - increased oxidation during friction and weaker strain hardening.

3.2 Nickel

Table 1 gives the grain size, microhardness and friction coefficient. The grain size ranged from 8 to 28 nm. The microhardness of Ni with a grain size of 8 and 10 nm was close, which is unusual for the general dependence on the grain size It was previously shown that in this size range there is a deviation from the Hall-Petch law, not only quantitatively, but also in sign. This may be due to a change in the deformation mechanism from purely dislocation slip in polycrystalline nickel to intergranular slippage in a nanocrystalline sample.

Table 1. Grain size, microhardness and friction coefficient of electrodeposited nickel.

Grain size, nm	Microhardness, MPa	Friction coefficient
61000	2690	0.62
28	5720	0.55
10	7240	0.29
8	6890	0.16

Figure 3 shows the dependence of the friction coefficient on the number of cycles for various grain sizes. In all cases, this value increases in the first period (up to 500 cycles) of cycling, and then reaches a stable saturation stage (presented in the table). An extremely low friction coefficient (k = 0.16) was obtained for nickel with a grain size of 8 nm. In contrast, nickel with the largest grain size of 61 microns showed the maximum value of the coefficient of friction.

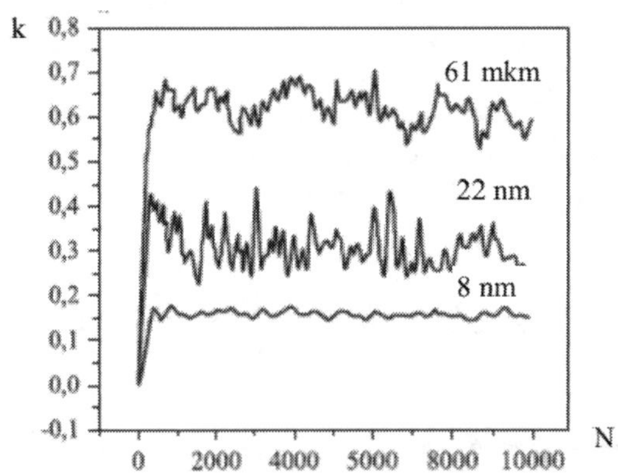

Fig. 3. Friction coefficient dependence for Ni on the number of cycles and grain sizes [6].

Thus, the study of nanocrystalline nickel showed that the higher microhardness of the electrodeposited sample, associated with the smallest grain size, is the main reason for the decrease in the friction coefficient.

3.3 Titanium

Figure 4 shows the temperature dependences μ_a of titanium in three states at different loads. At all temperatures above room temperature, CG titanium has a higher friction coefficient than UFG or NC titanium. In addition, it increases with temperature for CG titanium and, as a rule, decreases for UFG titanium (except for the state of ECAP + CR at 480 N and ECAP Ti at 960 N.

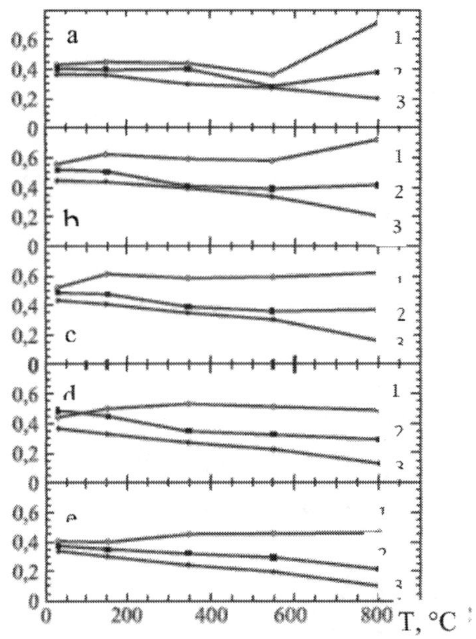

Fig. 4. Temperature dependences of the adhesive component of the friction coefficient in CG (1), UFG (2) and NC (3) titanium at loads: a - 480 H; b - 960 H; c - 1440 H; d - 1920 H; e - 2400 H [7].

Thus, the formation of UFG or NC structures in titanium reduces the adhesive component of friction μ_a and the tendency of titanium to stick and set. It is also important that μ_a decreases with increasing temperature and grain size. As subsequent studies of mechanical behaviour have shown, this effect is associated with multiple hardening of titanium as a result of structure refinement. A similar effect of reducing friction and wear during the structure refinement to the nanometer range was observed in titanium alloys [10].

3.4 Ceramic composite

Table 2 shows the microhardness, wear and friction characteristics for the Al2O3 / 0.5% G composite. With increasing dispersion of the matrix phase, microhardness naturally increases, and tribological characteristics decrease. This is especially pronounced in the wear characteristics, which are almost halved when the submicron matrix is replaced by a nanocrystalline one.

Table 2. Wear and friction coefficient of Al_2O_3 composites.

Dispersity	Wear (10^{-3} mm^3)	Wear rate Kv (10^{-6} $mm^3/N{\cdot}m$)	Friction coefficient
submicron	6.38	7.97±0.05	0.61±0.05
nanocrystalline	3.54	4.42±0.05	0.53±0.05

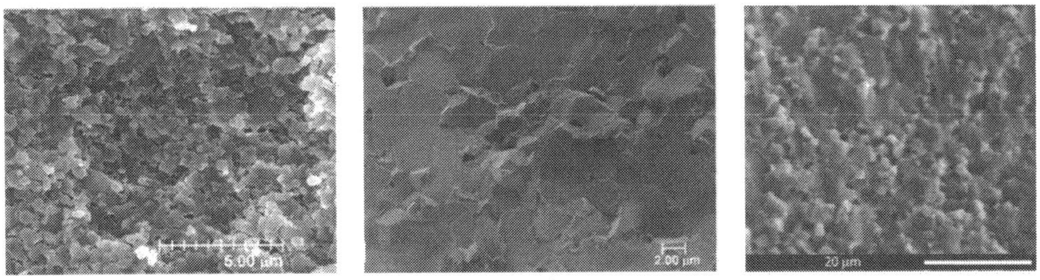

a b c

Fig. 5. Fracture surface of sintered nanoceramic: a - without graphene; b, c - with graphene 0.5% (large and low magnification)

An analysis of the fracture surface patterns of nanoceramic (Fig. 5) suggests that the addition of even minimum graphene content can reduce porosity and thereby contribute to an increase in wear resistance.

3.5 Friction and wear models

To calculate the adhesion and friction forces acting between the contacting surfaces of solids, methods of continuous contact mechanics based on the linear theory of elasticity are used. The surfaces are assumed to be smooth and even; their atomic structure is not taken into account. For example, the force required to tear off a surface with a radius of curvature R from a flat substrate (Fig. 6) is found by the simple formula $F = 3pRg$, where p is a dimensionless coefficient that takes into account surface roughness and elastic compression at the contact point; g is the surface energy.

This expression can be generalized to other geometries by replacing R with the corresponding characteristic length. In some cases, the continuum theory gives surprisingly accurate results, and sometimes diverges from experiment tenfold. The reason for this is that the surfaces of real solids in most cases are very different from their idealized mathematical images. Their roughness significantly affects the strength of adhesion.

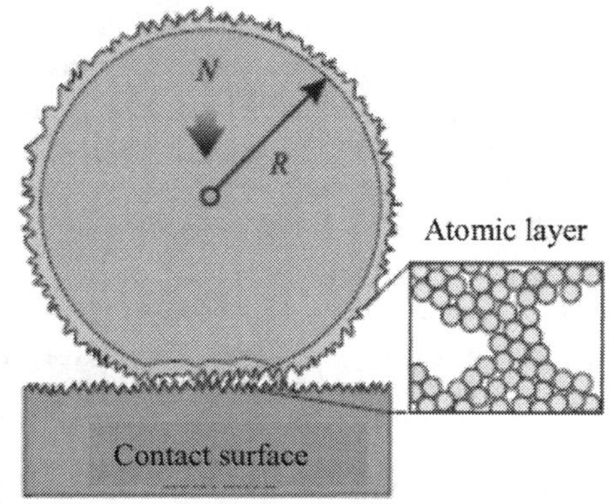

Fig. 6. Mechanical contact of two solids: continual (dashed line) and real surface shape [11].

In addition, there has been a recent surge in interest in micro- and even nanoscale electromechanical devices. They often fail to work due to unwanted adhesion. A continuum models, in principle, are not suitable for calculating their characteristics. All this suggests the need to develop a new, microscopic theory of contact mechanical phenomena.
A step in this direction was taken in [12], the authors of which are from Johns Hopkins Univ. (USA) used the molecular dynamics method to test the limits of applicability of the macroscopic description of contacting surfaces. They studied the contacts between a flat (001) FCC crystal substrate and three different types of cylindrical surfaces having the same radius of curvature (Fig. 7): (a) a curved crystal lattice with an atomically smooth surface, (b) the surface of a sample cut from amorphous solid, and (c) the "stepped" surface of the sample cut from the crystal. In cases (b) and (c), the surface roughness did not exceed one effective atomic diameter (one average interatomic distance). First, the dependences of the displacement d, the contact radius a, and the static friction force F on the magnitude of the pressing force N directed along the normal to the substrate were calculated. It turned out that for all types of surfaces the calculated dependences d (N) are in excellent agreement with the predictions of the continuum model. For the dependences a (N), the qualitative agreement between the numerical results and the analytics is preserved, but the quantitative discrepancy reaches 100%.

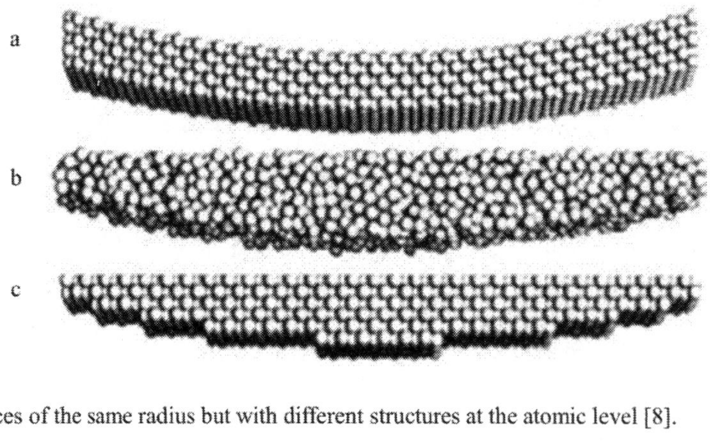

Fig. 7. Surfaces of the same radius but with different structures at the atomic level [8].

First of all, this concerns the "stepped" surface, for which a increases with increasing N not "monotonously", but "jumps". The continuum approximation can give a significantly reduced contact area, especially for small N. If for a smooth and "stepped" crystalline surfaces the friction force F within (10 - 20) % coincides with its value expected from macroscopic approach, then for an amorphous sample it turned out to be ~ 5 times less. The pressure distribution over the contact area is also very sensitive to the surface structure at the atomic level and in some cases qualitatively differs even with the same roughness, quantified in terms of the standard deviation of the atoms from the ideally smooth surface. Thus, the roughness is an average value that does not allow us to unambiguously predict the mechanical characteristics of the contact without detailing the specific type of atomic disorder, which creates the same roughness.

The results obtained are of both fundamental and practical importance. They not only make it possible to understand the physical nature of contact phenomena better, but also pave the way for "surface engineering", which will make it possible to make contacts with the required macroscopic properties by proper microprocessing of the contacting surfaces.

So far, only the limiting case of small strains has been more or less studied, which corresponds to the contacts of metals or ceramics. In addition, the description of the interaction of "real surfaces" in the "real time" mode requires the mandatory consideration of non-stationary phenomena. These and many other questions that have not yet been resolved is the subject of further theoretical and experimental research.

4 Conclusion

The tribological behavior of the pure metals and ceramic composites in the nanocrystalline and traditional coarse-grained states is significantly different. The friction coefficient and the amount of wear in nanocrystalline materials are minimal and are due to the maximum value of microhardness, which in turn depends on the degree of structure refinement. The transition to nanocrystalline contact surfaces can be carried out within the framework of models based on the surface relief at the atomic level.

The work has been supported by Russian Science Foundation [grant № 16-19-10213].

References

1. Y.T. Zhu, Z. L. Wang, T.G. Langdon, J. Mater Sci 421401(2007)
2. E.N.Hahn, M.A. Mater. Sci.Eng.,A **646** 101(2015)
3. S. Hassan, S. Sharma, B. Kumar, Inter. Ref. J. Eng. Sci. **6**, 66 (2017)
4. N. Gao, C.T. Wang, R. J. K. Wood, T.G. Langdon, J. Mater. Sci.47 4779 (2012)
5. Z. Han, L. Lu, K. Lu, Tribol. Lett., **21**, 47 (2006)
6. B. Basu, R. Balasubramaniam, R. Mishra, Mater. Sci. Eng., A **373** 370 (2004)
7. V. V. Stolyarov, L. Sh. Shuster, M. Sh. Migranov, R.Z. Valiev, Y.T. Zhu, Mater. Sci. Eng., A **371**, 313 (2004)
8. L. Sh. Shuster, Adhesive interaction of a cutting tool with the processed material, M: Mashinostroenie, **96** (1988) *in Russian*
9. G. S. Fox-Rabinovich, A. J. Kovalev, L. Sh. Shuster, Yu. F. Bokiy, G. K. Dosbayeva, D. L. Wainstein, V. P. Mishina, Wear **206** 214 (1997)
10. S. Chertovskikh, L. Shuster, V. Stolyarov, Friction and wear, **1**, 80 (2005)
11. J. N. Israelachvili, Nature, **435**, 893 (2005)
12. B. Luan, M. O. Robbins, Nature, **435**, 929 (2005)

https://doi.org/10.1051/matecconf/202031506002

The growth mechanism of ordered nanoporous alumina

Marat Makhmud-Akhunov[1,*], *Artem* Adamovich[1] , and *Igor* Yavtushenko[1]

[1]Ulyanovsk State University, Ulyanovsk, 432017 Leo Tolstoy Street, 42, Russia

Abstract. The paper scrutinizes the peculiarities of porous alumina growth during electrochemical anodic oxidation in the orthophosphoric acid solution at constant potential of 100 Volt. On the basis of atomic-force microscopy data the paper presents consecutive stages of ordering pores in the oxide with the formation of hexagonal dense package after two-stage anodic treatment.

1 Introduction

Oxide films on the surface of metals and semiconductors are widely used as electric isolation, corrosion-inhibiting covering, and intermediate buffer layers during hermetic metal brazing with different glasses, etc. Dielectric films bear the functional load in semiconductor condensers, transistors of the type metal-oxide-semiconductor (MOS), and etc.

Nowadays much attention is given to nanoporous oxides [1, 2], which are anode-structured in oxygen-containing electrolytes. Such structures often have nonstandard physical, optical and electric properties, including high photosensitivity [3] and catalytic selectivity, which gives an opportunity of practical use of nanoporous structures as extremely sensitive sensors of gas environment components [4]. The priority of such structures is noticeable in photocatalysis [5], in production technique of lithium-ion batteries of extended capacity [6], in solar cells of new generation [7]. However, despite practical success in the sphere of getting porous oxide structures, the description of mechanism and pores formation kinetics is not sufficiently presented; most of works are focused on the description of methods and techniques of getting highly-ordered porous structures, on the modification of their properties and on the application area without a detailed description of the ordering mechanism. Within the scope of the research this paper will present consecutive stages of the growth of porous anodic alumina, having an ordered honeycomb structure.

2 Experiment

During the research we used polycrystalline Al samples of a cylindrical shape ($S = 0.2\ cm^2$), produced from cast especially pure Al of brand A999. The internal stress

* Corresponding author: maratmau@mail.ru

relieving was carried out by two-hour annealing of the researched samples at temperature T = 450 °C. The samples were poured with epoxy and kept at room temperature during 24 hours. The working surface of each sample consecutively underwent grinding and polishing till the smooth state with further chemical polishing in the solution H_3PO_4:HNO_3 = 85:15 at 85 °C during 5 min [8]. For the neutralization of the acid residue on the surface the sample was kept for 5 min in 0.1 M solution NaOH and was washed properly with distilled water, after that the sample was placed in the cell. The structure of the cell is presented in [9].

To implement two-stage anodic treatment the selective removal of the "sacrificial" oxide layer after the first stage of oxidation was done in the mixture of 20 gr/l CrO_3 and 35 ml/l H_3PO_4 at 85 °C during 30 min.

The experiments on porous alumina were carried out under controlled potential conditions at U=100 V with the lower location of the researched electrode under plumbeous cathode. As an electrolyte 0.75 M solution of orthophosphoric acid was used. The kinetics of the phenomena in progress and the parameters of the oxide layers were judged by the data of atomic-force microscopy (Solver P47-PRO).

3 Results and discussion

Due to the little resistance of the metal under research, at the moment of turning on the outer potential the rupture of the natural oxide layer happens, and the high current goes through the sample (~ 850 mA). To exclude the rupture the increase of the anodic potential was done with the rate of 1V/s till the necessary value.

The supplied voltage U, sent to the electrodes, is redistributed unevenly between anode and cathode, and is under prevailing control of anode-polarized Al electrode, screened by the oxide. The presence of surface imperfection cannot help influencing the redistribution of electric field force lines. The main active centers of current localization can be the boundaries of the grains, dislocation and mechanical defects like protrusions and microscratchings, remaining after the mechanical polishing of the sample. However, the used annealed coarse-grained aluminum allows doing atomic-force analysis within the limits of one single-crystal block, excluding the influence of grain boundaries. Preliminary done chemical polishing of the sample due to corrosion of the surface layer weakens the influence of the rest imperfections. However, the presence of some concentration of non-corroded imperfections leads to the uneven distribution of electric field force lines, and consequently, to the uneven growth of the oxide film in the shape of the islands (Figure 1-a).

The formation of the islands is going on continuously because, having reached critical thickness on the prevailing imperfections and roughness, the current is redistributed to other areas with less resistance. This raises the resultant resistance of the aluminum electrode and lowers the current going through the system. Growing, the oxide islands interface between each other through a thin barrier layer, forcing the current to become localized in the gap of the joined protrusions. The rise of the electrolyte chemical activity nearby the anode [10] leads to the local dissolution of the oxide layer. Cavities, filled with the electrolyte, and transforming according to the oxide film growth into pores, are formed in the film. At the regular arrangement of growing hemispherical oxide islands they form a hexagonal dense package. Then each island will be surrounded by six interstices (Figure 1-a), at which the current localization takes place, and consequently the pores formation and corrosion of the support occur. During the oxidation process the expansion and deepening of the corroded zone under each island take place. At the moment of six nearest chemical milling zones contact the tangential growth of the oxide stops, and only deepening of the oxide into the support goes on. As a result, the corroded zones under the oxide film get the shape of hexagons and form a dense package (Figure 1-b). The offered model of the anodic dissolution of the aluminum support prognosticates the presence of a non-corroded zone in Al support straight under the oxide island.

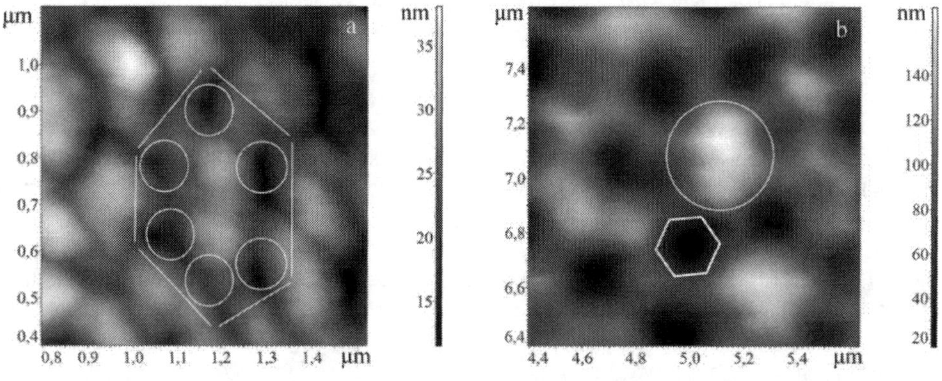

a b

Fig. 1. AFM picture of oxide film surface with the indication of interstices (a) and of Al support under the film with the indication of the non-corroded zone and of cavities in the shape of hexagons (b) after primary anodic treatment at U=100 V during 30 min.

As it is known, during electrochemical processes, besides the growth of the oxide film, the electrolyte warming up nearby the reaction zone and gas bubbles formation take place. All that violates the regular film growth, and distortions in the regular arrangement of the islands are observed in the oxide. As a result, a typical for this symmetry imperfection of pentagons formation is observed (Figure 2).

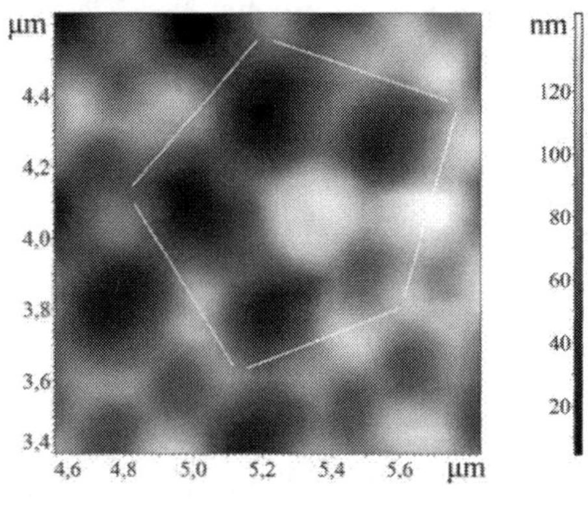

Fig. 2. AFM picture of Al support surface under the oxide film with the breach of the regular arrangement of the corroded areas after the primary anodic treatment at U = 100 V during 30 min.

It is significant that the oxide film topology at the second anodic treatment is set by the relief of the corroded Al support surface (Figure 1-b). The current localization mainly occurs on the protrusions, in this case, on the vertices of the hexagons (Figure 1-b), as a result, each pore turns out to be surrounded by six islands (Figure 3-a). Then the non-corroded zone in the support, formed at the primary anodic treatment, (Figure 1-b) is located in the centre of the nearest six oxide islands interface (Figure 3-b). This leads to the current localization exactly in it, i.e. to its corrosion. As a result, after more prolonged oxidation the oxide film surface will be covered with the pores of hexagonal symmetry without imperfection in the package.

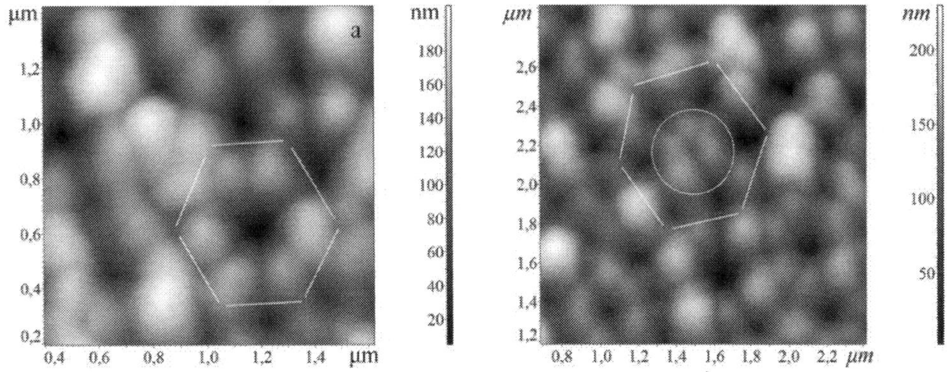

a b

Fig. 3. AFM picture of Al surface after the second anodic treatment at U=100 V during 90 min. a is the shape of the oxide cell ; b is the dense package with the indication of oxide islands above the non-corroded zone in the support (See Fig. 1-b).

4 Conclusion

Thereby, this paper scrutinized the peculiarities of electrochemical formation of porous alumina. Using the method of atomic-force microscopy, we showed the mechanism of pores honeycomb structure formation in the oxide film during two-stage anodic treatment in the self-organized process of the electric field force lines redistribution. The research also showed that the formation of island oxide layer at Stage One of oxidation leads to the incomplete corrosion of Al support with the presence of untouched areas. The second oxidation of the modified Al surface after the primary treatment sets the relief of the growing oxide layer with the formation of the ordered structure.

This work is supported by the Russian Science Foundation (project no. 19-71-10063).

References

1. Y. Guo, Li. Zhang, M. Han, X. Wang, J. Xie, L. Deng, R., Soc. Open Sci. **5**, 3 (2018)
2. M. Fukuhara, T. Kuroda, F. Hasegawa, Sci. Rep. 6 (2016)
3. A. Hierro-Rodriguez, P. Rocha-Rodrigues, F. Valdés-Bango, et al., J. Phys. D: Appl. Phys. **48**, 45 (2015)
4. V. Galstyan, E. Comini, G. Faglia, G. Sberveglieri, Sensors **13**, 11 (2013)
5. B. Niu, X. Wang, K. Wu, X. He, R.Zang, Materials **11**, 10 (2018)
6. T. Song, H. Han, H. Choi, et al., Nano Res. **7**, 4 (2014)
7. T. Balasundaram, K. Raja, Int. J. Chemtech Res. **7**, 1 (2014)
8. S. Jin, V. Pham, J. Dickerson, M. Brochu, J. Electrochem. Soc. **162**, 11 (2015)
9. A. Orlov, I. Yavtushenko, M. Makhmud-Akhunov, Phys. Solid State **57**, 8 (2015)
10. U.R. Dotel, M.O. Sydnes, H.Urkedal, T. Hemmingsen, Math. Models Methods Appl. Sci. **8**, 11 (2018)

Kinetics of liquid-phase shear exfoliation of graphite in synthetic oils

Zaman A A Alhilo[1], *Vladimir* Pershin[2*], and *Aleksey* Osipov[3]

[1]Post-graduate student from the Republic of Iraq, Tambov State Technical University, 106 Sovetskaya Street, 392000, Tambov, Russian Federation
[2]Department of Technology and Methods of Nanoproducts Manufacturing, Tambov State Technical University, 106 Sovetskaya Street, 392000, Tambov, Russian Federation
[3]LLC "NanoTechCenter", 51 Sovetskaya Street, 392000, Tambov, Russian Federation

Abstract. Lubricants modified with nanosized particles are of great interest to science and industry, since they have much better tribological characteristics compared to traditional lubricants. One of the most promising nanoparticles is graphene, which has an extremely low coefficient of friction, is very wear-resistant and environmentally friendly. Today, the main problem for the development of a new nano-lubricant is the creation of an environmentally friendly and cheap technology for the industrial production of graphene suspensions or graphene concentrates for the modification of traditional lubricants. The article describes the process of liquid-phase shear exfoliation of graphite in a rotary apparatus with moving blades, in synthetic oils. The kinetic dependences of the exfoliation process, i.e. the dependence of the concentration of graphene nanostructures in suspension over time at different values of the process parameters. It has been experimentally proved that using a stator-rotor mixer with moving blades, it is possible to obtain graphene nanostructure concentrations of at least 2 mg / ml.

1 Introduction

Graphite has been used as an antifriction material for several centuries, but the production of graphene and graphene nanostructures (low-layer graphene and multilayer graphene) has greatly increased the interest of scientists and engineers in this material [1, 2]. It has been experimentally established that the addition of graphene nanostructures to lubricants reduces not only the coefficient friction and wear, but also at times increases the ultimate load on friction pairs. The experimental results show that the addition of 0.4 wt.% graphene nanostructures to lithium greases gives good antifriction properties and wear resistance [3, 4]. Processing graphite lubricant in a ball planetary mill and converting a portion of graphite into low-layer and multi-layer graphene reduced the friction coefficient by 2.2–2.5 times [6]. Improving the tribological characteristics of lubricants by adding low-layer and multi-layer graphene is one of the most promising ways to reduce friction and wear in various machines and mechanisms. The results of our studies confirm the promise of this

*Correspondingauthor:pershin_home@mail.ru

area. The main obstacle to the transition from laboratory miserable volumes to large-tonnage industrial technologies is the lack of environmentally friendly and cheap methods for producing low-layer and multi-layer graphene, which can be used to modify lubricants. An analysis of the use of nanomaterials to improve the operational characteristics of lubricants showed that one of the most promising is liquid-phase shear exfoliation of graphite. One of the first full-scale studies was conducted to obtain aqueous suspensions of graphene nanostructures using shear exfoliation in a stator-rotor mixer [6]. The results of these studies quickly found practical application in the modification of concrete with graphene nanostructures [7]. Our group improved the design of the stator-rotor of the mixer and began to use moving blades, which fundamentally changed the mechanism of shear exfoliation. If in the works [6, 7] the exfoliation of graphite particles was carried out in a very thin (less than 0.1 mm) high-speed fluid flow, in our design shear effects on particles in graphite are transmitted directly from the blade, which glides along the inner surface of the stator. To move to the industrial scale of production of suspensions, which contain low-layer and multi-layer graphene, it is necessary to know the kinetics of the process, i.e. how the concentration of graphene nanostructures changes over time if the geometric and operational parameters of the process change. It is these problems that are investigated in this article.

2 Experiment

We conducted experimental studies on the installation, which is shown in figure 1. The main unit is a rotary apparatus with movable blades, similar to the one that we used earlier [8].

Fig. 1. Experimental setup for exfoliation of graphite in oil.

In contrast to this design, this stator was equipped with upper and lower covers. In addition, there were only two holes in the stator. One in the upper part of the stator and through this hole from the tank, which was located above the apparatus, the suspension was discharged into the apparatus by a pump with a fixed flow rate. The second hole is located in the lower part of the stator and through this hole the suspension exits the apparatus into a container.

After all the suspension from the upper tank passed through the rotor apparatus, the rotor rotation drive was turned off and the suspension treatment cycle ended. The suspension was pumped from the lower tank to the upper tank and the recycle cycle was repeated. This scheme, unlike the one used in the works [6, 7], guaranteed the processing of the entire suspension in one cycle. As a rotor rotation drive, we used a Black & Decker milling machine with a power of 1200 watts and a rotation speed of up to 27,000 rpm. A control unit for rotation speed and fixing of power consumption was specially designed and manufactured. This made it possible to control the process and record all process parameters on a personal computer.

As source materials we used: industrial oil I20A; organosilicon liquid; polyaliphine oil; crystalline graphite powder GSM-2. During the experiments, the following was changed: the concentration of graphite in the initial suspension from 1 to 5 weight percent; rotor speed from 6000 to 20,000 rpm; suspension flow rate, i.e. the volume that passed through the rotary apparatus per unit time from 3 l / min to 15 l / m.

After every 30 treatment cycles, samples were taken from the suspension and the concentration of graphene nanostructures in the suspension were determined. In the first series of experiments, a mixture of oil and graphite with different concentrations was used as the initial suspension. In the second series of experiments, the initial mixture with a graphite concentration of 10% was processed in a rod drum mill for 1 to 2 hours, then pure oil was added to the mixture so that the concentration of graphite and the resulting graphene nanostructures was from 1 to 5% and then the suspension was started to be processed on a rotary apparatus. Thus, in the second series of experiments, graphene nanostructures were already in the initial suspension, and graphite particles were mechanically activated.

3 Results and discussion

Figure 2 shows the characteristic dependence of the change in the concentration of graphene structures in suspension on the number of treatment cycles. It should be especially noted that since the suspension was passed through a rotary apparatus, the number of treatment cycles could be clearly fixed. The graph shows that after 60-90 cycles, the concentration does not increase. Given this fact, in further studies, the concentration of graphite in the initial suspension was increased and the yy was processed from 20 to 30 cycles.

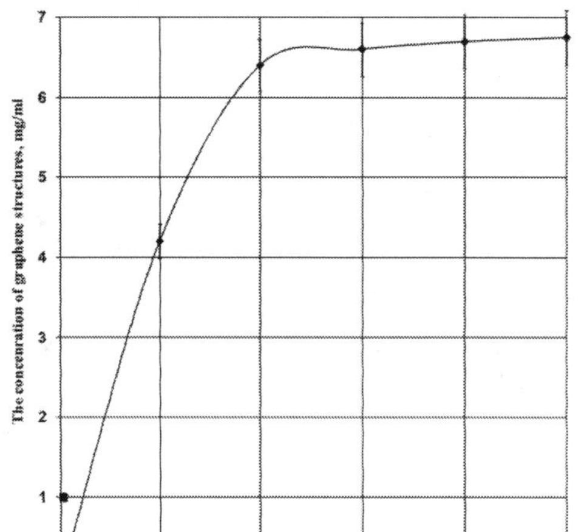

Fig. 2. The dependence of the concentration of graphene structures in suspension on the number of cycles

Figure 3 shows the dependence of the change in the optical density coefficient on the number of treatment cycles for the initial suspension (curve 1) and the suspension pretreated in a core drum mill. On the graphs, you can see that curve 2 is slightly steeper than curve 1. After pre-treatment in a rotary bar drum mill, the maximum concentration is reached faster and is 1.5 times higher than for untreated suspension (curve 1).

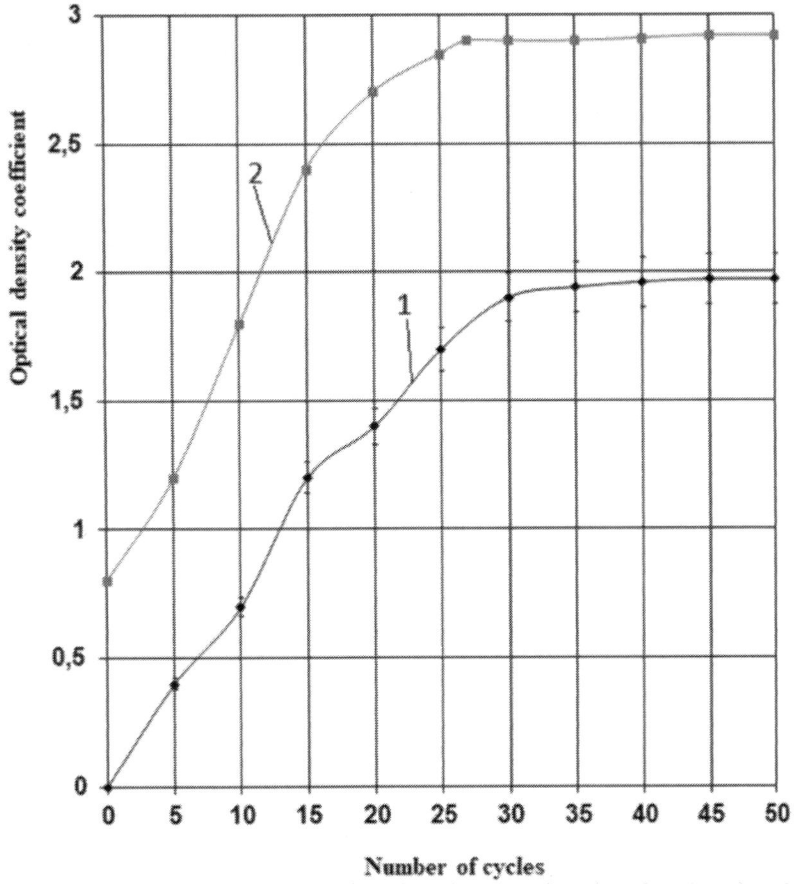

Fig. 3. Dependence of the change in the optical density coefficient on the number of treatment cycles

Figure 4 shows the characteristic dependence of the change in the optical coefficient on the number of cycles, with periodic addition of pure oil to the suspension. On the graphs, you can see that when processing a mechanically activated suspension, curve 2 is steeper than curve 1 and faster reaches the maximum optical coefficient, and therefore the maximum concentration of graphene nanostructures in the suspension. It is very important to note that despite the fact that the optical coefficient at 20 and 40 processing cycles is almost the same, the liquid volume is 2 times larger and this means that the mass of graphene particles is also 2 times larger. Of course, it must be borne in mind that after adding pure oil not only the volume has increased, but also the time during which one processing cycle is carried out. However, if the volume increased by 2 times, then the time of one cycle increased by a maximum of 1.5 times, because after adding oil, the viscosity of the suspension decreased and, consequently, the volumetric productivity of the rotor apparatus increased. However, this is a new problem that requires additional research. In this article, we did not set ourselves such a task. At the moment, we can only say that the productivity of the rotary apparatus with the periodic addition of pure oil, in terms of dry graphene nanostructures, increased by 33%. Indeed, if we received 10g in 10 minutes, and began to receive 20 grams in 15 minutes, then the increase in productivity by 33%.

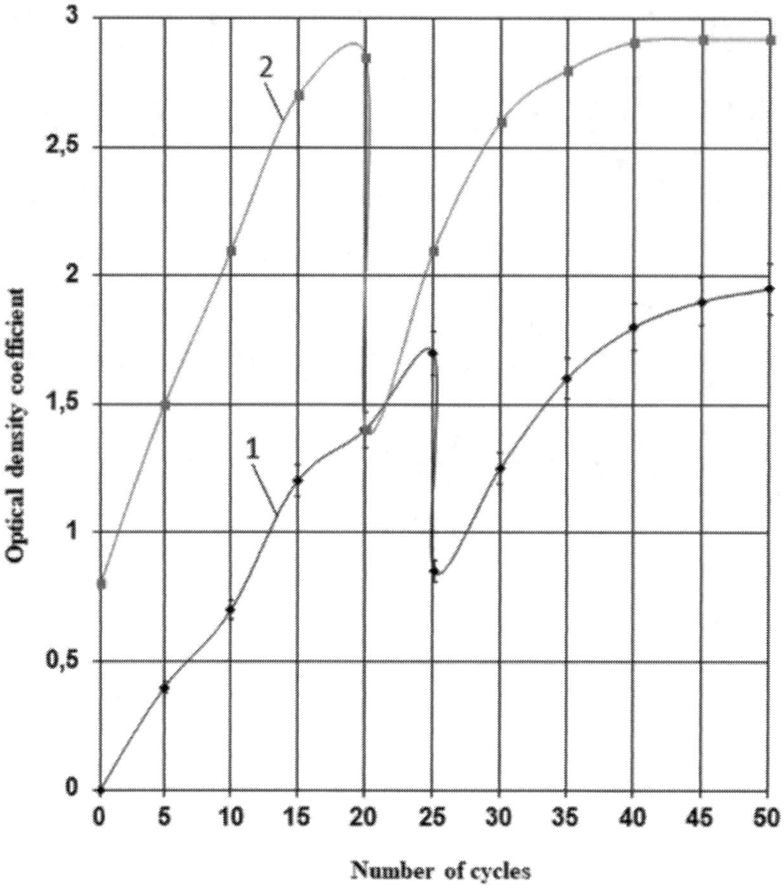

Fig. 4. Dependence of the change in the optical coefficient on the number of cycles, with periodic addition of pure oil to the suspension

4 Conclusion

As a result of experimental studies, it has been established that the concentration of graphene structures in the finished suspension is directly proportional to: the number of processing cycles; rotor speed; the concentration of graphite in the original suspension. In addition, it was found that to increase the efficiency of the exfoliation process, the fraction of graphite that could be converted into graphene structures can be achieved by periodically reducing the concentration of graphene structures in the suspension being processed, by adding pure liquid to this suspension. The technology for producing graphene structures directly in oil by shear exfoliation of graphite has good prospects for industrial implementation.

The reported study was funded by RFBR, project number 19-38-90159.

References

1. M.A.M. Souza, L.C. Pardini, Materia, **21(3)**, 706-713 (2016)
2. D. Berman, S.A. Deshmukh, Adv. Funct. Mater., **24(42)**, 6640-6646 (2014)

3. A.A. Gomaa, M.I. Khashaba, J. Multidiscip. Eng. Sci. Technol., **3(8)**, 5459-5465 (2016)
4. V.F. Pershin, K.A. Ovchinnikov, A.A. Alsilo, R.A. Stolyarov, N.R. Memetov, Nanotechnol. Russ. **13(5-6)**, 344-348 (2018)
5. D.A.Y. Al-Saadi, V.F. Pershin, B.N. Salimov, S.A. Montaev, J. Frict. Wear **38(5)**, 355-358 (2017)
6. K.R. Paton, E. Varrla, C. Backes, R.J. Smith, U. Khan, A. O'Neill, C. Boland, M. Lotya, O.M. Istrate, P. King, T. Higgins, S. Barwich, P. May, P. Puczkarski, I. Ahmed, M. Moebius, H. Pettersson, E. Long, J. Coelho, S.E. O'Brien, E.K. McGuire, B.M. Sanchez, G.S. Duesberg, N. McEvoy, T.J. Pennycook, C. Downing, A. Crossley, V. Nicolosi, J.N. Coleman, Nat. mater. **13(6)**, 3944 (2014)
7. D. Dimov, I. Amit, O. Gorrie, M.D. Barnes, N.J. Townsend, A.I. Neves, F. Withers, S. Russo, M.F. Craciun, Adv. Funct. Mater. **28**, 1705183 (2018)
8. K.A. Al-Shiblawi, A.A. Pasko, V.F. Pershin, Transactions TSTU, **24(4)**, 717-726 (2018)

MATEC Web of Conferences **315**, 06004 (2020) https://doi.org/10.1051/matecconf/202031506004
ICMSSTE 2020

Rotor apparatus performance in the production of graphene-containing suspensions

Ruaa A. Al-Jarah[1], *Andrey* Baranov[2*], and *Anastasiya* Pasko[2]

[1]Post-graduate student from the Republic of Iraq, Tambov State Technical University, 106 Sovetskaya Street, 392000, Tambov, Russian Federation
[2]Department of Technology and Methods of Nanoproducts Manufacturing, Tambov State Technical University, 106 Sovetskaya Street, 392000, Tambov, Russian Federation

Abstract. A new design of the rotor apparatus and the order of the experiments are proposed, which allows you to change individual parameters and study their influence on the intensity of the process. It has been theoretically and experimentally proved that with a change in the volume of the suspension passing through the rotor apparatus per unit time, the performance on graphene structures does not change. It is necessary to increase productivity either by increasing the rotor speed, or by increasing the concentration of graphite in the initial suspension.

1 Introduction

Graphene is one of the most promising materials of our time due to its unique properties [1]. One of the most remarkable advantages of graphene is that even in very small quantities it can greatly improve the performance properties of structural and functional materials. For example, adding only 0.07 weight percent of few-layer graphene to concrete increases the compressive and bending strengths by 146 and 79 percent, respectively [2]. In addition, a reduction in water permeability of almost 400% compared with conventional concrete makes this new composite material ideal for construction in areas prone to flooding. It has been experimentally established that the three-point bending strength of the graphene-modified samples (graphene concentration of about 0.05 wt.% with respect to cement) is about 1.5 times greater than that of the pristine concrete, whereas the water absorption of the 2DCM-modified concrete decreases by 3 times. Using the method for modifying the epoxy resin with the 2DCM the Charpy impact strength is increased by 25 %. The tensile and bending strength increase by 10%. The test results of the repaired beams after breaking at three-point bending, as well as on the adhesion of the reinforcing fabric to the concrete, revealed a synergistic effect while using the 2DCM-modified concrete and epoxy resin [3]. The review [4] details the methods for modifying epoxy resin with few-layer graphene.

Few-layer and multilayer graphene is used to improve the tribological characteristics of lubricants [5-8]. The paper [5] describes the results of a study of the graphene concentration effect on the viscosity of base oil for preparing plastic lubricants. It was established that the

*Corresponding author: barr2@yandex.ru

viscosity of the graphene masterbatch reaches the viscosity of the plastic lubricant at 9-11 wt% for few-layered graphene and 5-8 wt% for multi-layered one. It was found that if graphite is peeled off in the base oil used for the production of lubricant, the friction coefficient and the diameter of the wear spot are reduced by 35 and 25%, respectively, due to the modification of the lubricant with 0.12 wt.% few-layer graphene [6]. Modification with 0.1% multilayer graphene complex calcium grease, the diameter of the wear spot decreased by 50%, the scoring index increased 2.9 times, and the bearing capacity increased 3.8 times [7].

As a result of processing graphite lubricant in a planetary mill, the friction coefficient decreased by 3 times [9]. For the industrial production of graphene concentrates, which can modify plastic lubricants, a technology is proposed that is implemented in a drum rod mill [10-12]. Graphene nanoparticles can be obtained from expanded graphite compounds by mechanochemical synthesis [13].

The problems of the modification of lubricants by nanoparticles, including graphene, are considered in detail in a review [14].

2 Experiments

The authors [2] used the Silverson model L5M mixer, which generates high shear due to the closely spaced (0.1 mm) rotor / stator combination. The experiments were carried out as follows. A fixed amount of crystalline graphite (powder) was poured into 4.5 liters of water, which was in a cylindrical container. To prevent aggregation of graphene plates, N-methyl-2-pyrrolidone (NMP) or sodium cholate (NaC) was added to water. The stator and rotor were placed in a container and the rotor began to rotate at a speed 4500 rpm. Suspension due to centrifugal forces, which creates a rapidly rotating rotor, is ejected through holes in the side surface of the stator. In the zone between the rotor and the stator, a vacuum (vacuum) is created and the suspension is sucked in through the upper and lower ends of the stator. Thus, the mixer works like a pump. After centrifugation, these suspensions contained a large number of high-quality graphene nanosheets, including some monolayers. It was experimentally established that exfoliation takes place only if the gap between the stator and rotor blades is less than 0.1 mm and the shear rate is more than 10^4 s^{-1}.

In our opinion, this device design has several disadvantages. First of all, the fluid motion is random in nature. Despite the fact that there is a circulation circuit there is no guarantee that all particles of graphite, for a fixed period of time, will pass through the holes in the stator the same number of times. Changing the position of the mixer in the tank we obtained different results, in particular, different concentrations of graphene nanostructures in suspension after the same centrifugation regimes. In addition, during prolonged use, the blades and the inner surface of the stator are abraded and the gap between them increases. So, after 700 hours of operation, the exfoliation of graphite particles practically stopped, probably because the gap became more than 0.1 mm.

The design of a rotary apparatus with moving blades was previously proposed [15], and experiments were performed as follows. We prepared 3 ... 6 liters of oil suspension of crystalline fuel GSM-2 with a concentration of 2.5-10%. The suspension was poured into a cylindrical container, a rotary apparatus was installed in this container, and processing of the suspension was started. Every 5 cycles of suspension processing, samples were taken to determine the concentration of graphene structures.

During rotation of the rotor, the centrifugal forces acted on the suspension, which ejected the suspension through the holes in the stator of the apparatus. In the zone between the rotor and the stator a rarefied state (vacuum) was formed and new portions of the suspension were absorbed into this zone. In fact, the rotary apparatus worked like a pump.

https://doi.org/10.1051/matecconf/202031506004

The suspension in the container moved along a closed circulation loop periodically passing through the apparatus where graphite exfoliated. The dependence of changes in the concentration of graphene structures (low-layer and multilayer graphene) in suspension on the main parameters was experimentally obtained: the concentration of graphite in the initial suspension is c_O; rotor speed; stator inner radius R; the number of blades is n; processing time t. The concentration of graphene structures depends almost linearly on all these parameters:

$$c = kc_L\left[(c_O / c_{LO}) \times (R / R_L) \times (\omega / \omega_L) \times (n / n_L) \times (t / t_L)\ \right] \tag{1}$$

where c is the concentration of graphene structures in the designed apparatus, the index L – refers to a laboratory setup.

Thus, in order to calculate the concentration of graphene structures that can be obtained in the designed rotary apparatus, it is necessary to experimentally determine the numerical value of the concentration c_L in a laboratory setup and then calculate the value of c from dependence (1).

In the transition from laboratory installations to industrial apparatus, it is advisable to switch from a batch process to a continuous one. Figure 1 shows a diagram of the rotary apparatus that we used in this work. A rotor *2* is located in the cylindrical stator *1*. The suspension is fed into the apparatus through the nozzle *3* and leaves the apparatus through the nozzle *4*. The rotor *2* has radial grooves in which the movable blades *5* are mounted. When the rotor rotates, under the influence of centrifugal forces, the blades *5* are pressed against the inner surface stator *1* and slide on it without a gap. A roof *6* was installed on top of the stator *1*, and a bottom *7* was installed on the bottom.

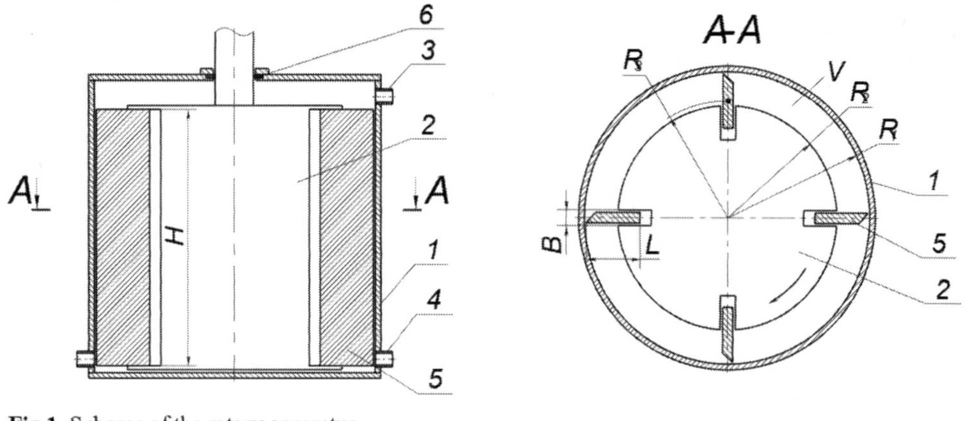

Fig.1. Scheme of the rotary apparatus.

The general installation diagram, which is shown in figure 2, was also changed.

From 3 to 5 liters of liquid was poured into a container *1* and crystalline graphite powder. The concentration of graphite in the initial suspension ranged from 2.5 to 20%. The liquid used was water or synthetic oil. The paddle mixer installed in the container *1* prevented the deposition of graphite to the bottom. After preparing the initial suspension, the valve V_1 was opened and the suspension was supplied to the rotary apparatus *2* using pump P_1. After all the suspension from the container *1* passes through the rotary apparatus and moves into the container *3*, we closed the valve V_1, turned off the pump P_1 and the rotary apparatus *2*. One cycle of processing the suspension was completed.

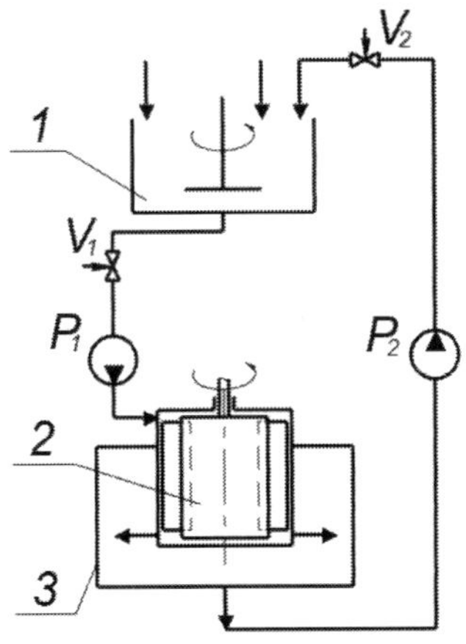

Fig. 2. Schematic diagram of the installation for liquid-phase shear exfoliation of graphite.

Then we opened the valve, turned on the pump and the suspension passed from container *3* to container *1*. After turning off the pump and closing the valve, the device is ready for the next suspension treatment cycle.

Next, open the valve, turn on the pump and move the suspension from container *3* to container *1*. After turning off the pump and closing the valve, the unit is ready for the next suspension treatment cycle.

In the experiments, we changed the flow rate of the suspension from 5 to 10 liters per minute, and the rotor speed from 5000 to 15000 rpm.

Every 30 cycles of suspension processing, samples were taken to determine the concentration of graphene structures.

3 Results and discussion

Figure 3 shows the dependence of the concentration of graphene structures on the number of suspension processing cycles. The rotor speed was 10,000 rpm, the graphite concentration in the initial suspension was 50 mg / ml, suspension volume was 5 liters and pump *P* was 5 liters per minute.

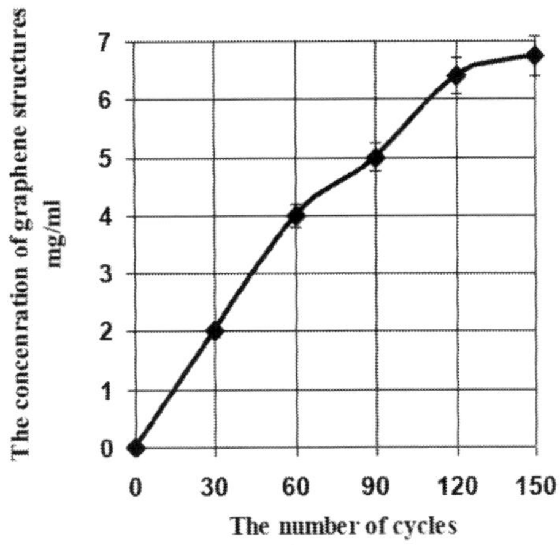

Fig. 3. Dependence of the concentration of graphene structures on the number of suspension processing cycles.

Comparing the results obtained earlier with the results presented in figure 2, we saw that previously the maximum concentration of graphene structures in oil was achieved in 25 cycles [16], and in the present work for 150 cycles. Assuming that such a big difference arose because of different experimental methods, we calculated the time of one cycle for these two options for exfoliation. In the first case, the time of one cycle was taken equal to 300 s, and in our case, the time of one cycle (the time during which the entire suspension passes through the rotary apparatus) is 60 s. Thus, in [16], the maximum concentration of graphene structures in the suspension was achieved in 125 s, and in our case, in 90 s. It is likely that such a reduction in processing time occurred because in our case the entire suspension passes through the apparatus only once. With a periodic process, it is possible that part of the suspension does not enter the apparatus during one cycle.

It was experimentally established that with an increase or decrease in the rotor speed, the change in the concentration of graphene structures in one processing cycle did not change. In order to explain this fact, we determine the length of the path that a graphite particle travels in a rotary apparatus in one exfoliation cycle. First of all, we calculated the volume between the rotor and the stator (Figure 1):

$$V = \pi H(R_1^2 - R_2^2) = 3.14 \cdot 70 \cdot (20^2 - 18.5^2) = 12693 \text{ mm}^3. \quad (2)$$

Given that the pump capacity is equal to 5,000,000 mm^3, the time of one cycle will be equal to:

$$t = V / Q = 126993 / 5000000 = 0.015 \text{ s}. \quad (3)$$

The length L_1 of the sliding path of the particle along the inner surface of the rotor will be equal to:

$$L_1 = 2\pi R_1 nt, \quad (4)$$

where n is the rotor speed of revolutions per second.

Particular attention should be paid to the fact that if we increase productivity Q, then the residence time t of particles in the apparatus will be proportionally reduced, i.e. in the exfoliation zone. As a result, the increase in the concentration of graphene structures in the suspension will decrease. However, at the same time, the time of one cycle will be proportionally reduced. In the final result, an increase in the concentration of graphene structures per unit time will not change. If, at a constant value of productivity, the rotor speed is increased, the path length in one cycle will increase and, as a result, the concentration of graphene structures will increase.

In the periodic mode, it was not taken into account that with an increase in the rotor speed, the volume of the suspension increases, which passes through the rotor apparatus per unit time, which leads to a decrease in the time for which one exfoliation cycle takes place. The presented methodology for the exfoliation process (figure 2) allows you to independently change the process parameters and find the real influence of these parameters on performance.

4 Conclusion

A new design of the rotor apparatus and the order of the experiments are proposed, which allows you to change individual parameters and study their influence on the intensity of the process. It has been theoretically and experimentally proved that with a change in the volume of the suspension passing through the rotor apparatus per unit time, the performance on graphene structures does not change. It is necessary to increase productivity either by increasing the rotor speed, or by increasing the concentration of graphite in the initial suspension.

References

1. A.P. Kauling, A.T. Seefeldt, D.P. Pisoni, R.C. Pradeep, R. Bentini, R.V.B. Oliveira, K.S. Novoselov, A.H.C. Neto, Adv. Mater. **30**, 03784 (2018)
2. D. Dimov, I. Amit, O. Gorrie, M.D. Barnes, N.J. Townsend, A.I. Neves, F. Withers, S. Russo, M.F. Craciun, Adv. Funct. Mater. **28**, 1705183 (2018)
3. V. Pershin, K.A. Al-Shiblawi, A.M.R. Al-Mashhadani, A. Pasko, D. Melekhin, IOP Conf. Ser-Mat. Sci. **693(1)**, 012046 (2019)
4. K.A. Al-Shiblawi, V.F. Pershin, V.P. Yartsev, Mater. Sci. Forum **4**, 42-53 (2018)
5. V. Pershin, K. Ovchinnikov, Z. Al-Hilo, N. Memetov, A. Tkachev, E. Galunin, AIP Conf. Proc. **2041**, 020016 (2018)
6. V. Pershin, G. Zhumagalieva, A. Tkachev, A. Pasko, N. Memetov, J. Phys.Conf. Ser. **1278(1)**, 012025 (2019)
7. V.F. Pershin, K.A. Ovchinnikov, A.A. Alsilo, R.A. Stolyarov, N.R. Memetov, Nanotechnol. Russ. **13(5-6)**, 344-348 (2018)
8. V.F. Pershin, M.N. Krasnyanskiy, Z.A.A. Alhilo, A.M.R. Al-Mashhadani, A.A. Baranov, A.A. Osipov, IOP Conf. Ser.-Mat. Sci. **693(1)**, 012023 (2019)
9. D.A.Y. Al-Saadi, V.F. Pershin, B.N. Salimov, S.A. Montaev, J. Frict. Wear **38(5)**, 355-358 (2017)
10. G. Zhumagalieva, V. Pershin, A. Tkachev, A. Vorobiev, A. Pasko, E. Galunin, AIP Conf. Proc. **2041**, 020010 (2018)
11. V.F. Pershin, G.B. Zhumagaliyeva, A.G Tkachev, A.A. Pasko, A.M. Vorobyev, IOP Conf. Ser-Mat. Sci. **693(1)**, 012035 (2019)
12. A. Tkachev, G. Zhumagalieva, Z.A.A.A. Ali, N. Memetov, E. Galunin, V. Pershin, *Modification of Frost-Resistant Plastic Lubricants Using Few- and Multi-Layered Graphene*, in Proceedings of the 4th World Congress on Recent Advances in Nanotechnology (RAN'19), 14-16 April 2019, Rome, Italy (2019)
13. A.V. Melezhik, V.F. Pershin, N.R. Memetov, A.G. Tkachev, Nanotechnol. Russ. **11(7-8)**, 421-429 (2016)
14. I. Ali, Al A. Basheer, A. Kucherova, N. Memetov, T. Pasko, K. Ovchinnikov, V. Pershin, D. Kuznetsov, E. Galunin, V. Grachev, A. Tkachev, J Mol. Liq. **279 (1)**, 251-266 (2019)
15. K.A. Al-Shiblawi, A.A. Pasko, V.F. Pershin, Transactions TSTU, **24(4)**, 717-726 (2018)
16. V.F. Pershin, A.G. Tkachev, R.A. Al-Jarah, T.V. Pasko, A.A. Osipov, IOP Conf. Ser.-Mat. Sci. **693(1)**, 012007 (2019)

Environmentally friendly technology for the modification of lubricants with graphene nanostructures

Zaman A A Alhilo[1*], *Gaukhar* Zhumagalieva[2], and *Tatiana* Pasko[3]

[1]Post-graduate student from the Republic of Iraq, Tambov State Technical University, 106 Sovetskaya Street, 392000, Tambov, Russian Federation
[2]Zhangir Khan West Kazakhstan Agrarian-Technical University, Zhangir Khan Str., 51, Uralsk, 090009, Republic of Kazakhstan
[3]Department of Technology and Methods of Nanoproducts Manufacturing, Tambov State Technical University, 106 Sovetskaya Street, 392000, Tambov, Russian Federation

Abstract. Nanomaterials, and in particular graphene, significantly increase the basic tribological characteristics of lubricants. One of the most effective ways to quickly use laboratory results on an industrial scale is to modify traditional lubricants with graphene concentrates. The research results showed that in order to solve this problem successfully, it is necessary to create graphene concentrate and distribute it evenly over the total volume of the lubricant. It is necessary that the concentrate be prepared on the same basis as the lubricant. In the present work, previous studies on the mechanical activation and exfoliation of graphite in a core drum mill, liquid-phase shear exfoliation in a rotary apparatus with moving blades, and the homogenization of graphene nanostructures in lubricants using a rotary mixer are combined. Changes were made to the design of the main components, taking into account the use of this technology in the industry. The presented technology is environmentally friendly, since only oil and graphite are used and there are no harmful wastes that must be disposed of.

1 Introduction

Currently, lubricants have high requirements regarding antifriction and anti-wear characteristics in friction units of many mechanisms and machines operating at variable and elevated speeds, pressures and under various temperature conditions [1]. The formation of composite lubricants with desired properties has become the main direction of development of modern tribology. The anisotropic nanofriction force on MoS_2 as a function of thickness was studied by a calibrated atomic force microscopy (AFM) in ambient air. The observed anisotropic nanofriction on MoS_2 with a thickness of 4.18 nm was attributed to the combination of the lattice orientation and puckering effect. The anisotropic nanofriction on

*Corresponding author: tpasko@yandex.ru

MoS_2 with a thickness of 1.49 nm was mainly due to the puckering effect dominating the nanofriction. As the thickness of MoS_2 increased, the magnitude of nanofriction decreased, and the anisotropy ratio of nanofriction increased. The structural anisotropy coupling with thickness plays an important role in the anisotropic nanofriction on MoS_2. This study could provide the clue for the design of MoS_2 and other two-dimensional materials in lubrication applications.

Graphite is traditionally used as an antifriction material [2]. A new explosion of interest in materials that can be obtained from graphite occurred after graphene was obtained and its unique physical and mechanical characteristics were determined. It was experimentally established that one layer of graphene grown on steel withstands 6400 slip cycles, while low-layer graphene (consisting of 3-4 layers) withstands 47,000 cycles. In addition, it was shown that these graphene layers completely stop wear, despite the difficult sliding conditions, including high contact pressures (≈0.5 GPa), which are usually observed in small-scale trials [3]. The experimental results demonstrate that the addition of 0.4 wt.% Graphene nanofiller to lithium lubricants gives good antifriction properties and wear resistance indicators [4]. Theories of friction and graphite wear were analyzed in [5]. The combined action of graphite and graphene in grease can reduce the friction coefficient by 2.2-2.5 times [6]. Adding graphene concentrate to a standard lubricant is one of the promising ways to increase the tribological characteristics of lubricants.

Earlier, we analyzed the effect of carbon nanomaterials on the tribological characteristics of lubricants [7]. In spite of good development and progress in lubricants, billions US dollars are being lost in many industries and motor vehicles globally due to mechanisms frictions. Therefore, there is a great need to develop more efficient and effective lubricants. The review paper considers the application of carbon nanostructures, such as carbon nanotubes and graphene, as nano-additives to lubricants in order to improve their tribological characteristics i.e. to decrease friction and wear. Some important aspects of nanoparticles in modifying lubricants are size, form, structure, surface functionalization, concentration, physical and chemical properties of surfaces, which are outlined in this article. It was observed that the adding of nano-additives to lubricants; even at low concentrations; decreased the friction and wear coefficients considerably. Moreover, carbon nanotubes and graphene are environmentally friendly and safe. Thus, all these factors allow positive assessment of the prospects of using the above mentioned nanomaterials to modify lubricants.

The review [5] compares the classical theory of friction and wear with the modern one. The advantages and disadvantages of each of these theories are analyzed. In addition, the graphology tribology, the prospects of its use are considered and shown.

In recent decades, much attention has been paid to studies of the effect of nanoscale fillers and, in particular, graphene nanostructures on the performance of lubricants. Two-dimension graphene-like covalent-organic frameworks (GCF) nano-sheets were successfully fabricated to evaluate the tribological behavior as lubricant additive in comparison to GO and rGO in PAO 10 oil [8]. The mechanical strength of covalent bond structure in-plane ensures carrying capacity and durability, and the ultrathin layered feature reduces shear stress due to slide effect. Eventually, the excellent reducing friction (up to 53.5%) and anti-wear (up to 95.4%) were achieved at an ultralow dosage of 0.008 wt%. By contrast, dissociation of C–O linkages of rGO and GO induced by tribo-stress and tribo-heat deteriorates its stability and durability.

Graphite oxide (GO) with N doping is synthesized from the waste carbon source. The GO products obtained at 1200–1400 °C are utilized as nano-additive in synthetic engine oil [9]. The GO is produced in the form of layered structure with Ca compound incorporation and functional groups. GO addition has improved the friction (2.8–21.1% reduction) and wear (8.2–18.5% reduction in worn volume) properties due to its easy shear force of layered structure.

Natural crystalline graphite of GSM-2 grade (ash content <0.5%) was used in the study [10]. The oxidative intercalation of graphite was carried out using ammonium persulfate (pure for analysis grade) dissolved in sulfuric acid containing 5% free sulfur trioxide. Intercalated graphite was subjected to cold expansion at 40°C for 3 h. The apparent volume of graphite is 270–280 cm3/g of the initial graphite. In order to prepare a paste to be used for mechanical treatment, carbamide (on the basis of 0.5 mole per 1 mole of sulfuric acid contained in EGC) was added to EGC. The reaction mixture self-heated to 80°C and a thick gray-green paste was formed. Glycerol (100 g per 5 g of the initial graphite) was then added to this paste. The resulting paste was treated in a stainless steel cylinder-shaped container with an internal diameter of 60.0 mm and a steel rotor 59.3 mm in diameter rotating inside the container. The net time of treatment was 30 min. The GNPs in the form of an aqueous paste containing 3–5% of dry substance were obtained.

GNPs are successfully produced in rod drum mills [11, 12] and rotary apparatuses (stator-rotor mixer) [13, 14].

In our opinion, it is better to use graphene nanoparticles (GNP) to modify lubricants than graphene oxide because the technology for producing GNP is environmentally friendly. It should be noted that they are used to modify not only lubricants, but also other materials, such as epoxy [15, 16].

It is very important that using the method of liquid-phase shear exfoliation it is possible to produce GNP directly in synthetic oils, and on an industrial scale [17, 18]. In addition, graphene-modified lubricants can operate at low temperatures [19, 20].

This article proposes a technology for the modification of lubricants with graphene nanostructures obtained by shear exfoliation of graphite, which can be implemented on an industrial scale.

2 Technology for the production of graphene structures and the modification of lubricants

Figure 1 shows a scheme of the modification of lubricants by graphene nanostructures.

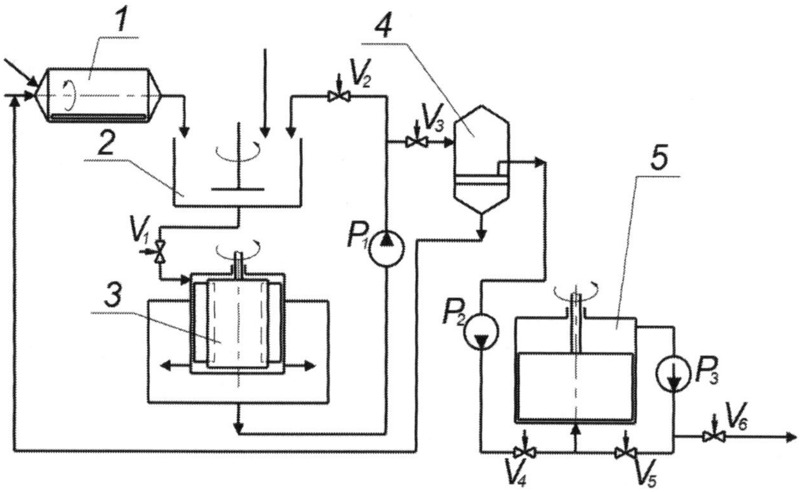

Fig. 1. Functional diagram of the production of graphene nanostructures and the modification of lubricants: *1* – rod drum mill; *2* – stirred tank; *3* – rotary apparatus (exfoliator); *4* – filter; *5* – rotary mixer; *6* – rotary homogenizer; P_1, P_2, P_3 – pumps; V_1, V_2, V_3, V_4, V_5, V_6 – valves.

This technology is implemented as follows. Pre-prepare a mixture of synthetic oil and crystalline graphite powder and load them into a rod drum mill *1*. Mass concentration of graphite is from 10 to 20%. In the mill, graphite is mechanically activated and partially exfoliated, i.e. stratification of graphite particles into graphene nanostructures. The term "graphene nanostructures" should be understood as a mixture of particles consisting of a different number of graphene layers. Graphene nanostructures contain both single-layer graphene and low-layer and even multi-layer graphene. However, all particles have at least one of the sizes less than 100 nanometers. The duration of processing the mixture in the mill, depending on the desired result and the type of synthetic oil, can vary from 1 to 5 hours. A little later we will discuss the features of processing the mixture in the mill. After mechanical activation is complete, the mixture is reloaded into a container *2* and either pure synthetic oil or clarified suspensions from the filter *4* are added. After adding oil, the concentration of graphite in the suspension should be 1-2%. Open the crane *1* and turn on the rotary drive of the exfoliator *3*. The design of the exfoliator is similar to the one used in the work. The exfoliator consists of a cylindrical stator inside which a rotor with movable blades is installed. The stator is closed by covers at the top and bottom. The suspension is fed into the hole, which is located at the top of the stator, and exits through the hole at the bottom of the stator. Under the action of centrifugal forces, the movable rotor blades are pressed against the inner surface of the stator and constantly slide along this surface. Graphite particles and graphene nanoparticles that are in suspension are pressed against the inner surface of the stator because their density is greater than the density of synthetic oil. Thus, the particles fall between the stationary stator and the moving blade and dissolve under the action of shear forces, i.e. two particles are formed from one particle. This is the mechanism of shear exfoliation in a rotary apparatus with moving blades. From the exfoliator *3*, the suspension is pumped into the container *2* and then fed back to the exfoliator. Thus, the exfoliator operates in a cyclic mode. Processing the suspension is from 1 to 5 hours. Exfoliation lasts from 1 to 5 hours. At the end of exfoliation, close the valve *2* and open the valve *3*. The suspension is fed to the filter. The liquid part of the suspension, which contains a certain amount of low-graphene graphene, is returned to the beginning of the process and is used either to prepare the initial mixture for the mill, or to dilute the mixture after processing in the mill in a container with a mixer *2*. The precipitate from the filter *4* is fed into the mixer *5*, which is also fed lubricant to be modified. From the mixer *5*, a mixture of lubricant and graphene nanostructures is supplied to the rotary homogenizer *6* by a pump P_2. If the valves are closed and the valve is open, the mixture is processed in the homogenizer in a closed cycle for from 0.5 to 1 hour, i.e. the mixture goes through the garlic homogenizer many times. In a homogenizer, graphene nanostructures are not only uniformly distributed over the entire volume of the lubricant, but also exfoliate due to the high velocity gradient in a very thin layer of lubricant. After processing the mixture in the homogenizer, close the valve V_5, open the valve V_6 and unload the finished product – lubricant modified with graphene nanostructures.

3 Discussion

First of all, it should be noted that the proposed technology is an almost continuous production. Two nodes are limiting by the processing time: a mill; rotary apparatus. These nodes can work simultaneously and the processing time of the material in each of them is the same from 1 to 2 hours. Thus, the period of time when processing of the material does not occur is the time the material moves from the mill 1 to the tank *2*.

Now consider the features of processing a mixture of graphite with synthetic oils in the mill *1*. Figure 2 shows the cross section of a drum mill. Grinding rods *2* are located inside the drum *1*, which are interconnected by a flexible connection *3* and rigidly fixed relative to the base of the mill *4*. Since the drum *1* rotates at a speed greater than critical, that is, the centrifugal forces that act on the mixture are greater than the forces of gravity, the mixture forms a thin ring *5*, which rotates together with the drum *1*. The particles of graphite together fall into the contact zone between the rotating shell of the drum and the fixed rods. Shear forces begin to act on these particles and the particles exfoliate. Since these effects occur repeatedly, graphite particles gradually turn into particles of multilayer graphene, then into particles of few-layer graphene, and ultimately into single-layer graphene.

The results of our experiments showed that the preliminary mechanical activation of a mixture of graphite and oil in a core drum mill significantly increases the intensity of subsequent exfoliation, therefore it is not advisable to process the mixture in a drum mill for more than 2 hours, since a high concentration of graphene nanostructures in the oil can be obtained much faster in a rotary apparatus. It should be noted that with an increase in the concentration of graphene nanostructures in oil, its viscosity increases [19] and therefore the productivity of the rotor apparatus decreases.

Significant changes have also been made to the design of the rotary mixer compared to the design that we used earlier. The rotor height was increased, and therefore the homogenization of a mixture of lubricants with graphene nanostructures and exfoliation of graphene nanostructures is carried out not only when moving along a spiral path in the horizontal plane, but also when moving along a helical line in the cylindrical contact zone of the stator and rotor.

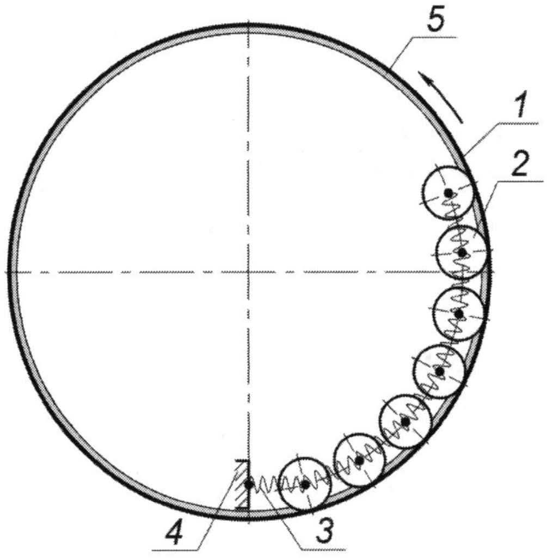

Fig. 2. Cross section of a drum mill: *1* – drum; *2* – rod; *3* – flexible connection; *4* – mill base; *5* – mixture of graphite and oil.

3 Conclusion

In the present work, previous studies on the mechanical activation and exfoliation of graphite in a core drum mill, liquid-phase shear exfoliation in a rotary apparatus with moving blades, and the homogenization of graphene nanostructures in lubricants using a rotary mixer are combined. Changes were made to the design of the main components, taking into account the use of this technology in the industry. The presented technology is environmentally friendly, since only oil and graphite are used and there are no harmful wastes that must be disposed of.

The reported study was funded by RFBR, project number 19-38-90159.

References

1. X. Cao, X. Gan, H. Lang, K. Yu, Sh. Ding, Y. Peng, W. Yi, Tribol. Int. **134**, 308-316 (2019)
2. M. Schmitt, S. Bistac, Kh. Jradi, J. Phys. Conf. Ser. **61(1)**, 1032-1036 (2007)
3. D. Berman, S.A. Deshmukh, Adv. Funct. Mater. **24(42)**, 6640-6646 (2014)
4. A.A. Gomaa, M.I. Khashaba, J. Multidiscip. Eng. Sci. Technol. **3(8)**, 5459-5465 (2016)
5. S. Thakur, K.K. Thakur, H. Singh, AGU Int. J. Eng. Technol. **4**, 62-69 (2017)
6. D.A.Y. Al-Saadi, V.F. Pershin, B.N. Salimov, S.A. Montaev, J. Frict. Wear **38(5)**, 355-358 (2017)
7. I. Ali, Al A. Basheer, A. Kucherova, N. Memetov, T. Pasko, K. Ovchinnikov, V. Pershin, D. Kuznetsov, E. Galunin, V. Grachev, A. Tkachev, J Mol. Liq. **279 (1)**, 251-266 (2019)
8. P. Wen, Y. Lei, W. Li, M. Fan, Tribol. Int. **143**, 106051 (2020)
9. B. Sivakumar, N. Ranjan, S. Ramaprabhu, M. Kamaraj, Tribol. Int. **142**, 105990 (2020)
10. A.V. Melezhik, V.F. Pershin, N.R. Memetov, A.G. Tkachev, Nanotechnol. Russ. **7-8**, 421-429 (2016)
11. G. Zhumagalieva, V. Pershin, A. Tkachev, A. Vorobiev, A. Pasko, E. Galunin, AIP Conf. Proc. **2041**, 020010 (2018)
12. V.F. Pershin, G.B. Zhumagaliyeva, A.G. Tkachev, A.A. Pasko, A.M. Vorobyev, IOP Conf. Ser-Mat. Sci. **693(1)**, 012035 (2019)
13. K.A. Al-Shiblawi, V.F. Pershin, V.P. Jarcev, T.V. Pasko, AIP Conf. Proc. **2041**, 020015 (2018)
14. V. Pershin, K.A. Al-Shiblawi, A.M.R. Al-Mashhadani, A. Pasko, D. Melekhin, IOP Conf. Ser-Mat. Sci. **693(1)**, 012046 (2019)
15. K.A. Al-Shiblawi, V.F. Pershin, V.P. Yartsev, Mater. Sci. Forum **2**, 68-78 (2018)
16. K.A. Al-Shiblawi, V.F. Pershin, V.P. Yartsev, Mater. Sci. Forum **4**, 42-53 (2018)
17. V.F. Pershin, M.N. Krasnyanskiy, Z.A.A. Alhilo, A.M.R. Al-Mashhadani, A.A. Baranov, A.A. Osipov, IOP Conf. Ser.-Mat. Sci. **693(1)**, 012023 (2019)
18. V.F. Pershin, K.A. Ovchinnikov, A.A. Alsilo, R.A. Stolyarov, N.R. Memetov, Nanotechnol. Russ. **13(5-6)**, 344-348 (2018)
19. V. Pershin, K. Ovchinnikov, Z. Al-Hilo, N. Memetov, A. Tkachev, E. Galunin, AIP Conf. Proc. **2041**, 020016 (2018)
20. V. Pershin, G. Zhumagalieva, A. Tkachev, A. Pasko, N. Memetov, AIP Conf. Proc. **2101**, 020011 (2019)

Improving functional and environmental performance of Portland cement-based materials by graphene nanostructures

Vladimir Pershin[1*], *Ali* Mashhadani[2], *Denis* Melekhin[3], and *Aleksey* Osipov[1]

[1]Department of Technology and Methods of Nanoproducts Manufacturing, Tambov State Technical University, 106 Sovetskaya Street, 392000, Tambov, Russian Federation
[2]Post-graduate student from the Republic of Iraq, Tambov State Technical University, 106 Sovetskaya Street, 392000, Tambov, Russian Federation
[3]Architecture Faculty, Czech Technical University in Prague, 1903/4 Zikova, 166 36 Prague 6 Czech Republic

Abstract. Nanoparticles can accelerate cement hydration due to their high activity compacting microstructure, and consequently increase the strength. For the industrial production of concrete modified with nanoscale particles, it is necessary to develop an environmentally friendly and cheap technology for the industrial production of nanoparticles. The article presents the results of increasing the operational characteristics of concrete by adding graphene nanostructures obtained by liquid-phase shear exfoliation of graphite. This environmentally friendly technology requires energy costs, at least ten times less than technologies that use ultrasonic treatment of the suspension. Our industrial partner set a specific task: to find the minimum concentration of graphene structures in relation to cement, at which the compressive strength of cement-sand mix increases by 30%. It was experimentally established that at concentrations of graphene nanostructures with respect to cement not more than 0.02%, the compressive strength increases by at least 30%.

1 Introduction

Concrete is the most used material in the global construction industry, but so far there are a number of problems: insufficient strength, especially when bending; high absorption of moisture and therefore not sufficient durability; high carbon dioxide emissions from cement production. Of course, there are other disadvantages of concrete, but in our opinion these are the three main disadvantages that need to be urgently addressed. There is a wide consensus in the research community that concrete should be developed on a nanoscale, where its chemical and physiomechanical properties can really be improved [1]. Nanoparticles can accelerate cement hydration due to their high activity, densifying the microstructure and, therefore, increasing strength and durability [2]. By introducing

*Correspondingauthor:pershin.home@mail.ru

engineering nanomaterials, such as nano-SiO_2, nano-Fe_2O_3, nano-Al_2O_3, nano-CuO, nano-TiO_2, it is possible to increase the mechanical properties and durability of concrete [2].

The effect of nanometakaolin (NMK) on the mechanical properties and microstructure of Portland cement mortar was investigated [3]. The NMK was prepared by thermal activation of kaolin clay for 2 hours at 750°C. Based on experimental studies, the authors made the following conclusions:

• Compressive and tensile strength of the cement mortars with NMK is higher than that of the plain cement mortar with the same w/b ratio.

• The enhancement of tensile strength was 49% above control mortar; whereas the enhancement of compressive strength was 7% at 8% NMK replacement.

• The NMK in cement mortar acts as a nano-fiber due to its morphology.

• The SEM observations confirmed that the NMK was not only acting as filler, but also as an activator to promote hydration process.

Many articles are devoted to studies of the effect of graphene oxide and graphene on concrete performance [4-10]. Graphene oxide (GO), which has been successfully used in cement composites as an admixture, can improve their mechanical and durability properties [4]. This study examined and elucidated the effect of GO on the hydration products of a cement system in detail. The hydration precursors, morphology, composition, and chemical bonds of cement pastes were studied by SEM/BSE, XRD, LA-ICP-MS, and 29Si/27Al MAS-NMR, respectively. The experimental results indicated that the 28-day compressive strength of cement pastes (w/c ¼ 0.35) was increased by 29% when admixing 0.02 wt% GO. There existed chemical reactions between the admixed GO and the cement hydration products. The improved strength actually was not only attributed to the improved hydration degree of cement, but also the newly found tobermorite-like hydrates and jennite-like hydrates due to the Ca ions consumption by the negatively charged GO. The polymerization of hydration products was also improved due to the addition of GO. The incorporation of 0.16 wt % GO into the cement matrix can enhance the flexural strength of the material by 11.62% due to the higher hydration degree, nano-filler effect and cracking-bridging effect [5].

The rate of hydration accelerated and the cumulative hydration heat increased with the increasing proportions of GO in the composite [6]. GO dosages about 0.02 and 0.04 wt% of cement in the composites resulted the maximum enhancement of compressive and flexural strength by 83 and 26%, respectively, compared to the control mix (0 wt % GO). The microstructural investigation shows that GO enhanced the hydration of calcium hydroxide (CH) and calcium silicate hydrate (C-S-H) during the nucleation and growth stages, filled pores, bridged micro-cracks and created interlocking between the cement hydration products. Collectively, these effects ultimately improved the mechanical properties of the composites. Also, in this process, the 0.02 and 0.04 wt% GO cement composite increased the electrical resistivity by 11.5%, and decreased the absorptivity by 29%, respectively, both of which improved the overall performance of the composite.

The study [7] examines and compares the performance of two specific forms of graphene nanomaterials in the cement based composite, namely graphene oxide (GO) and reduced graphene oxide (rGO). The effect of GO and rGO on workability, early age hydration, microstructure, mechanical and transport properties was determined. The XRD and TGA quantification of the hydration products shows that GO composites have a greater content of $Ca(OH)_2$ and C-S-H compared to rGO composites measured at 1, 7 and 28 days. Meso and gel pores (<2 nm) related to C-SH show increasing trend in the GO and rGO composites compared to the control mix in dynamic vapour sorption (DVS) analysis. Compared to the control mix, the 28-day compressive strength increased by 10% and 15% in 0.06 wt% of GO and rGO in the composites. However, the highest flexural strength was increased in the 0.04 wt% of GO and 0.06 wt% rGO composite, and the enhancement was 75.7% and 33.7%, respectively, compared to the control mix.

The effects of graphene oxide (GO) on the hydration kinetics, pore structure, mechanical properties, and the structure of calcium silicate hydrate (C-S-H) gels were systematically investigated by combinatorial techniques [8]. GO can accelerate the cement hydration, refine the pore structure and increase the polymerization degree of C-S-H gels due to the nucleation effects of GO. The specific surface area of cement paste and the number of gel pores were reduced with the addition of GO. A 3D network structure of GO modified C-S-H gels was originally proposed based on the obtained results that GO could intercalate into the interlayer space of C-S-H gels through ionic bonding with Ca^{2+} and fill in the gel pores. A small fraction of GO (0.022 wt%) increased the 28 d compressive strength by 16.31%–25.60% at a various water to cement ratios, indicating that GO is a potential nano-reinforcing material for cement composites.

Acid attack on cement concrete results in the development of a degraded layer surrounding the unaffected material, which causes a deterioration of mechanical properties. The effect of reduced graphene oxide, alumina and silica nanoparticles on the deterioration characteristics of 28-day cured cementitious pastes after storage in 0.5 moL/L HNO3 solution for a period of 56 days is reported [9]. The porosimetry results showed that the presence of reduced graphene oxide and nano alumina decreased the amount of capillary pores (10 nm–10 μm) by up to 46% and 51% than the control paste. Overall, the results suggest that the presence of these nanomaterials refined the pore structure of the cementitious matrix and thereby increased the resistance to leaching of calcium ions from the binder phases exposed to aggressive aqueous media.

To further delineate the collective effects of GNPs and GONPs, particularly their contents, on the performance of concrete, a comprehensive investigation is carried out in the study [10]. Mortar samples, characterized by different water-cement ratios and containing different amounts of GNPs and GONPs, are experimentally tested to systematically probe their strength, electrical resistance and piezoresistive reactions. With a proper water-cement ratio, a small percentage of GNPs and GONPs can significantly improve the concrete strength, while the content of GNPs needs to be substantially increased to achieve low electrical resistance and accurate piezoresistive reaction.

This article presents the results of the influence of GNP on the strength of a cement-sand mixture, since the strength of concrete mainly depends on the strength of this mixture. In our opinion, the use of GNP for concrete modification on an industrial scale has better prospects than civil defense, for the following reasons: GNP production technology is environmentally friendly; no industrial waste; less energy consumption than in the production of GO. Moreover, our industrial partner set a specific task: to find the minimum concentration of graphene structures in relation to cement, at which the strength of cement-sand mix increases by 30%.

2 Materials and experimental procedure

2.1 Materials

To obtain few- layer graphene in this work, we used graphite grade GSM-2 according to GOST 18191-78 (Russian Standard) "Special low-ash graphite. Technical conditions" According to physic-chemical indicators, it meets the requirements indicated in the table. The choice of material is justified by its use at NanoTechCenter LLC (Tambov, Russia) for producing few-layer graphene by liquid-phase exfoliation by ultrasonic treatment of graphite oxide according to the Hamers-Ofemann method [5].

When preparing the suspension, water was used in accordance with GOST 23732-2011 (Russian standard) "Water for concrete and mortar. Technical conditions. " To prevent agglomeration of graphene nanostructures, the surfactant OP-7 was used according to GOST 8433-81 "Substances auxiliary OP-7 and OP-10. Technical conditions. " The concentration of OP-7 was 0.3%, by weight of water. In addition, concrete modifiers were used as surfactants according to GOST R 56178-2014 "Organic-mineral modifiers MB type for concrete, mortar and dry mixes. Technical conditions".

2.2 Experimental setup

Many methods for producing graphene flakes are known [11], but we used the method of liquid-phase exfoliation of graphite [1, 12].

The apparatus consists of a cylindrical stator, a rotor and a rotor rotation drive. In contrast to the apparatus used in [1, 12], radial grooves in which movable blades were mounted were cut in the rotor. In addition, a cover was installed on the stator; holes were placed only in the upper part of the cylindrical shell. The blades were installed in such a way that under the action of centrifugal forces that arose during the rotation of the rotor, the blades moved in the radial direction and glided along the inner surface of the stator with almost no gap. The inner diameter of the stator is 40 mm, and the height is 50 mm. The rotor speed varied from 5,000 to 20,000 rpm.

Changes were made to our design for the following reasons. The authors of [1] believe that exfoliation occurs mainly when particles passing through holes in the stator are exposed to blades. The necessary conditions for the exfoliation of graphite particles are: the gap between the stator and the rotor blades should be no more than 0.1 mm; the local shear rate in the zone of contact of the blades with the stator should not be less than 10 s. In our design, exfoliation occurs in the gap between the blade and the inner surface of the stator. The cover provides a stable circulation loop for the movement of the suspension, which enters the processing zone only through the lower end of the stator, and exits through the holes. The location of the holes in the upper part of the stator provides the maximum residence time of particles in the contact zone of the blades with the inner surface of the stator.

Centrifugal forces act on the fluid and particles located in the area between the stator, the rotating rotor and adjacent blades. Under the action of these forces, the particles are pressed against the inner surface of the stator and the peripheral speed of their motion decreases. When a particle enters the zone between the stator and the moving blade, shear forces begin to act, as a result of which inside the particle, in addition to normal stresses, tangential stresses arise, which leads to a shift of the graphene layers relative to each other. Suppose that a particle consists of five graphene layers. As a result of a shift, two are formed from one particle. In this case, the following options are possible: 1 layer + 4 layers; 2 layers + 3 layers. Apparently, the probabilities of each of these options are the same.

2.3 Methodology of the experiments

Investigations of the exfoliation process were carried out as follows. Prepared 5 litters of an aqueous suspension of crystalline graphite GSM-2 with a concentration of 5 to 20%, poured it into a cylindrical container, installed the apparatus and began processing the suspension. Every 10 minutes, the process was stopped and 100 ml of suspension was taken. The suspension was centrifuged, the precipitate was removed, and the concentration of graphene nanostructures in the remaining suspension was determined. After analysis, the sample was poured into a container and the exfoliation process was continued. The process was continued until the increase in the concentration of graphene structures ceased. During the experiments, the rotor speed was changed: 5000; 10000; 12000; 15,000; 20,000 rpm At the end of the exfoliation process, the suspension was centrifuged and the precipitate was removed. The purified suspension was used as mixing water in the preparation of concrete samples for compression and bending tests. These tests were carried out in accordance with GOST 310.4-81 "Cements. Methods of bending and compression strength determination". For testing, 40×40×160 mm beams were prepared. Three-point bending tests were carried out on the machine, and compression tests on the press.

The concentration of graphene structures ranged from 0 to 0.05% after 0.01% (relative to cement). Thus, we included few- layer graphene in a mixture of cement and sand, mixing water-based graphene dispersions with ordinary Portland cement (OPC) and fine dry sand. The cement / sand / water ratio was 1 / 2.7 / 0.53. The fresh mixture was poured into standard steel molds measuring 40 × 40 × 160 mm, subjected to vibration to remove air, removed from the steel molds after 24 hours and stored in a water tank for curing. Beams (Fig. 1) were tested for three-point bending (Fig. 2) after 7, 14 and 28 days of curing.

Fig. 1. Beams after curing.

a) b)

Fig. 2. Tests of beams for three-point bending: a – before testing; b – after testing.

In addition, cubes were made from the halves of the beams after bending tests and tested for compression.

3 Results and discussion

Figure 3 shows the characteristic changes in bending strength for control beams and beams made of cement-sand mixture, modified with few-layer graphene. In this case, the concentration of few-layer graphene in water ranged from 0.25 to 0.3 mg / ml. Considering that the water-cement ratio was 0.53, the concentration of few-layer graphene in the cement-sand mixture ranged from 0.013 to 0.016%. As you can see, the increase in bending strength during curing (7, 14 and 21 days) is almost linear. Ultimately (after 28 days), the strength of the beams made of cement-sand mixture modified with few-layer graphene increased by 14%, compared to control samples. This result agrees well with the data presented in the papers. As you can see, the increase in bending strength during the curing process (7, 14 and 28 days) is almost linear. Ultimately (after 28 days), the strength of the beams made of cement-sand mixture modified with few-layer graphene increased by 14%, compared to control samples. This result agrees well with the data presented in [1, 5].

Figure 4 shows the change in compressive strength of control samples modified with graphene. You can see that the strength increases linearly, and ultimately for modified samples, the compressive strength is 33% higher than the strength of the control samples. Earlier [13], we experimentally found that when the concentration of few-layer graphene with respect to cement was 0.05%, the compressive strength increased by about 60%.

Thus, with an increase in the concentration of few-layer graphene by 3.5 times (from 0.014 to 0.05%), the strength increases only 2 times.

It should be noted that in a number of experiments using a suspension with the same concentration of few-layer graphene we obtained an increase in strength by 50%. Such a large discrepancy in the results (33 and 50%) can probably be explained by the different quality of few-layer graphene, more precisely the average number of graphene layers in the particles that were in suspension after centrifugation. Indeed, to obtain better results, we processed the suspension in a rotary apparatus for a larger number of cycles and centrifuged for a longer time.

For industrial use of concrete modified with few-layer graphene, it is necessary to solve the problem of accurately determining the concentration of this graphene in suspension and the average number of layers. In our opinion, the main importance is not the concentration of few-graphene, but the number of particles per unit volume, since it is the particles that are the centers of formation of complexes on which the strength of concrete depends. In one liter of suspension, which contains particles with two graphene layers, there will be twice as many of these particles compared to a suspension in which particles consist of four graphene layers.

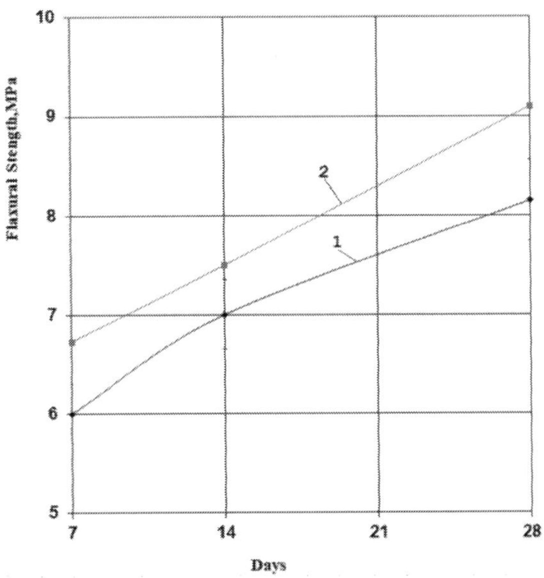

Fig. 3. Change in bending strength during curing: 1 – control images; 2 – modified with few-layer graphene.

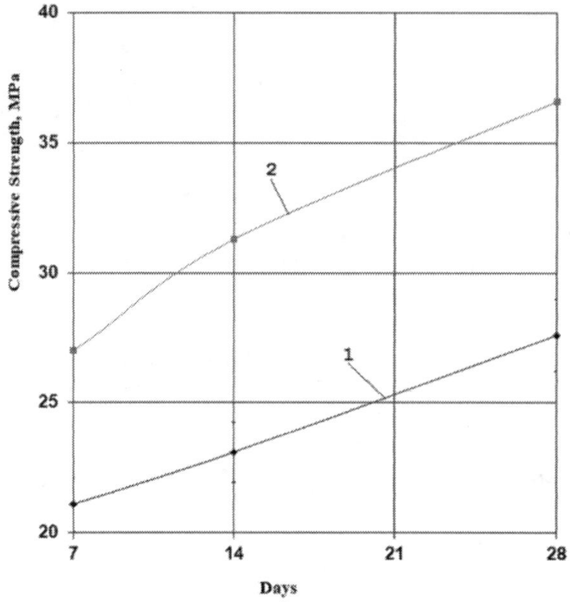

Fig. 4. Change in compressive strength in the curing process: 1 – control samples; 2 – modified with few-layer graphene.

4 Conclusion

It was experimentally established that the bending strength at three points of samples modified with graphene (graphene concentration of about 0.015 wt.% With respect to cement) is about 16% higher than that of the original concrete, while the compressive strength is 33% higher. There is reason to assume that the main role in increasing the strength of concrete is played not by the mass concentration of few-layer graphene, but by the amount of its particles per unit volume. This assumption requires further detailed studies, because a decrease in the mass concentration of few-layer graphene in concrete, while maintaining its strength action, opens up a new path to the large-scale industrial use of environmentally friendly and inexpensive technology to improve the performance of concrete.

References

1. D. Dimov, I. Amit, O. Gorrie, M.D. Barnes, N.J. Townsend, A.I. Neves, F. Withers, S. Russo, M.F. Craciun, Adv. Funct. Mater. **28**, 1705183 (2018)
2. D. Nivethitha, S.R. Srividhya, S. Dharmar, Int. J. Sci. Research, **5 (1)**, 913-916 (2016)
3. M.S. Morsy, S.H. Alsayed, M.A. Aqel, Int. J. Civ. & Environ. Eng. **10 (01)**, 23-27 (2010)
4. G. Xu, S. Du, J. He, X. Shi, Carbon **148**, 141-150 (2019)
5. D. Hou, Z. Lu, X. Li, H. Ma, Z. Li, Carbon **115**, 188-208 (2017)
6. T.S. Qureshi, D.K. Panesar, B. Sidhureddy, A. Chen, P.C. Wood, Composites B **159**, 248-258 (2019)
7. T.S. Qureshi, D.K. Panesar, Constr. Build. Mater. **206**, 71-83 (2019)

8. L. Zhao, X. Guo, Y. Liu, Y. Zhao, Z. Chen, Y. Zhang, L. Guo, X. Shu, J. Liu, Constr. Build. Mater. **190**, 150-163 (2018)
9. M. Muthua, M. Santhanam, Cement Concrete Comp. **91**, 118-137 (2018)
10. Q. Liu, Q. Xu, Q. Yu, R. Gao, T. Tong, Constr. Build. Mater. **127**, 565-576 (2016)
11. A.P. Kauling, A.T. Seefeldt, D.P. Pisoni, R.C. Pradeep, R. Bentini, R.V.B. Oliveira, K.S. Novoselov, A.H.C. Neto, Adv. Mater. **30**, 03784 (2018)
12. K.R. Paton, E. Varrla, C. Backes, R.J. Smith, U. Khan, A. O'Neill, C. Boland, M. Lotya, O.M. Istrate, P. King, T. Higgins, S. Barwich, P. May, P. Puczkarski, I. Ahmed, M. Moebius, H. Pettersson, E. Long, J. Coelho, S.E. O'Brien, E.K. McGuire, B.M. Sanchez, G.S. Duesberg, N. McEvoy, T.J. Pennycook, C. Downing, A. Crossley, V. Nicolosi, J.N. Coleman, Nat. mater. **13(6)**, 3944 (2014)
13. V. Pershin, K.A. Al-Shiblawi, A.M.R. Al-Mashhadani, A. Pasko, D. Melekhin, IOP Conf. Ser-Mat. Sci. **693(1)**, 012046 (2019)

Analysis of structure-forming role of phosphogypsum in the production of nonfired cementless building composites

Olga Kukina[1,*], *Andrey* Eremin[1], and *Dmitriy* Shuvaev[1]

[1] FSBEI of Higher Education "Voronezh State Technical University", Moscovskiy prospect, 14, Voronezh, 394026, Russia

Abstract. The working hypothesis is the development of the theory of disperse system aggregative stability considering the condensation processes of structure-forming of nonfired cementless building materials on the basis of phosphogypsum. The results of phisicomechanical and physicochemical analyses of Uvarovo chemical plant phosphogypsum are presented. To define the properties of phosphogypsum and phosphogypsum-based composite modern methods of analysis were applied with the use of the following equipment: the universal electromechanical test system Instron, the automatic diffractometer PANalytical EMPYREAN, the device for synchronic thermal analysis. The results of the research showed that the first endoeffects of Uvarovo chemical plant phosphogypsum have the dehydration energy of more than 200 joule/g, and they do not get lost within the time; this proves of the cementing properties of phosphogypsum. Due to the analysis of the differential scanning calorimetry results of phosphogypsum, pressed at 5 Mpa, it was found out that at the models compression the dehydration energy increases, and the correlation between the changes of water films' thickness and dehydration energy can be observed. With the decrease of water films' thickness we can observe the increase of dehydration energy with the simultaneous increase of the structure density and its transformation into a monolithic structure.

1 Introduction

The problem of phosphogypsum utilization is one of the most significant ecological problems all over the world [1-4]. This problem is usually solved by developing new technologies in the construction sector in the areas of road building, production of gypsum and anhydrite cementing, the filler in the building mixes. The economic effieciency of the immediate use of phosphogypsum and the products of its processing instead of traditional natural raw material in the construction sector in different areas has the following indices: road building – 5.4 %; the production of gypsum cementing – 2.5 %; the production of anhydrite cementing for construction – 3.9 %; the producion of the filler – 21 - 53 % [5].

* Corresponding author: u00136@vgasu.vrn.ru

We may single out several ways of using phosphogypsum: as one of the cementing component in the production; while manufacturing the fillers [6], in the production of gypsum plasterboard without using paper and fiber [7], in the manufacture of wall goods, for instance, bricks by the kilning technology [8]; while manufacturing loess [9], in road building; in nonfired technologies of obtaining wall goods and pavement tiles. Nonradioactive rare-earth metals, which are of a great demand on the market and are used in the war and semiconductor industries, are extracted from phosphogypsum [10-13]. An autoclave way of processing phosphogypsum into the gypseous cementing is known; it happens when phosphogypsum of the moisture up to 55 % from the sulfuric acid production workshop is supplied to the receiving bunker of the workshop where it is processed into the gypseous cementing. The obtained gypseous cementing can reach the durability of 40 – 50 MPa, and water needs are 30 – 36 % [14]. From waste phosphogypsum it is possible to get water-resistant anhydrite cementings by kilning it with the additions of alumosilicious materials, which reach the compression strength of dry models up to 10-15 MPa in the mortars with sand [15]. To get mortars from phosphogypsum with high-durability performances they apply grinding in the vibration grinding mill [16, 17]. Composite water-resistant gypseous cementing is also got by combined grinding of fine-ground mortar of dross and gypsum of calcium sulfate dihydrate [18]. Phosphogypsum can be used in the production of building materials as a basis for crystallization structures, but not for the cementing, thus offering the challenge of its large-scale use in the industry [19, 20]. It will be impossible to bring the gypsum dihydrate particles together to the required distance by the casting technology, that is why we propose our own method of molding such systems. There is a technology of producing decorative tiles from phosphogypsum by pressing aqueous pastes with a simultaneous water derivation [21, 22]. A cheaper and less resource-demanding method is known. Pressing of harsh mixes, containing 70-80% of phosphogypsum, 20-30% of gypsum, anhydrate or sulfate-cindery cementing under pressure of 5-20 MPa gives the opportunity to manufacture products with the ultimate strength under compression of 5 Mpa and more, which also provides an opportunity to use widely the obtained mix for production of blocks [23, 24].

As a working hypothesis the authors of the paper offer the next theoretical propositions, following from the theory of the aggregative stability of colloidal systems. While mixing the components of the raw meals, the water is absorbed in the form of films on the hydrophilic particles. The interaction forces of electrical nature and the disjoining pressure quantity are of the highest values. The maximum durability characteristics of the obtained materials are got if the water films' thickness is 10^{-9}–10^{-7} m. The total energy of the composite interacting particles also comprises induction interaction and dispersion forces of intermolecular interaction [12, 13].

Besides, during the semidry molding, used in the nonfired technology of producing building materials, jointing of the particles at their contact results in the relaxation of the charged particles to the boundary zone and structure formation with the minimum junction energy in the particles contact zone. The energy stability of the contacting particles zone and the material density depend on the amount of lattice vacancies in the crystal volume:

$$N_v = Ne^{\left[-\left(\frac{E_v+P_v}{KT}\right)\right]}, \qquad (1)$$

where N_v is the amount of point imperfections (lattice vacancies in the crystal); N is the total amount of sites; E_v is the thermal energy of the crystal activation; P_v is the pressure against the crystal; K is Boltzmann constant; T is temperature [25].

At the temperature of 65…70 °C and the change of aqueous extracts pH from 2 to 8 water films change their thickness dramatically. At the same temperatures the water penetrability increases dramatically, too.

Dispersed materials with the hydration energy more than 200 joule/g can develop cementing properties [26 – 29].

The aim of the given paper is to determine the structure-forming role of phosphogypsum while producing nonfired cementless building composites. To achieve the aim the authors of the paper set the following tasks: to analyze physicomechanical and physicochemical properties of phosphogypsum as a dispersed material and artificial stone, to determine its cementing properties, considering the thickness of water films, and energy stability of the contacting particles zone during the condensation.

2 Materials and Methods

The phosphogypsum, produced by Uvarovo chemical plant, was used for the research. The physicomechanical properties of the materials were estimated in the Center of common use in Voronezh State Technical University according to the requirements of the State Standards SS 23789-79, SS 8736-93, SS 5802-86, SS 310.4.81, SS 22688-77, and SS 125-79 with the help of the universal electromechanical test system Instron 5982 with an error of ± 0.5 % when being loaded. The radiography phase analysis was done by the automatic diffractometer PANalytical EMPYREAN. Differential scanning calorimetry DSC was done by the synchronic thermal analysis device STA 449 F5 A-0082-M (NETSCH, Germany) with the software NETSCHProteusco, with the integrated reliable monolithic system of weights and extremely sensitive DSC- and DTA-sensors in the modern construction of STA 449 F3 Jupiter®.

The analysis of the materials' microstructure was done by a scanning electronic microscope JSM-6380LV. The pictures present the microstructure of the chips' surface of the models. The micron marker in the pictures allows evaluating the grain dimensions and intergranular thicknesses of the water films.

3 Results

Chemical and granulometric composition of the phosphogypsum produced by the mentioned above enterprises is presented in Tables 1 and 2.

Table 1. Chemical composition of the phosphogypsum (mass %).

Enterprise	Total moisture	pH	P_2O_5	F	CaO	SiO_2	Fe_2O_3	SO_3
Uvarovo chemical plant	15.50–20.0	7.15–8.02	6–7	–	27–28	1–2	2–3	40–42

Table 2. The results of the grain-size analysis of the waste phosphogypsum of Uvarovo chemical plant.

The name of the residue	The sieve residues, mass %					Pass through the sieve with 0.16 net, mass %
	2.5	1	0.63	0.315	0.16	
The phosphogypsum of Uvarovo chemical plant						
Particular	13.78	8.74	9.0	17.86	30.82	19.80
Total	13.78	22.52	31.52	49.38	80.20	-

Figure 1 presents the diffraction pattern of the waste phosphogypsum model from Uvarovo chemical plant, having the following phases:

a) semiacquatic gypsum $CaSO_4 \cdot 0.5\ H_2O$ with a hexagonal crystal lattice a = 6.93 Å; b = 6.93 Å; c = 6.34 Å with interplanar spacings d = 6.01; 3.47; 3.00; 2.80; 1.84; 1.69 Å;

b) two-acquatic gypsum $CaSO_4 \cdot 2 H_2O$ with a monoclinic crystal lattice a = 7.61; 6.28 Å; b = 15.20 Å; c = 6.52 Å with interplanar spacings d = 4.28; 3.06; 2.87; 2.68;2.08 Å.

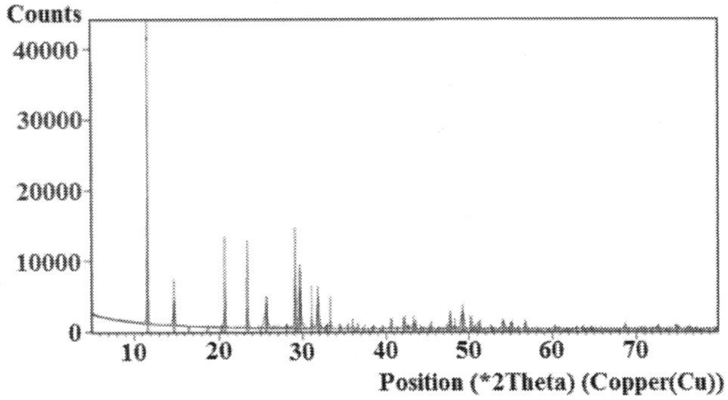

Fig.1. The diffraction pattern of the waste phosphogypsum of Uvarovo chemical plant.

Figure 2 presents a microphotograph of the waste phosphgypsum model from Uvarovo chemical plant.

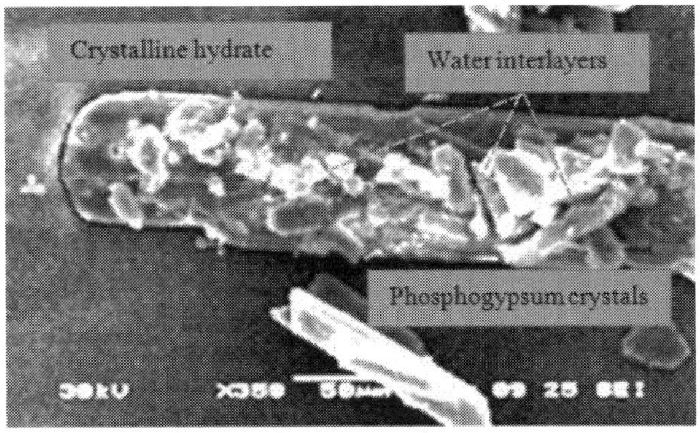

Fig.2. The microphotograph of the waste phosphogypsum model from Uvarovo chemical plant.

Figure 3 presents the data of the differential scanning calorimetry of the waste phosphogypsum model from Uvarovo chemical plant.

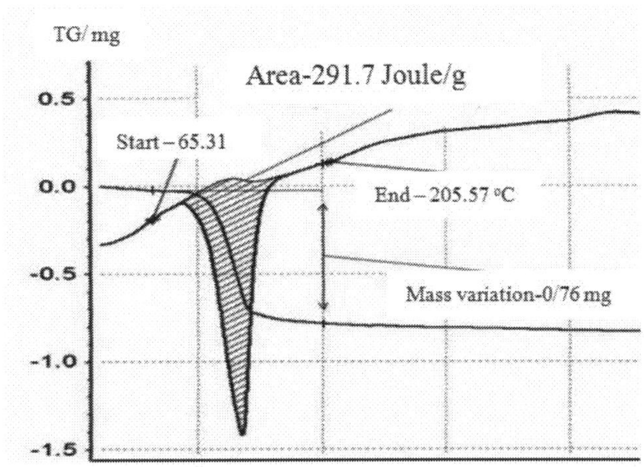

Fig. 3. The result of the differential scanning calorimetry of the waste phosphogypsum from Uvarovo chemical plant (the dehydration energy is 291.7 joule/g).

4 Discussions

The research was carried out to scrutinize physicomechanical, physicochemical and cementing properties of phosphogypsum as well as to obtain nonfired cementless composite building material on its basis. Phosphogypsum is a dispersed powder on the surface of which both adsorptive and crystallized water can be kept.

The prisms of a typical ionic crystalline hydrate with the dimensions from 100 to 500 micro lengthwise and from 10 to 100 micro in width can be seen clearly. Besides, water interlayers inside these crystalline hydrates with the thickness from 10 to 1000 nm can be seen well as the distances between the crystalline hydrate particles are too big (the interstices between them exceed 100 nm), and the number of the contacts is small (Fig. 2).

The results of the differential scanning calorimetry show that the first endoeffects of Uvarovo chemical plant phosphogypsum have the dehydration energy of more than 200 joule/g, and they do not lose it within the time; this proves of the cementing properties of phosphogypsum [12-13].

By analyzing the results of the differential scanning calorimetry of phosphogypsum, pressed at 5 Mpa and at the age of 1, 3, 24 days, it was found out that at the compression, leading to tightening, the models demonstrate the dehydration energy increase, and the correlation between the changes of water films thickness and dehydration energy can be observed. With the decrease of water films' thickness we can watch the increase of dehydration energy with the simultaneous increase of the structure density and its transformation into a monolithic structure (Fig. 4).

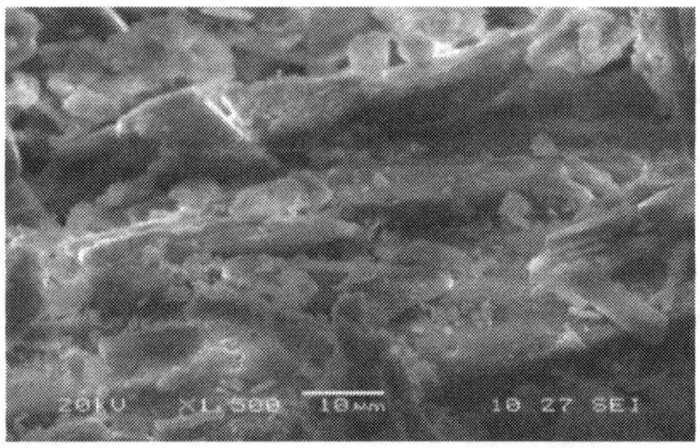

Fig. 4. The microphotograph of the pressed waste phosphogypsum (× 1500).

Nonfired cementless composite building material was got by semidry molding from phosphogypsum with the addition of sand and slack lime. The results of researching the physicomechanical properties of the products, manufactured from the composite of phosphogypsum (60 %), sand (30%), lime (10 %) under the laboratory conditions on the compression equipment "Condor", are presented in Table 3.

Table 3. The physicomechanical properties of the products, manufactured from the composite of phosphogypsum (60 %), sand (30%), lime (10 %) under the laboratory conditions on the compression equipment "Condor".

Property	Value
Average strength, MPa	3 … 5
Water absorption	0.23
Softening coefficient	0.63
Density, kg/m^3	1700

The prime cost of producing 1 m^3 wall gypseous materials, depeding on the equipment efficiency, varies from 900 to 1950 rouble/m^3, which is 2 or 3 times cheaper than the counterparts.

5 Conclusion

According to the data of the radiography phase analysis there were no new formations found in the structure of the lime-sandy phosphogypsum material. That is why we may assert that the durability of the materials, got by the nonfired technology, is obtained due to the self-organization of the structure, in which the particles of the dispersed material (phosphogypsum) with the dehydration energy of more than 200 joule/g play the main structure-forming role.

The components with nanodimensional particles, having thermodynamic instability, form crystalline contacts with the nanodimensional thicknesses of the water films. The decrease of porous space (the growth of compression pressure, the decrease of the lime mass fraction, the increase of clay components mass fraction, the increase of the contacts' amount between crystals and the decrease of lattice vacancies amount) contributes to the growth of the durability and water resistance of the composite according to the propositions suggested by the authors of the paper.

References

1. J. Yang, L. Ma, H. Liu, Z. Guo, Q. Dai, W. Zhang, K. Bounkhong, Chemosphere, **248**, 125979 (2020)
2. S.N. Zolotukhin, O.B. Kukina, D.I. Shuvaev, M.L. Zolotukhin, *Scientific support of Voronezh region: Collected papers of the winners in the contest of VSTU students' and post-graduate students' scientific-research works in the priority spheres of science and technology development*, 50-51 (2019)
3. D. Nizevičienė, D. Vaičiukynienė, B. Michalik, M. Bonczyk, V. Vaitkevičius, V. Jusas, Constr. Build. Mater., **180**, 134-142 (2018)
4. A.M. Rashad. J. Clean. Prod., **166**, 732-743 (2017)
5. V.V. Ivanitsky, et al., *Phosphogypsum and its use* (Moscow : Chemistry, 1990)
6. D.S. Liu, C.Q. Wang, X.D. Mei, C. Zhang, Environ. Sci. Pollut., **26** (29), 30533-30539 (2019)
7. J. Zhou, X. Li, Y. Zhao, Z. Shu, Y. Wang, Y. Zhang, X. Shen, Constr. Build. Mater., **243**, 118091 (2020)
8. Y. Rakhila, A. Mestari, S. Azmi, A. Elmchaouri, J. Chem., **11** (4), 1552-1563 (2018)
9. K. Gu, B. Chen, Constr. Build. Mater., **231,** 117195 (2020)
10. O. Cheremisina, V. Sergeev, A. Fedorov, D. Alferova, E. Lukyantseva, *International Multidisciplinary Scientific Geo Conference Surveying Geology and Mining Ecology Managemen - SGEM,* **19**, 903-910 (2019)
11. L.A. Podolyanets, Opcion, **34**, 1491-1508 (2018)
12. S.N. Zolotukhin, O.B. Kukina, D.E. Barabash, *IOP Conf. Series: Mater. sci. eng.*, **687**, 022028 (2019)
13. S. Zolotukhin, O. Kukina, V. Mishchenko, S. Larionov, Adv. Intell. Syst. Comput., **983**, 339-351 (2019)
14. A.F. Tubolkin, *Waste-free production* (Leningrad: Knowledge, 1980)
15. A.V. Volzhensky, *Mineral cementing agents: tutorial for higher educational establishments* (4th Impression, revised and enlarged edition, Moscow: Stroyizdat, 1986)
16. A.V. Terekhov, V.P. Varlamov, Building Materials, **2**, 22-24 (1985)

17. Iu.B. Potapov, Building materials, **7**, 37-39 (2003)
18. M.V. Zuev, S.A. Mamaev, M.A. Mikheenkov, A.I. Stepanov, *Composite water-resistant gypsum cementing,* Bul. 3 (Patent of the Russian Federation 2505504, MPK C04B28/14, C04B7/14, patent holder Limited Liability Company «EVEREST»: 201212993/03, publ. 27.01.2014)
19. V.B. Petropavlovskaya, A.F. Burianov, T. B. Novichenkova, Building materials, **7**, 8-9 (2006)
20. V.B. Petropavlovskaya, V.V. Belov, A.F. Burianov, Building materials, **12**, 46-47 (2007)
21. I.M. Lyashkevich, G.S. Raptunovich, A.F. Polak, *Proceedings of Higher Educational Establishments. Series Building and Architecture*, **12**, 60-63 (1985)
22. A.F. Polak, et al., *Proceedings of Higher Educational Establishments. Series Building and Architecture*, **10**, 55-59 (1987)
23. R.N. Mirsaev et al., Building materials, **7**, 4-6 (2010)
24. R.N. Mirsaev et al., Building materials, **5**, 55 (2004)
25. V.I .Solomatov, S.F. Korenkova, Iu.V. Sidorenko *Proceedings of Higher Educational Establishments,* **2 – 3**, 38- 44 (2001)
26. A.S. Bagdasarov, Building materials, **1**, 13-17 (2002)
27. A.T. Baranov, *Ash concrete. Porous and dense* (Ed. by a member of the USSR Academy of Science prof. S. A. Mironov, Moscow : Stroyizdat, 1960)
28. V.P. Berezovsky, R. I. Kontor, M.A. Matusevich, Building materials, **2-2**, 30 (1964)
29. P.I. Bozhenov, *Complex use of mineral raw materials and ecology* (Publishing House ASV, Moscow, 1994)

The deformability of the stress-strained concrete strips between the cracks during the short-term dynamic loading

Zaur Galyautdinov[1,*], *Oleg* Kumpyak[1], and *Daud* Galyautdinov[1]

[1]Tomsk State University of Architecture and Building, 634003 Tomsk, Russia

Annotation. The formation of non-intersecting cracks in stress-strained ferroconcrete elements leads to separation of concrete strips between the cracks. The results of the experimental research indicate a significant decrease of the durability and deformability of the stress-strained concrete strips between cracks both under static and short-term dynamic loading. At the same time physico-mechanical properties depend on the straining deformations and rebars' inclination angle towards the cracks. The existing theoretical and experimental results evaluate only the durability of the concrete strips between the cracks. The current paper presents the results of experimental and theoretical studies on the dynamic deformability of the stress-strained discs between the cracks. The statistic analysis of the experimental data is done; on the basis of the analysis we suggest the dependencies, characterizing the deformability of the concrete strips during the short-term dynamic load depending on the level of the straining deformations and rebars' inclination angle towards the cracks.

1. Introduction

Single dynamic loadings on buildings and constructions make special demands of ensuring durability of separate constructions and the building as a whole [1 –3]. The trustworthiness of the construction deflected mode evaluation is determined, in the first place, by the concrete deformation model [4–6]. Modern models allow considering the main peculiarities of the dynamic deformation of concrete, reinforcement and ferroconcrete at all the stages of the construction work [7–13]. At the same time there is a range of issues, requiring an additional study, especially under high-speed loading conditions.

In flat-strained ferroconcrete constructions cracks appear in the areas of the main tensile stresses and, in general case, may be skew-oriented towards the bars of the mesh reinforcement. After appearing of a crack, the concrete cannot take tensile loads in the direction perpendicular to the crack direction; that is why all the tensile loads along this direction are transferred through the reinforcement. The loads in the orthogonal direction are taken by the concrete strips between the cracks.

The researches of the durability of concrete strips between cracks are presented in the works of our country and foreign scientists [14, 15, 16, 17]. According to the results of the

*The corresponding author: gazr@yandex.ru

experimental researches, we considered the influence of the samples' deformation level in the direction perpendicular to the crack and the inclination angle of the rebars towards the cracks. The scientists proposed the formulae for durability of the concrete strips between cracks under static [16, 17] and short-term dynamic loads [18, 19, 20]. At the same time, the issues of the deformability of the concrete strips between cracks under intense dynamic loads have not been scrutinized yet by now.

2. The experimental research

The current research presents the results of the experimental-theoretical analysis of the deformability of the stress-strained concrete strips between cracks under short-term dynamic load. We studied both the influence of the sample's deformation level and the inclination angle of the rebars towards the cracks' direction. The experimental researches were carried out on the ferroconcrete discs with fixed cracks (**Ffig. 1**).

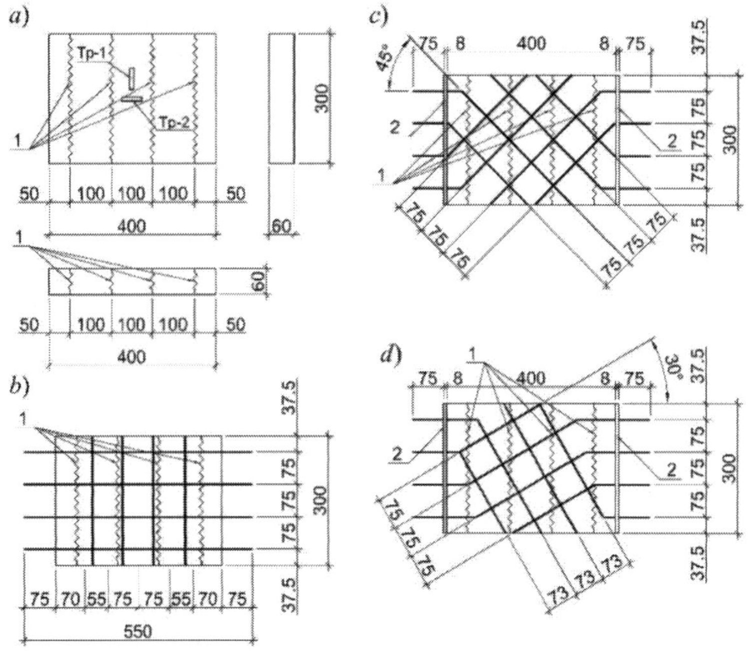

Fig. 1. The general view of the samples and the strain resistors placing on the concrete (*a*) and the scheme of samples' reinforcement at different rebars' inclination angles towards the cracks (*b*, *c*, *d*): 1 – cracks; 2 – metal plates.

The general amount of the tested samples is 72, including: 23 samples at $\alpha = 90°$; 28 – at $\alpha = 45°$ and 21 – at $\alpha = 30°$. During concrete casting celluloid inserts were placed in the samples; then during the testing the cracks, dividing concrete blocks, appeared along the inserts. The rebars were placed at an angle towards the crack. The rebars' inclination angle towards the crack α was taken as equal to 90°, 45° and 30°. In the samples with $\alpha = 30°$ and $\alpha = 45°$ metal plates were placed on the outer edges (position 2, **fig. 1**); the plates prevented the mutual shift of the rebars. The main dimensions and schemes of samples' reinforcement are presented in Fig. 1. All the tested samples were produced from the heavy concrete of class B15. The mesh reinforcements were made from the reinforcement of class A400 with a diameter 8 mm.

During the tests the load was applied to the sample in two mutually orthogonal directions. The sample was strained along one direction, and was compressed along the other one (Fig. 2). The tensile stresses were fully transmitted through the reinforcement. The compressive stresses were transmitted to the concrete blocks between the cracks (Fig. 2). The tests were carried out at two stages; first, the strain was created up to the necessary level, and then the compression – till the destruction. During the strain there appeared the cracks, dividing the sample into concrete blocks, connected with each other by the reinforcement.

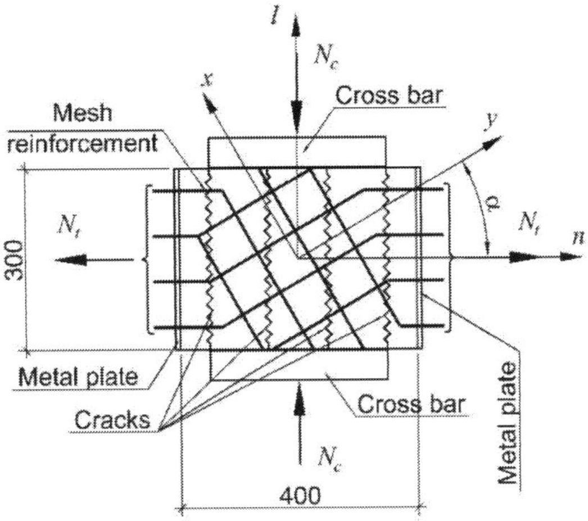

Fig. 2. The general view of the samples and the strain resistors placing on the concrete (*a*) and the schemes of samples' reinforcement at different rebars' inclination angles towards the cracks (*b*, *c*, *d*): 1 – cracks; 2 – metal plates.

The tensile load on the sample was created by the hydraulic jack. The dynamic compression stresses were created due to the energy of the falling weight. The height of the weight falling and its size was chosen providing the total exhaustion of concrete strips bearing capacity between cracks at single load action.

Figure 3 presents the data on the deformability of concrete strips between cracks at different levels of samples tension ε_l, for the accepted rebars' inclination angles. According to the research results, it is obvious that the deformability of concrete strips between cracks decreases with the growth of the deformation level perpendicular to the crack ε_l for all the values of the rebars' inclination angle towards the crack α . Along with this, in contrast to the concrete strips' durability between cracks [17, 18, 19, 20], the influence of the angle α on their deformability is of a less degree, the level of ultimate strains ε_{Rp} lowering at different angles α is practically identical. The most intensive lowering of the concrete strips' deformability is observed in the interval $\varepsilon_l = 100 \cdot 10^{-5} ... 400 \cdot 10^{-5}$, at $\varepsilon_l > 400 \cdot 10^{-5}$ the deformations ε_{Rp} remain almost invariable. According to the experimental research data, the deformability of concrete strips at short-term dynamic load, depending on the mentioned above factors, decreases twice.

3. The theoretical analysis

The experimental research of the deformability of the concrete stripes between cracks allowed evaluating qualitatively the influence of the studied factors. The quantitative evaluation of deformations ε_l influence at different rebars' inclination angles α on the deformability of the concrete strips between cracks is done by mathematic statistical methods [21, 22].

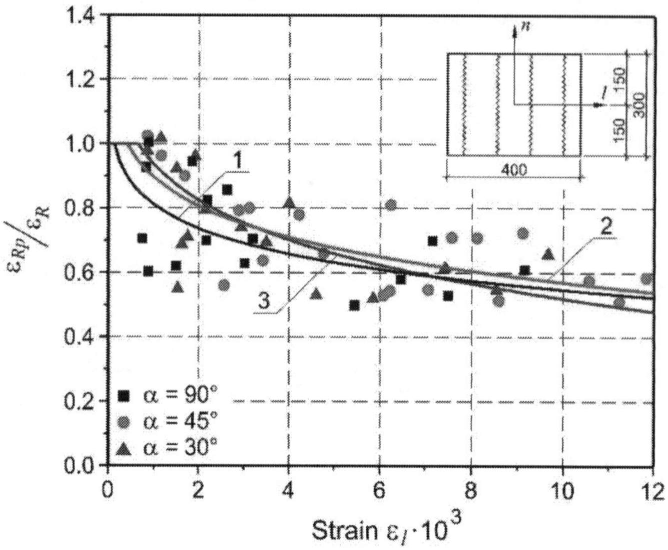

Fig. 3. The change of the deformability of the concrete strips between cracks under a short-term dynamic load depending on the inclination angle α and deformation level ε_l : 1 – according to the formula (1); 2 – according to the formula (2); 3 – according to the formula (3);■, •, ▲ – the experimental data.

The experimental data results indicate non-linear change of the concrete strips deformability depending on deformations' level ε_l (**fig. 3**). At the initial stage the transformation of the data to the linear kind is done, then the evaluation of the data selection is done, including grave errors' sifting and checking the correspondence of the concrete strips' durability between cracks to the normality test. The results of the selections' evaluation are presented in Fig. 4. The grave errors' sifting is done according to Student's test. The checking of the data distribution according to the normality law, done on the grounds of Kolmogorov-Smirnov test and Shapiro-Wilk test criteria, showed their correspondence to the normality distribution law (**fig. 4**, *b*, *d*,*f*).

The calculated correlation coefficients, characterizing the dependence of the deformability on the tensile strain level ε_l, equal to: at $\alpha = 90°$ – 0.551; at $\alpha = 45°$ – 0.652; at $\alpha = 30°$ – 0.798. This indicates an average and high degree of the studied values ties.

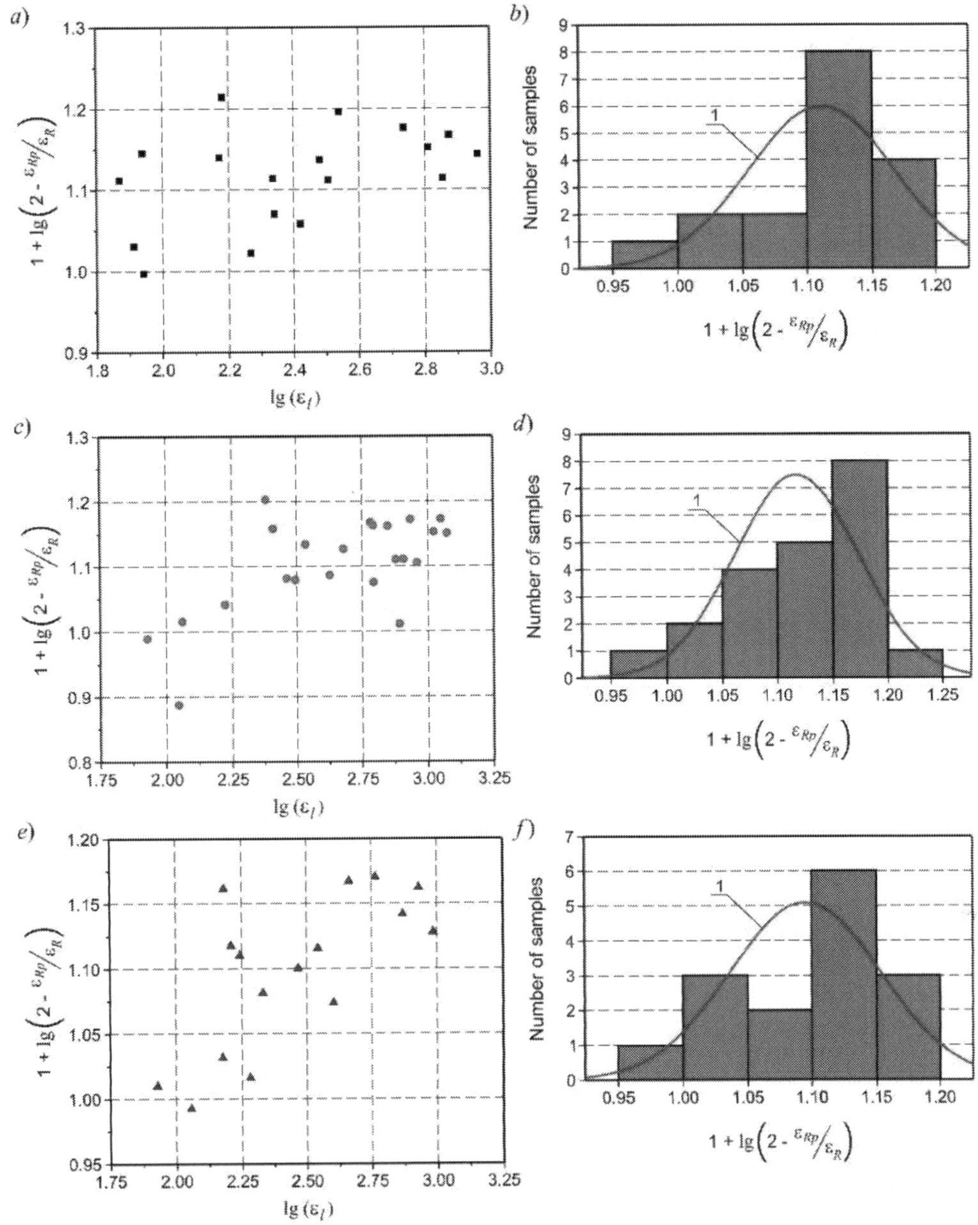

Fig. 4. The deformability change of the concrete stripes between cracks during short-term dynamic loading (*a, c, e*), and experimental and theoretical distribution of the studied values after bringing them to the linear kind (*b, d, f*):*a, b* – at $\alpha = 90°$;*c, d*– at $\alpha = 45°$; *e, f*– at $\alpha = 30°$; 1 – normal theoretical distribution.

The regression analysis is done according to the methodology presented in the research [21, 22]. By the least-squares method we got the regression equations which, considering reverse conversion, are of the following kind:

- at $\alpha = 90°$

$$\gamma_{bp\varepsilon} = \varepsilon_{Rp} / \varepsilon_R = 2 - 10^{-0.0979+0.0867\lg(\varepsilon_l)}, \quad (1)$$

- at $\alpha = 45°$

$$\gamma_{bp\varepsilon} = \varepsilon_{Rp} / \varepsilon_R = 2 - 10^{-0.1808+0.1121\lg(\varepsilon_l)}, \quad (2)$$

- at $\alpha = 30°$

$$\gamma_{bp\varepsilon} = \varepsilon_{Rp} / \varepsilon_R = 2 - 10^{-0.259+0.1431\lg(\varepsilon_l)}. \quad (3)$$

The evaluation of the obtained expressions significance is done on the basis of F-test criterion which equals: at $\alpha = 90°$ – 6.253; at $\alpha = 45°$ – 14.09; at $\alpha = 30°$ – 22.85. The obtained values are higher than the tabular values corresponding to 5% of probability being equal to 4.543, 4.381 and 4.667 respectively [21, 22]. Thereby, the presented dependencies allow determining the deformability of the concrete strips between cracks with confidential probability of 0.95, laid in regulatory acts.

The analysis of the obtained theoretical dependencies, presented in Fig. 3, shows that the deformations ε_l influence the deformability of the concrete strips samples. The deformations of the concrete strips between cracks, corresponding to ultimate stresses at different inclination angles of the rebars α are comparable; i.e. the value of the angle α almost does not influence the deformability of the samples.

4. Conclusions

The results of the experimental-theoretical researches indicate a considerable influence of the samples deformation level in the direction perpendicular to the crack ε_l on the deformability of concrete strips between cracks. The rebars' inclination angle towards cracks α has an insignificant influence on the deformations of the concrete strips ε_{Rp}. On the basis of the experimental research data elaboration we offered the dependencies for the determination of the deformability of the concrete strips between cracks. The evaluation of the given formulae indicates a sufficient reliability of the calculated values of the deformations ε_{Rp} of concrete strips.

The reported study was funded by RFBR, project number 19-31-90097

References

1. G.I. Odnokopilov, D.Ju. Sarkisov, E.A. Butuzov, Proceedings of Tomsk Polytechnical University. Engineering of Geological Resources **329**, 122–135 (2018)
2. G.I. Odnokopilov, D.Ju.Sarkisov, Proceedings of Tomsk Polytechnical University. Engineering of Geological Resources **328**, 85–95 (2017)
3. G.I. Odnokopilov, O.G. Kumpyak, Z.R. Galyautdinov, D.R. Galyautdinov, Proceedings of Tomsk Polytechnical University. Engineering of Geological Resources **330**,110 – 125 (2019)
4. N.I.Karpenko *Obshhie modeli mehaniki zhelezobetona* [General models of reinforced concrete mechanics] (Strojizdat, Moscow, 1996) (in Russian)
5. N.I.Karpenko, Structural Mechanics and Analysis of Constructions**1**, 28 – 32 (1984)
6. Z.R. Galyautdinov, News of higher educational institutions. Construction **10**, 93 – 98 (2010)
7. J. Jones, C. Wua, D.J. Oehlers, A.S. Whittaker, W. Sunc, S. Marksa, R. Coppola, Engineering Structures **31**,2825 – 2832 (2009)

8. B. Chiaia, O. Kumpyak, L. Placidi, V. Maksimov, Engineering Structures **96**, 88 – 99 (2015)
9. R. John, S.P. Shah, Impact: Eff. Fast Transient Loadings: Proc 1st Int. Conf. Fast Transient Loadings **8**, 37 – 65 (1987)
10. Y. Lu, K. Xu, International Journal of Solids and Structures **41**, 131 – 143 (2004)
11. T. Ngo, P. Mendis, Electronic Journal of Structural Engineering 8, 80 – 89 (2008)
12. V.S. Plevkov, A.I. Malganov, I.V. Baldin, P.V. Stukov, K.V. Kalachev, D.Ju. Sarkisov, 8th Korea-Russia International Symposium on Science and Technology, KORUS, 342 – 344 (2004)
13. Z.R. Galyautdinov, AIP Conference Proceedings **1800**, 010001 (2017); doi: 10.1063/1.4973016
14. A. Belarbi, L.-X. Zhang, T.T.C. Hsu, Sociedad Mexicana de Ingenieria Sismica in World Conference On Earthquake Engineering, 1 – 8 (1996)
15. B.Y. Bahn, T.T.C. Hsu, ACI Structural Journal **95**, 178 – 193 (1998)
16. F.J. Vecchio, M.P.Collins, ACI Journal **83**, 219 – 231 (1986)
17. A. F. Yaremenko, V. S. Gapshenko, Concrete and Reinforced Ferroconcrete 12, 23 – 24 (1986)
18. Z.R. Galyautdinov, MATEC Web of Conferences **143**, 01013 (2018); doi: 10.1051/1.4973016
19. Z.R. Galyautdinov, Bulletin of Tomsk State University of Architecture and Building **6**,97 – 105 (2017)
20. O.G. Kumpyak, Z.R. Galyautdinov, D.N. Kokorin *Prochnost' ideformativnost' zhelezobetonnyh konstrukcij na podatlivyh oporah pri kratkovremennom dinamicheskom nagruzhenii* [Strength and deformability of reinforced concrete structures on yielding supports under short-term dynamic load] (TSUAB Publ.,Tomsk, 2016) (in Russian)
21. E.N.L'vovskij*Statisticheskiemetodypostroenijaempiricheskihformu* /[Statistical methods of empirical formulas] (VysshayaShkola Publishers,Moscow, 1988) (in Russian)
22. R. Vadzinsky *Statisticheskie vychisleniya v srede Excel. Biblioteka pol'zovatelya* [Statistical calculations in Excel environement. User's library].– (Peter, StPetersurg, 2008) (in Russian)

The study of the ferroconcrete beams with thrust on the foil bearings during short-term dynamic loading

Oleg Kumpyak[1], and *Daud* Galyautdinov[1*]

[1]Tomsk State University of Architecture and Building, 634003 Tomsk, Russia

Abstract. While designing and maintaining ferroconcrete beam constructions subjected to a short-term dynamic loading, it is necessary to consider the emergence of the thrust reaction; this leads to a considerable increase of durability and fracture strength. It is also necessary to consider the use of foil bearings as the power-intensity increase of the researched constructions. The aim of the experimental research is to expose the reasonability of using foil bearings in thrust constructions. The paper presents the results of the experimental research of ferroconcrete beam constructions with thrust on foil bearings during short-term dynamic loading. The influence of the thrust on the durability, deformability and fracture strength of ferroconcrete constructions during short-term dynamic loading, as well as joint use of foil bearings and limitation of support contour horizontal shift are studied. The results of the research indicate the positive effect of using foil bearings in the constructions with thrust.

1 Introduction

The existing analytical methods of calculating ferroconcrete constructions during a short-term dynamic loading are based on the preconditions which do not always fully consider the peculiarities of the actual operation of the construction. A considerable part of bendable constructions has connections in the bearing part; these connections prevent from horizontal shift of the bearing sections at the loading and lead to thrust reaction emergence. A group of scientists carried out experimental-theoretical researches which exposed a considerable positive effect of considering the thrust reaction to the deflected mode of bendable structures [1–5].

The thrust under static influence is studied sufficiently [1, 3, 5]. The results of the analytical calculations of the constructions on the effect of static loadings with thrust indicate the increase of their bearing ability depending on the percentage of the cross-section reinforcement (μ) and the yielding of the bearing contour (c) [3].

Under the dynamic influence the impact of the thrust on the constructions' resistance is ambiguous: on the one hand, the presence of the horizontal reaction increases the bearing

* The corresponding author: DaudG@yandex.ru

ability of the element, on the other hand, it decreases its deformability; this leads to the reduction of the deformation plastic phase. This issue is scrutinized in the works of N. N. Popov and B. S. Rostorguev [6], T. N. Vinogradova [3], and I. N. Tikhonov [5].

Dynamic loading on buildings and constructions makes special demands of providing survivability of separate constructions and of a building as a whole [7, 8, 9]. The use of foil bearings leads to the decrease of the resistance of buildings' constructions and buildings to the influence of dynamic loadings due to the energy absorption of the dynamic influence. The results of the experimental research of bendable constructions [9, 10, 11, 12] on foil bearings demonstrate high efficiency of their use under the influence of high intensity dynamic loadings. The maximum effect is observed during the deformation of bearings at the elastoplastic phase.

The issue of the optimization of foil bearings' parameters in statically indefinable systems under a short-term dynamic loading is a new task which is topical and which requires the further study.

2 Experiment

With the aim of experimental evaluation of the deflected mode of ferroconcrete beam constructions on foil bearings with thrust under short-term loading the following research programme was elaborated and implemented (Table 1).

Table 1. The programme of the experimental research.

Type of loading	Short-term dynamic				
Type of bearings	Rigid supports	Foil bearings			
Work stage of foil bearings		Elastic-plastic with hardening	Elastic-plastic		
Hardness of foil bearings, *l*, mm	-	50	100	160	180
Code of the sample	BDP-1	BDPFh-2	BDPFe-3	BDPFe-4	BDPFe-5
Amount	1	1	1	1	1

Notation conventions: «B» is a ferroconcrete beam; «D» is a short-term dynamic loading (drop of the load weighed 450 kg from the height 1150 mm); «P» is the presence of the thrust construction; «F» is a foil bearing (made from hollow pipes: the outside diameter is 32 mm, the inside diameter is 26.1 mm) working at the elastoplastic phase «e» or at the hardening phase «h»; «1…5» is the ordinal number of the construction.

The beams with the dimensions 1900x150x220 mm are designed and made. The concrete is heavy, of class B35. The beams are reinforced with spatial framework. The cold-deformed reinforcement Ø5 mm of class Bp500, set with a step of 50 mm in the close to bearing zone and 130 mm in the middle of the span, is used as a cross reinforcement; the hot-rolled reinforcement Ø 6 mm of class A240 in the compression area and Ø10 mm of class A500 in the tension area is used as a longitudinal reinforcement. To reinforce end areas of the beams nets with cells of 50×50 mm from cold-deformed reinforcement Ø5 mm of class Bp500 were used, 7 nets on each side, and angle bars 100×10 mm were set (Fig. 1).

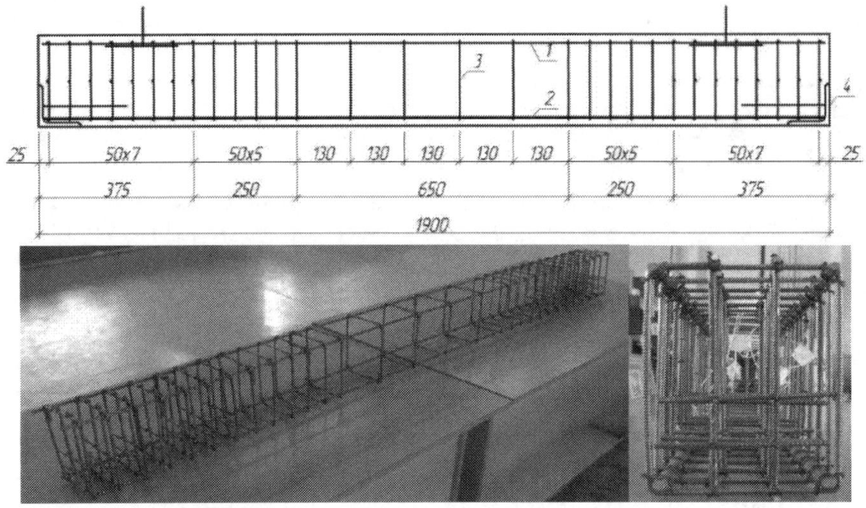

Fig. 1. The scheme of reinforcement of the researched ferroconcrete constructions: reinforcement Ø6 mm of class A240 (1); reinforcement Ø10 mm of class A500 (2); reinforcement Ø5 mm of class Bp500 (3); equilateral angle bar 100×10 mm (4).

For conducting the test a complex of measuring devices was placed on the samples: to determine the shift the inductive sensors of location Waycon of series RL150 were used. To measure the accelerations the accelerometers (DHE 100023) were used. The force-measuring tensoresistant sensor DST 4126 was used to determine the system reaction; and dynamometric bearings (Patent of RF for a useful model №161908) were used to determine the value of bearings' reactions. The devices were connected to the computers Mic-036R and Mic-400D.

Fig. 2. The stand for testing the construction under the conditions of the limited horizontal shift under a short-term dynamic loading (a photo, a three-dimensional model).

To implement the programme of the experimental research a stand was designed and manufactured (Fig. 2), the corresponding patent was obtained (Patent of RF for a useful model №147262). A short-term dynamic loading was created by means of a falling load weighed 450 kg from the height of 750 mm (Fig. 2).

The tested ferroconcrete beam is placed on the dynamometric bearings, providing hinged support. The bearings are fixed on the power floor, and then the initial longitudinal effort is created by the hydro jack through the system of traverse joint guide rods.

3 Results

By the results of the dynamic tests the graphic dependencies over the time are drawn up for the following parameters: the input force influence (Fig. 3); the construction shift without considering the mutilation of foil bearings (Fig. 4); the deformation of foil bearings (Fig. 5); the acceleration of the system (beam-yielding bearings) (Fig. 6).

Figure 3 shows that the presence of foil bearings leads to the decrease of the peak value of the input signal, while the time of reaching the maximum value is decreasing. This factor is explained as follows: the transition of foil bearings to the elastoplastic area of deformation happens later, for the obtained results in the interval of 0.007…0.011 s. The decrease of the deformability of statically indefinable beams is connected with the reduction of foil bearings rigidity (Fig. 4), and the more the decrease of the construction's deflection is, the more mutilation of foil bearings is. It is worth mentioning that the most efficient is the operation of foil bearings till the transition to the hardening phase, as in this case additional inertia forces emerge which lead to the growth of beams' deformation due to abrupt stop of bearings' mutilation.

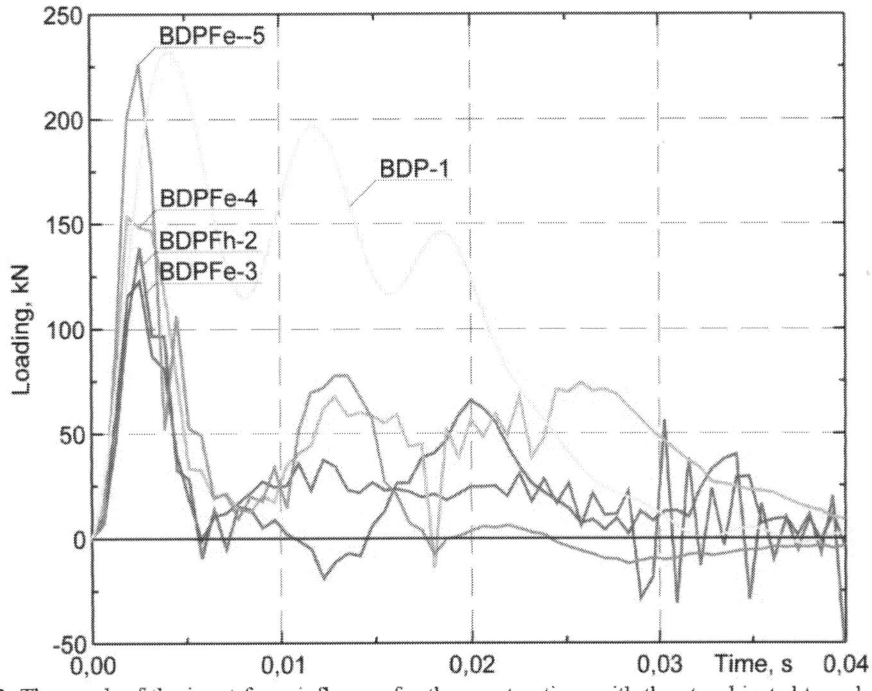

Fig. 3. The graph of the input force influence for the constructions with thrust, subjected to a short-term dynamic loading.

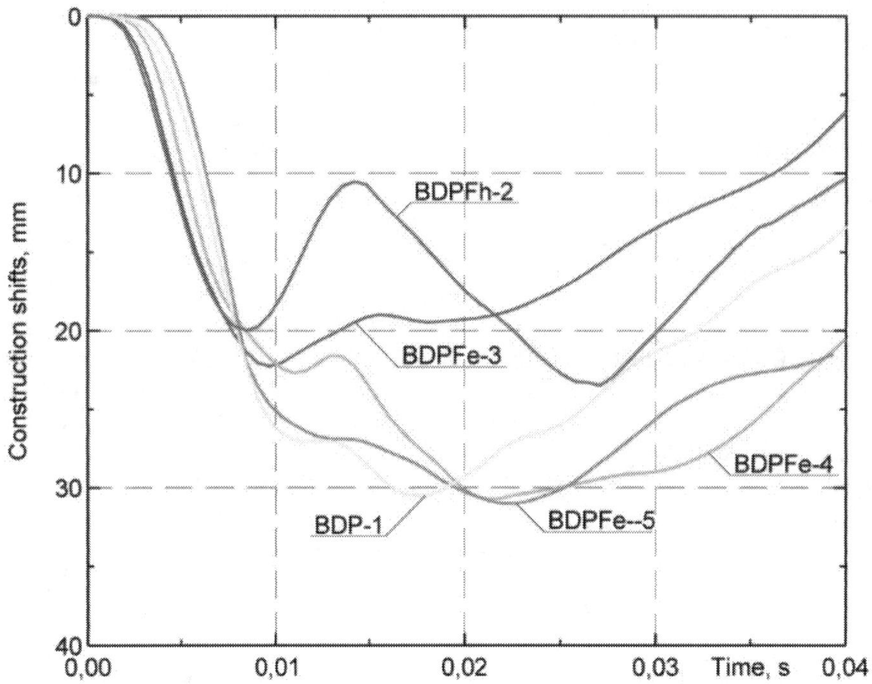

Fig. 4. The graph of the shift over time change without considering mutilation of foil bearings for the constructions with thrust, subjected to a short-term dynamic loading.

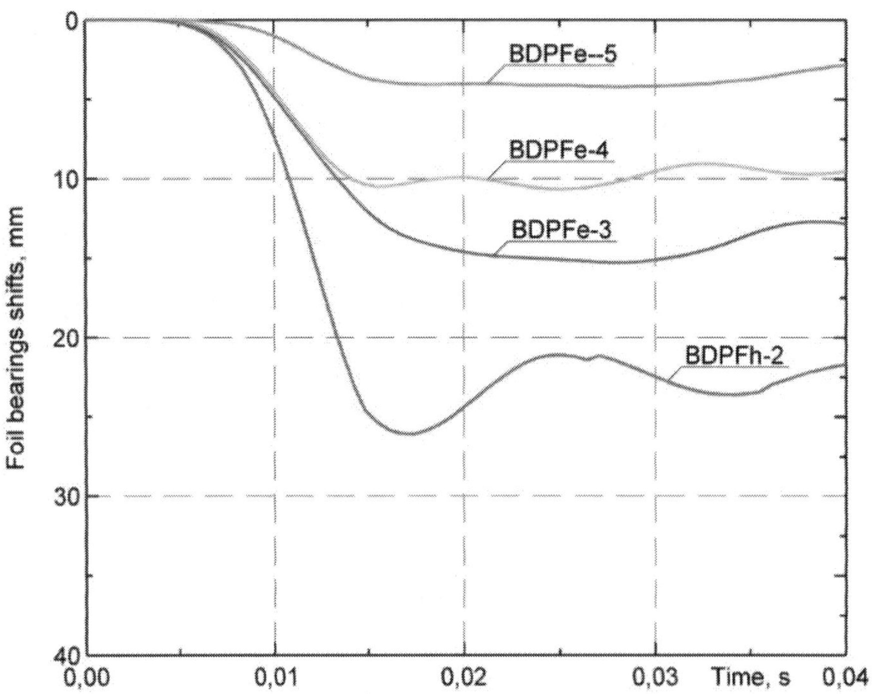

Fig. 5. The graph of the mutilation change of foil bearings for the constructions with thrust, subjected to a short-term dynamic loading.

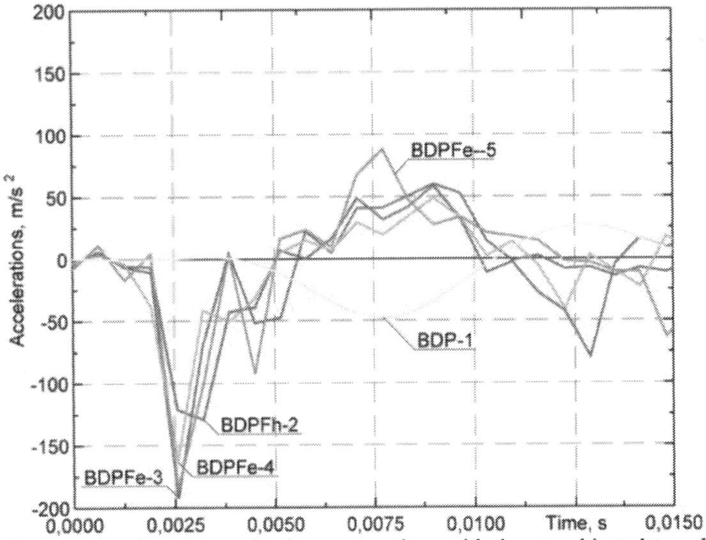

Fig. 6. The graph of acceleration change for the constructions with thrust, subjected to a short-term dynamic loading.

The analysis of the accelerations (Fig. 6) shows that the presence of foil bearings, working at the hardening phase for the beam BDPFh-2, leads to the additional growth of inertia, that is after a complete mutilation of the foil bearings (Fig. 5), the accelerations grow swiftly and aim at the peak values at the phase of elastic-plastic deformability.

4 Conclusion

By the results of the done experimental research it is ascertained that the use of foil bearings in the constructions, operating under the condition of the limited horizontal shift, leads to the decrease of deformability and fracture strength under the intensive dynamic loading. The increase of the system power-intensity (BDPFh-2…BDPFe-5) is observed. It is ascertained that the maximum efficiency of foil bearings is at the work in the elastoplastic area of deformation without transition to the hardening phase (BDPFe-3). The results of the experimental research also revealed hidden durability reserves of the beams on foil bearings with thrust, consideration of these parameters will allow increasing the accuracy of the calculations under short-term dynamic loading and their reliability while designing.

The reported study was funded by RFBR, project number 19-31-90097

References

1. A.A. Gvozdev, et al., *Novoe o prochnosti zhelezobetona* [Modern studies on reinforced concrete strength] (Strojizdat, Moscow, 1977) (in Russian)
2. L. Huynh, S. Foster, H. Valipour, R. Rendall, Construction and Building Materials **78**, 153-171 (2015)
3. T.N. Vinogradova, Vlijanie raspora na rabotu zhelezobetonnyh balochnyh konstrukcij pri kratkovremennyh dinamicheskih vozdejstvijah [The effect of thrust on performance of reinforced concrete beam structures under short-term dynamic load], PhD thesis (Moscow, 1977) (in Russian)

4. D.R. Galyautdinov, Proceedings of the 62nd Scientific University Conference of Students and Young Researchers, 21–24 (Tomsk, Russia, 26 April, 2016) (in Russian)
5. I.N. Tikhonov, Zhilishhnoe stroitel'stvo [Housing Construction] **2**, 40-45 (2013) (in Russian)
6. N.N. Popov, B.S. Rastorguev, Dinamicheskij raschet zhelezobetonnyh konstrukcij [Dynamic calculation of reinforced concrete structures] (Strojizdat, Moscow, 1974) (in Russian)
7. G.I. Odnokopilov, D.Ju. Sarkisov, E.A. Butuzov, Proceedings of Tomsk Politechnical University. Engineering of georesources 329, 122 – 135 (2018)
8. G. I. Odnokopilov, D.Ju. Sarkisov, Proceedings of Tomsk Politechnical University. Engineering of georesources 328, 85 – 95 (2017).
9. G.I. Odnokopilov, O.G. Kumpyak, Z.R. Galyautdinov, D.R. Galyautdinov, Proceedings of Tomsk Politechnical University. Engineering of georesources 330, 110 – 125 (2019)
10. O.G. Kumpyak, Z.R. Galyautdinov, D.N. Kokorin, *Prochnost' i deformativnost' zhelezobetonnyh konstrukcij na podatlivyh oporah pri kratkovremennom dinamicheskom nagruzhenii* [Strength and deformability of reinforced concrete structures on yielding supports under short-term dynamic load] (TSUAB Publ., Tomsk, 2016) (in Russian)
11. O.G. Kumpyak, Z.R. Galyautdinov. The calculation of ferroconcrete plates on foil bearings during a short-term dynamic loading // Proceedings of Tomsk State University of Architecture and Building. – Tomsk. – 2012. – № 2. – P. 107 – 111
12. O.G. Kumpyak, Z.R. Galyautdinov, D.N. Kokorin Strength of concrete structures under dynamic loading // AIP Conference Proceedings 1698, 070006 (2016); doi: 10.1063/1.4937876
13. A.R.A. Said, Povyshenie nesushhej sposobnosti zhelezobetonnyh konstrukcij pri vzryvnyh vozdejstvijah [Increase of bearing capacity of reinforced concrete structures under blast load], PhD thesis (Moscow, 1995) (in Russian)

Two-layer reinforced concrete panels with uniformly compressed sections

Dmitry Mailyan[1], *Alik* Blyagoz[2,*], *and Konstantin* Kretinin[2]

[1]Don State Technical University, 162 Sotsialisticheskayastr, Rostov-on-Don, 344022, Russia
[2]Federal State Budgetary Educational Institution of Higher Education «Kuban State Agrarian University named after I.T. Trubilin» 13, Kalininastr, Krasnodar, 350044, Russia

Abstract: The paper contains the issues related to the creation of a non-uniform prestress cross-section of reinforced concrete elements compressed during operation of working with a one-way eccentricity of the resultant longitudinal forces.

1. Introduction

Compressed reinforced concrete elements, as a rule, have a symmetrically arranged longitudinal reinforcement, so when it is pre-stressed, the symmetrical sections are uniformly compressed. This is reasonable because in most cases, compressed elements are subjected to alternating forces. Nevertheless, a number of compressed reinforced concrete elements in use operate with a one-way eccentricity of the resultant longitudinal forces. These include, for example, the racks of transport galleries, elevated roads, bridges, overpasses, etc. Moreover, the static scheme of such elements can be done so that the deflection moments along the entire length have the same non-changing sign. In these cases, it is advisable to create a non-uniform prestress section of the elements. Part of the external load will be used to offset the initial deflection created in this way, and the total axial eccentricity of the external longitudinal force *N* will be reduced by the value of the initial deflection f_p (Fig. 1), the boundaries of the effective use of prestressing in compressed elements are expanded. As a result of these factors, the load-bearing capacity of compressed elements increases, which makes it possible to obtain significant savings in valve consumption while maintaining its value unchanged [4,5].

2. Experimental program and research results

Non-uniform prestress section and the initial deflection of the element in cross-section and reinforcement are symmetric with respect to axis passing through center of gravity perpendicular to the force action plane, and these are formed by inequality of effort prestress in A_{sp} and A'_{sp}, i.e. at $\sigma_{sp}A_{sp} \neq \sigma'_{sp}A'_{sp}$. In case of an unsymmetrical arrangement of the rebars the deflection was formed even by the same value of forces in the rebars A_{sp} and A'_{sp}.

Compressed elements include wall panels along with columns, piles, and other structural elements. A progressive solution is the prestressed double-layer wall panels developed by the originators that combine load-bearing and enclosing functions. The use of such panels, whose height is equal to the height of a single-storey industrial- or agricultural building, allows you to abandon the frame solution, which increases prospects of a fully-prefabricated structure by elimination columns, half-timbered pillars, foundation beams, and other elements [8].

The connection between the layers of panels is provided by the adhesion between heavy and light concrete, which, as studies have shown, is quite sufficient and reliable. Concreting of panels is performed in a horizontal position - first, the first layer of heavy concrete is laid, and immediately after that, the second layer of light concrete is laid. After each layer is vibrated, there is some interpenetration of the concrete layers, which helps to increase the adhesion between them.

Pre-compression of the proposed panels, including non-uniform, allows to increase their rigidity and stability, which is especially important at a high height of buildings. The geometric flexibility (l_0 / h) of the panels can reach 30 or more. In addition, pre-compression increases the crack resistance and deformability of panels when operating loads with large eccentricities are applied, as well as when deflection moments are applied during transportation and installation of panels [6,7].

*E-mail: alfa-maikop@yandex.ru

With non-uniform pre-compression of panel sections, the initial deflection is formed to one side of the face with less compression (Fig. 1, b).

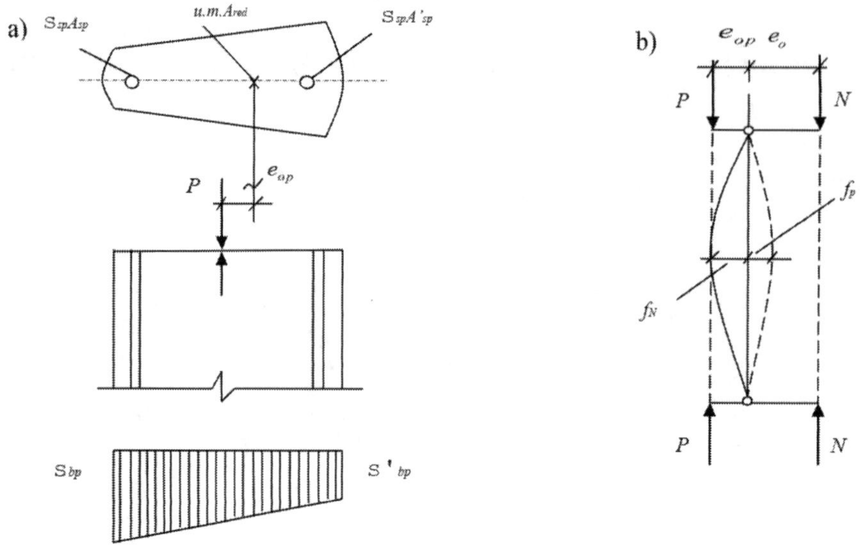

Fig.1. Compressed elements with non-uniformly prestressed sections: a - reinforcement of the section; b - diagram of the operating forces during pre-compression of the section of the element.

Misalignment of the force N from the axis passing through the center of gravity of the section decreases and becomes equal to $(e_0 + f_N - f_p)$, which leads to a significant reduction in the deflection of the bent panels.

However, this positive effect of non-uniform prestressing of cross sections occurs when the following condition is met:

$$e_0 > kf_p \tag{1}$$

where e_0 - is the axial misalignment of the longitudinal force measured from the axis passing through the center of gravity of the section close to the supports; f_p - is the deflection in the middle of the element caused by non-uniform compression; k - is a coefficient depending on the longitudinal force.

To determine a coefficient k, included in this condition, we make a differential equation of the elastic axis of the element (Fig. 2):

$$B(y''-1/r_p) = N(\Delta - y - e_0) \tag{2}$$

in which the left part is the product of stiffness of the cross sections B by the difference of curvatures from the external force $N$and the bend, and the right part is the moment from the force N in any cross section of the element in the deformed state.

We represent the solution of equation (2):

$$y = \frac{1}{\cos\frac{ml}{2}}\left(\frac{\xi}{m^2} - e_0\right)(1 - \cos mx), \tag{3}$$

where: $$m^2 = N/B \text{ and } \xi = 1/r_p = 8f_p/l^2 \tag{4}$$

Let us determine the misalignment e_0, at which the movement of the end of the rack does not exceed the deflection f_p [4,5]. To do this, in (3) we substitute the condition $y < f_p$ and having in mind (4) after the transformations we get:

$$e_0 > f_p\left(\frac{8B}{Nl^2} - \frac{\cos\frac{l}{2}\sqrt{\frac{N}{B}}}{1-\cos\frac{l}{2}\sqrt{\frac{N}{B}}}\right) \quad (5)$$

The coefficient k, equal to the expression in parentheses is a function of a free length of the column l, cross section stiffness B and longitudinal force N.

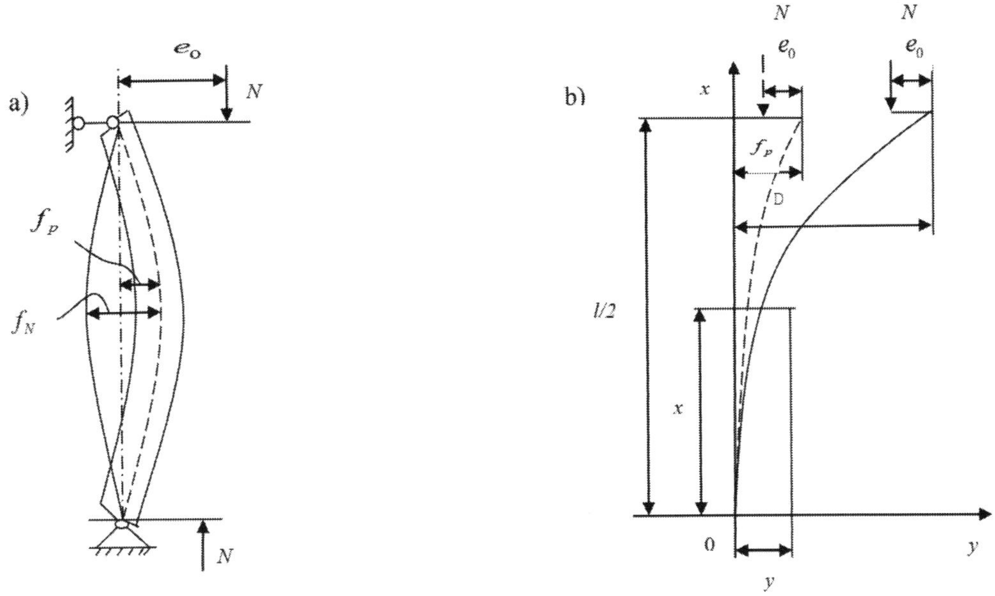

Fig.2. To determine the conditions for the feasibility of creating a non-uniform prestressing of concrete in flexible compressed elements; a - diagram of the panel with hinge supports; b - designations for eccentricity from the elastic axis.

Experiments conducted in the Russian State Social University [1,2] have shown that changes in these parameters over a wide range have little effect on the numerical values of the coefficients: with increase in N and l/B by 3-4 times k decreases from 0.85 up to 0.81 (Fig. 3).

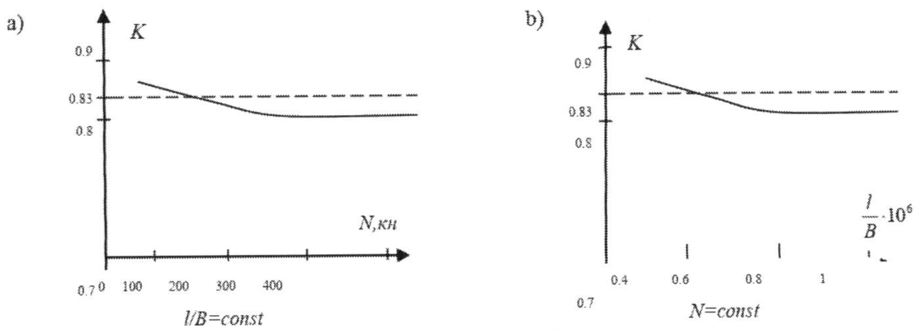

Fig.3. Dependence of coefficient K accordingly on N (a) and l_0/B (b) respectively to the data [1, 2].

If condition (5) is met, the non-uniform compression significantly increases the relative crack resistance and reduces the deflection of the compressed elements, and given that, the positive effect with an increase in the relative misalignment up to 0.8 or more is practically not reduced.

The results of experiments [1-3] with uniform $(\eta = \eta' = 0.34)$ and non-uniform $(\eta = 0.37; \eta' = 0.15)$ compression are shown in Fig. 4, where it can be seen that the non-uniform compression of sections at $e_0 > 0.83 f_p$ significantly increases the load-bearing capacity of the compressed elements in comparison with the uniform compression intensity equal to the maximum at the non-uniform. At $e_0/h = 0.25$ $(e_0 = 2.7 f_p)$ the bearing capacity of non-uniformly compressed elements is greater than that of uniformly compressed elements by 35%.

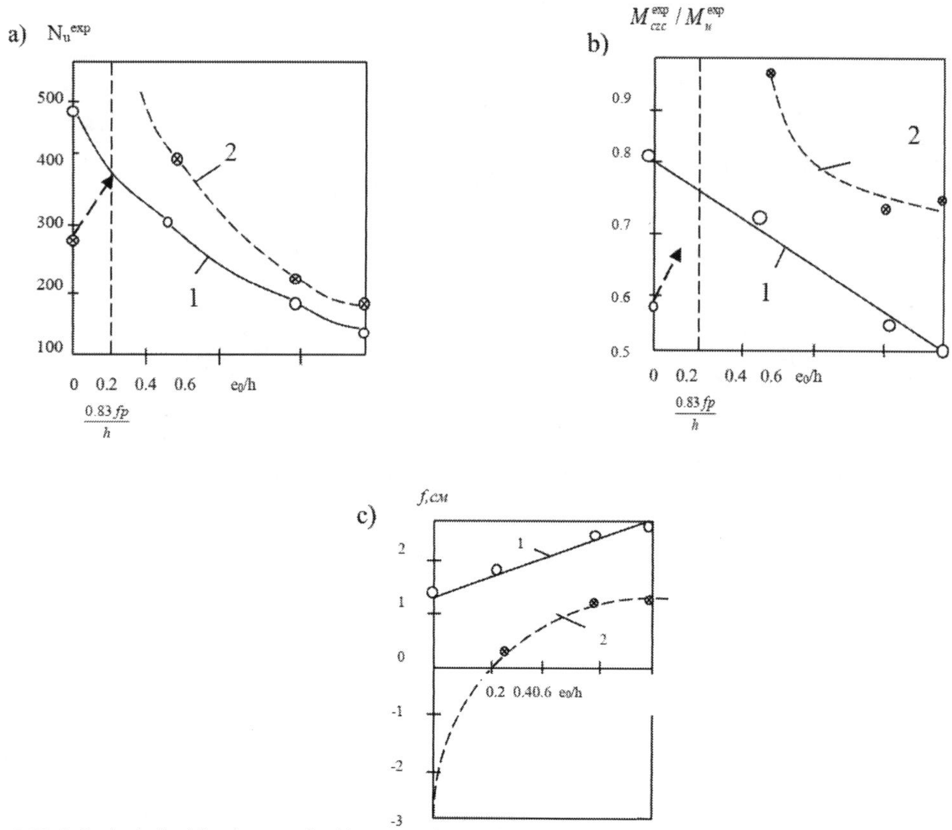

Fig.4. Variation in the load-bearing capacity (a), crack resistance (b) and deflections in the midsection at M=0.7Mu(c) with respect to the relative misalignment for the compressed elements with uniformly (1) and non-uniformly (2) pre-compressed sections according to data [1,2].

3. Conclusion

With an increase in the misalignment of the external longitudinal force, the load-bearing capacity of non-uniformly compressed elements approaches the load-bearing capacity of uniformly compressed elements and at $e_0 / h = 0.8$ becomes almost equal to the latter. The value of the boundary relative misalignment e_0^{rp} / h, at which the load capacity of uniformly and non-uniformly compressed elements becomes the same, depends on the level of compression of sections η and its difference η / η'. As they increase, the boundary value of the relative misalignment increases.

References

1. D. R. Mailyan Concrete and reinforced concrete **2** (1982)
2. D. R. Mailyan, N. Z. Sukhail *Questions of strength, deformability and crack resistance of reinforced concrete* (Russian Institute of Strategic Research, Rostov-on-Don, 1986)
3. V. F. Mayboroda, V. M. Karnyuk *Three-layer reinforced concrete structures* (Budivelnik,Kiev,1990)
4. A. Ryabukhin, A. S. Matsiy In-situ measurement of the anchor-pile displacement in the geotechnical conditions of Sochi *Proceedings of the 5th International Young Geotechnical Engineers' Conference (IYGEC 2013)* 576–578 (France. Paris: Ecole des PontsParisTech, 2013)
5. A.K. Ryabukhin, V.A. Lesnoy, D.Y. Kalashaov *IOP Conf. Ser.: Mat. Sci. and Eng.*. **698 (7),** 077003 (2019)
6. E. Bezuglova, S. Matsiy Engineering and geological grounds of landslide protection reliability of structures *Landslide Science and Practice: Risk Assessment, Management and Mitigation* **6**, 709-714 (2013)
7. E. Bezuglova, S. Matsiy, V. Podtelkov Landslide risk management at transport facilities *Landslides and Engineered Slopes. Experience, Theory and Practice* **2**, 405-409 (2016)

8. S.I. Matsiy, L.A. Sukhlyaeva *Geotechnics Fundamentals and Applications in Construction: New Materials, Structures, Technologies and Calculations* (Proceedings of the International Conference on Geotechnics Fundamentals and Applications in Construction: New Materials, Structures, Technologies and Calculations, GFAC 2019) 178-181 (2019)

Effective additive for foam concrete

Vladimir Belov[1, *], and *Pavel* Kuliaev[1]

[1]Tver State Technical University, Civil Construction Department, 22, Af. Nikitin Quay, Tver, 170026, Russia

Abstract. In search of a building material of the XXI-st century, architects, investors and builders are increasingly coming to one recently successfully approved solution - the use of foam concrete in construction, which, due to its exceptional characteristics, meets the multilateral requirements of the construction industry. The article presents the way to use technogenic wastes in the production of non-autoclave foam concrete.

1 Introduction

Foam concrete is characterized by high versatility, relative simplicity of technology, low production costs in the manufacture of products. This is owing to the fact that the production of concrete in a wide range of density values is possible on the same equipment by using natural sand as a filler. Due to the simplicity of the technology and the equipment used (when excluding grinding raw materials in ball mills from the technological cycle and autoclave processing), the cost of products is 1.5-2 times lower than the cost of the same products made of cellular aerated concrete.

The possibility of excluding foam concrete from the heat treatment technology ensures that it can be effectively used in monolithic construction. The density of foam concrete (using fine sand of natural dispersion as a filler) is usually within 600...1000 kg / m3, and the compressive strength is 1.5...3.5 MPa (concrete class 15...35). To obtain foam concrete with a lower average density with maintaining strength, ground sand is used, as well as hardening accelerators, reducing, water-reducing and other additives.

Sometimes, in order to reduce the density and eliminate the grinding operation, foam concrete is produced from cement binder without sand. This material is called foam cement. However, this concrete has essential shrinkage when drying, which reduces its quality indicators. All these difficulties prevent the foam concrete from widespread use altogether with a lack of of highly qualified specialists.

In the future, the most reasonable ways for small producers will be to use ready-made dry mixes for foam concrete, which will provide a significant improvement in its properties. In this case, the foam concrete mix will be prepared by professionals in advance, the materials of guaranteed quality being used. Manufacturers will have to put in the dry mixture some water according to the instructions. However, the practical implementation of this idea is a matter of the future.

[*] Corresponding author: vladim-bel@ya.ru

With regard to the most common products – wall blocks and stones made of foam concrete – practice has shown that they achieve an optimal ratio between the density and thermal conductivity of concrete, on the one hand, and its strength and frost resistance, on the other. The use of foam concrete as a material for lightweight elements and fencing structures allows us to meet modern requirements for heat efficiency of residential buildings. Houses made of foam concrete allow to save natural and energy resources. They are distinguished by high quality, energy efficiency, low construction costs; they have a long service life. At the same time, it is important to reduce the wall thickness while maintaining its necessary thermal resistance, which can be achieved by reducing the average density of the material.

Currently, finely-ground limestone with a nanoscale component is used to modify and improve concrete properties [1]. This is due to the ability of fine-ground limestone to interact with concrete clinker components at the physical and partially chemical levels, forming a monolith [2, 3]. The positive effect of carbonate fillers is explained by the fact that they have a chemical affinity with cement, which leads to the formation of a dense contact between them and the fusion of hydration products of cement with the filler during natural hardening. An important feature of limestone rocks is that they are not inert, but enter into an active physical and chemical interaction with clinker minerals of cement, participating in the formation of the structure of the cement stone.

In addition, it is known that finely ground limestone helps to reduce water segregation, water demand and delamination of the mixture, increase their water-holding capacity, plasticity, and also improves the water and frost resistance of the solution [4].

Physical and chemical compatibility of the nanosized component of finely-ground limestone with a binder and solid components of the concrete is technologically revealed in the following effects: powdering cement grains, micro-reinforcement of structure of cement stone and concrete, promoting the appearance of additional amount of cementitious formations. This positive effect of the carbonate microfiller is enhanced by a synergistic effect when optimizing the granulometry of the grain composition of the mineral part by implementing a model of the densest particle packages [5].

Thus, the use of a microdispersed limestone filler made of the crushed wastes of carbonate aggregates in combination with a superplasticizer as a complex additive in the manufacture of foam concrete in order to improve the operational and technological properties of this material, as well as increase its durability by forming a better microstructure of interstitial partitions is currently very actual for the construction industry.

Almost all types of natural-hardening foam concrete have low resistance to tensile stresses and increased brittleness, resulting in unwanted cut-chips and cracks. Non-autoclaved foam concrete is also characterized by high shrinkage, which leads to intensive cracking and even destruction of products. A radical way to eliminate these disadvantages is the dispersed reinforcement of foam concrete with fibers, which provides a significant improvement in the strength and deformative properties of the material, as well as increasing the operational reliability of products [6]. Dispersed-reinforced foam concrete is destroyed as a plastic material, without sudden opening of cracks and breakage into separate pieces. It is established that the putting of polymer fibers in foam concrete mixtures allows 2-2.5 times to increase the bending strength, and up to 1.5 times-the compressive strength. Fiber reinforcement completely eliminates the appearance and development of shrinkage cracks during the material hardening [7].

2 Materials and technics

In this paper, we studied the effect of a complex additive, including fine-grained limestone with a specific surface area of 310 m2/kg according to PSC-2, superplasticizer C-3, a fiber component in the form of the utilized textile cord obtained during the processing of old tires, as well as a hardening accelerator, on the properties of foam concrete obtained by "dry mineralization" [8].

Foam concrete of "dry mineralization" was obtained in the form of cubes with a rib size of 100 mm. The following basic raw materials were used: quartz sand with a grain size module; Portland cement of the PC-500 D0 class and foam agent FF-2000. The overdosed amount of water (based on w / solid=0.64) was mixed in a vertical mixer (1000 min-1) with the addition of 0.33 % of the foaming agent for 3 min, then with continuous mixing for 1 min (200 min-1), a dry mixture of cement, sand and a complex additive was evenly added to the foam in a ratio of 1: 0.18: 0.064, respectively, the resulting homogeneous molding mixture was poured into molds. The samples were tempered at a temperature of 60 °C for 3 days.

3 Test results

The analysis of the data obtained (table) shows that the use of the described additive with the same strength of foam concrete can reduce the average density of the material by 100 kg / m3 and the coefficient of thermal conductivity by 13.3 %.

Composition of "dry mineralization" foam concrete	Average density of samples, kg / m^3	Average compressive strength at the age of 3 days, MPa	Coefficient of thermal conductivity, W/(m*°C)
Without additives	600	1,2	0,15
With the addition	500	1,2	0,13

The introduction of a reinforcing fiber additive into foam concrete in an amount of 1-2 % increases its compressive strength by 20-44 % at the same density compared to foam concrete without a fiber additive. Further increase in the concentration of utilized textile cord is not possible, since this leads to the formation of lumps of tangled fibers. This reduces the strength of the concrete.

The positive effect of fiber additives on the properties of foam concrete is explained by the following reasons. On the surface of the fibers, evenly distributed in the volume of the mixture, there are particles of the solid phase. The first which move to the extended phase separation surfaces (fibers) are the smallest of them, whose energy is regulated by the volume concentration of the liquid phase. The thickness of water films firmly bound to the surface of the solid phase is proportional to the size of the particles. Therefore, a variatropic structure of the concrete mix is naturally formed on the surface of the dispersed reinforcement, the ability of which to accumulate defects in the form of cracks is limited by geometric parameters and the amount of fiber.

Cellular concretes are three-phase dispersed systems. In such systems, stability and cohesion are provided when the capillary and wedging pressures of the film water in the capillary meniscus zones are balanced, as well as by surface tension forces [9]. The most significant factor that increases the strength of cellular concrete at a given density is the "reinforcing" of air cells with cement particles [8]. This " reinforcing " is essentially a manifestation of the processes of capillary structure formation, similar to globulation in raw mixes of building materials, provided a kind of inversion-the replacement of large aggregate grains with air (in foam concrete) or gas (in aerated concrete) cells. Considering this, a parameter of the optimal structure of cellular concrete is the maximum efficiency coefficient (in MPa), determined by the formula

$$K_{eff} = R_{comp} / (\rho_o)^2, \quad (1)$$

where R_{comp} – compressive strength, MPa; ρ_0 – the average density of cellular concrete in the dry state (relative to the density of water).

Based on the planned experiments application and optimization of the obtained mathematical models, which present the dependence of the average density, compressive strength and efficiency coefficient on the composition factors, the foam concrete composition was chosen that meets the requirements of GOST (State standards) 25485 for structural and heat-insulating non-autoclaved concrete of the average density D500 and compressive strength class B1.

4 Summary and prospects

The use of technogenic wastes in the production of non-autoclave foam concrete, such as limestone crushed waste and utilized textile cord, contributes to improving the performance properties of the material, while solving the environmental problems of the region, expanding the raw material base for the production of construction materials, as well as reducing the cost of production.

References

1. V.G. Khozin, O.V. Khokhryakov, Carbonate cements of low water demand, Technologies of concrete, **11-12**, 25 (2009)
2. U. El Shamy, C. Denissen, Microscale energy dissipation mechanisms in cyclically-loaded granular soils, Geotechnical and Geological Engineering, **30(2)**, 343-361 (2012)
3. G. De Schutter, Influence of limestone filler as a mineral additive in self-compacting concrete, Belgium Magnel laboratory of concrete research, *36 Conference on Our World in concrete & Structures* (Singapore, **14-16**/08/, 2011)
4. I.A. Zinov, High-Strength concrete with the addition of micro-silica, Izvestiya vuzov. Construction and architecture, **4**, 55 (1990)
5. B. Moser, C. Pfeifer, Microstructure and Durability of Ultra-High Performance Concrete, *Proceedings of the Second International Symposium on Ultra High Performance Concrete Kassel* (Germany March, 05-07, 2008)
6. L.V. Morgun, Efficiency of application of fibropenobetona in modern construction, Construction materials, **3**, 16-17 (2002)
7. L.V. Morgun, V.N. Morgun, Influence of dispersed reinforcement on the aggregate stability of foam concrete mixtures, Construction materials, **1**, 33 (2003)
8. A.P. Merkin, Cellular concrete scientific and practical prerequisites for further development, Construction materials, **2**, 11-15 (1995)
9. V.V. Belov, Capillary autohesion in three-phase raw mixes, *Fundamental, research and applied research of the Russian Academy of architecture and construction Sciences on scientific support for the development of architecture, urban planning and the construction industry of the Russian Federation in 2018* (Russian Academy of architecture and construction Sciences, Moscow, 58-68, 2019)

Beneficiated ponded fly ash for concretes with high volume mineral additions

Nikolai Zaichenko[1,*], *Irina* Petrik[1], and *Liudmila* Zaichenko[1]

[1]Donbas National Academy of Civil Engineering and Architecture, 286123 Makeyevka, Ukraine

Abstract. The article presents the investigation results of effect of beneficiated ponded fly ash on the properties of cement pastes and concretes with high-level replacement of Portland cement. To improve the characteristics of ponded fly ash meeting technical requirements for replacing cement in concrete the triboelectrostatic beneficiation technology has been elaborated. This technology can produce low-carbon ash product (LOI = 2.52 % in this study) for the high replacing level of cement (45 %) in concretes. The beneficiated ponded fly ash has an improved granulometric and phase composition, a decreased content of unburned carbon that accelerates the hydration process of cement, increases the rheological properties of cement paste and the ability of air-entraining admixture to hold the required involved air. In a combination with the nanostructured-carbon-based plasticizing admixture the beneficiated ponded fly ash exhibits high rate of strength increase when is used in high-volume fly ash concretes.

1 Introduction

Fly ash is a by-product material being generated by thermal power plants from combustion of pulverized coal, and so may be regarded as a waste product [1, 2]. At thermal power plants, fly ash is presently collected or disposed by using either dry or wet systems. Being cheaper than any other manner of fly ash removal, it is widely used method, thus more than 65 % of fly ash is disposed in landfills and ash ponds [1].

The landfill of fly ash is the problem of all power plants because disposed waste product as a rule is not used in any practical application [3]. The American Coal Ash Association reported that utilization rates for freshly generated fly ash have increased considerably over recent years, with current rates near 60 %, however approximately 14 million tons of fly ash continue to be disposed of annually. While utilization rates in Europe have been much higher than in the US, considerable volumes of fly ash have also been stored in landfills [4]. It may cause substantial environmental hazards [5]. Thus, high disposal costs, increased interest in improving the environmental impacts are making fly ash utilization an increasingly attractive alternative to disposal [6].

The utilization of fly ash instead of dumping can be used on economic grounds as pozzolana for partial replacement of cement [7] on the one hand, and because of its

*Corresponding author: zaichenko_nikola@mail.ru

beneficial effects on the properties of concrete mixtures and concretes [8-10]. However, many variables affect the chemical composition, morphology, and relative phase proportions in fly ash [11]. As a result, the chemical composition in particular the loss-on-ignition (LOI) content could restrict the amount of fly ash application in cement concretes.

Fly ash particles have a wide size distribution ranging from molecular clusters with diameters of 2 nm up to particles with diameters of about 100 μm. The particles larger than 1 μm mostly contain unburned mineral materials [12]. In doing so, it was noted that loss-on-ignition increased with particle size of fly ash samples [13].Beneficiation of fly ash has been studied in recent years and several methods have been proposed to separate good quality cementitious material from the combustible one [14]. However, the electrostatic separation method of unburned carbon from specifically prepared fly ash has the greatest practical significance [15, 16]. A number of different types of triboelectrostatic separators are known. It was found that the quality of separation is dependent upon the nature of fly ash and the configuration of the separator used [17]. The plate triboelectrostatic separator has been used for this study.

Triboelectrostatic beneficiation of fly ash with high-unburned carbon content can produce low-carbon ash products having value as mineral additions and meeting technical requirements for replacing cement in concrete [18]. This effect is of paramount significance when the beneficiated fly ash will be used in concretes with high volume of mineral additions in particular high-volume fly ash concrete (HVFAC) which contains normally more than 50 % fly ash by mass of total cementitious materials [9]. Nevertheless, while HVFAC can be designed and produced to meet 28-d strength requirements and often exceeds the durability performance of conventional concretes, a persistent problem is the potentially long delay in setting time and hardening process at early curing age [19, 20]. To overcome these limitations, the next approaches have been proposed by various research teams: the addition of either calcium hydroxide powder (5 %) or rapid set cement (5-10 %) [19], silica fume (10 %) and lime (30 %) [21], the replacement of cement with 5 % of nano-limestone [22] or nano-SiO_2 [8], the use of ternary blended cements with limestone and Portland cement [7] or high-range water reducing admixtures [23] as well as internal curing to improve hardening process [20]. Belarusian scientists have developed a nanostructured-carbon-based plasticizing admixture, which makes it possible to produce concrete with exceptional physical-mechanical properties [24, 25]. Besides, the admixture provides an adequate compressive strength of concrete with addition of fly ash both at early and project ages of hardening. However, the effect of this admixture on the properties of HVFAC has not been investigated. On the other hand, no experimental data regarding concretes with high content of ponded fly ash beneficiated by triboelectrostatic separation technique have previously been reported. Thus, the aim and the objectives of this study are the following: i) to investigate the properties of ponded fly ash treated in the triboelectrostatic separator, ii) to determine the properties of HVFAC with the nanostructured-carbon-based plasticizing admixture and the ponded fly ash beneficiated by triboelectrostatic separation technique.

2 Experimental

2.1 Materials

Ordinary Portland cement (OPC) CEM I 42.5 N conforming to EN 197-1 and ponded fly ash (PFA) with Blaine specific surface area of 365 and 290 m^2/kg, respectively, were used as cementitious materials. The value of loss-on-ignition of PFA is 6.92 %. Quartz sand (fine aggregate FA) with fineness modulus of 2.2 and specific gravity of 2.68 as well as crushed granite (coarse aggregate CA) of 10 mm maximum size and specific gravity of 2.69 were used for producing normal weight concrete.

The nanostructured-carbon-based plasticizing admixture "ART Concrete-K" as an opaque dark brown liquid (superplasticizer based on a polycarboxylate polymer and nanostructured carbon material are in the ratio 100 : 0.09...0.11) has the following properties: dry content 37 %, density 1.1 ± 0.1 g/cm^3, pH 7.0 ±1.0. SikaAer®PRO-100 admixture was used as a high-efficiency air-entraining agent (AEA) (density 0.995-1.005 g/cm^3, pH 10.0-11.0).

The formulations of cement-fly ash pastes made with different content of fly ash collected from anode and cathode plates of electrostatic separator as well as with reference ponded fly ash are given in Table 1. The water to binder ratio was different to evaluate the constant value of standard consistency of the cement-fly ash pastes.

Table 1. The formulations of the cement-fly ash pastes.

Formulation index		Composition					w/b ratio
		OPC		PFA		Water, ml	
		g	%	g	%		
1	Control	690	100	0	0	172.5	0.25
2	PFA-anode	586	85	104	15	165.6	0.24
3	PFA-anode	380	55	310	45	138.0	0.20
4	PFA-cathode	586	85	104	15	179.4	0.26
5	PFA-cathode	380	55	310	45	193.2	0.28
6	PFA-reference	586	85	104	15	200.1	0.29
7	PFA-reference	380	55	310	45	220.8	0.32

2.2 Experimental procedure

2.2.1 Triboelectrostatic separation beneficiation technology

Ponded fly ash material was dried at 105 ± 5 °C and then screened at 315 μm as a precursor to further beneficiation. Fly ash particles are mostly composed of silica and alumina, which have a higher work function (SiO_2 = 5.0 eV, Al_2O_3 = 4.7 eV) than carbon (C = 4.0 eV). Therefore, when they contact to each other or the surface of copper tribocharger (Cu = 4.38 eV), carbon particles will be charged positively and fly ash particles will be charged negatively due to the differences in the work function values [6, 15].

The charged particles enter the separation chamber (Fig. 1) containing the electrodes made of two copper plates attached to its opposite sidewalls. Under the DC electric field generated by a high voltage power supply, the charged particles are deflected towards the electrodes according to their polarity. The electric field ranged from -2 kV/cm to +2 kV/cm.

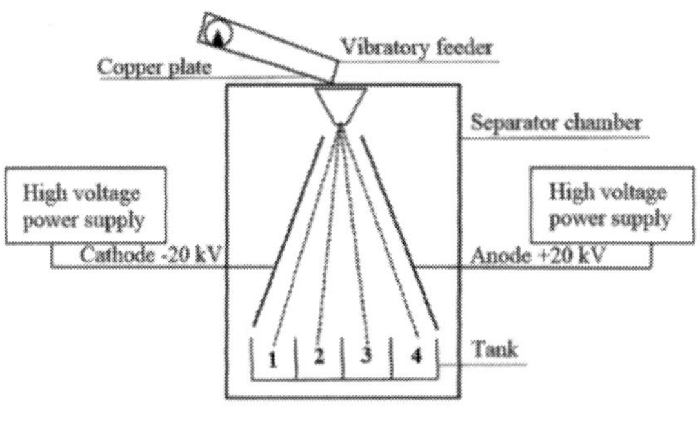

Fig. 1. Schematics of the experimental setup

2.2.2 Methods

The particle size distribution (PSD) of ponded fly ash was determined by the ANALYSETTE 22 Compact laser diffraction particle size analyser.

Loss-on-ignition measurements of fly ash were performed on 1-g samples pre-dried in a laboratory oven at130 °C for 2 h in accordance with ISO 1171-97. After drying, cooling and reweighing, the samples were placed into an air-ventilated laboratory oven at 815 ± 10 °C for 2 h to burn out the residual carbon. The loss in mass at 815 °C was reported as the LOI, and this is assumed to be the weight of unburned carbon in the original sample.

The Foam Index Test (FIT) was originally designed to evaluate the influence of cementitious materials on the air-entraining capacity of concrete [26]. At first, 20 g of Portland cement were placed in a 125 ml glass jar. Then 50 ml of water were added to the jar which was caped and shaken for a minute. Diluted AEA solution (50 ml AEA : 50 ml water) was added in small increments of 2 to 5 drops at a time. After each addition, the jar was shaken vigorously for 15 seconds. The stability of the foam was observed. The minimum amount of diluted AEA needed to produce a foam that is stable (bubbles exist over the entire surface) for 45 seconds is the Foam Index of the cement mixture. After that, the steps 1 through 4 using different proportions of cement and fly ash to develop the foam index of the cement and fly ash mixture were repeated [27].

Diffraction studies of the ponded fly ash samples were conducted with the help of 26 ARLX'TRA X-ray diffractometer (ARL, Switzerland) using CuK_{α} radiation (λ = 1.54056 Å) at a voltage of 40 kV and a current of 20 mA. The scanning was carried in 4 to 70 degrees 2-theta range with a step of 0.02 degrees. X-ray phase analysis of the cement-fly ash pastes was performed on the X-ray diffractometer DRON-4-07. The diffraction patterns were recorded at a wavelength of λ = 1.54178 Å in a step-by-step mode (2θ = 15-75° with a step of 0.1° and exposure time of 5 s) at 27 kV, 15 mA.

Vicat apparatus mold was used to determine the standard consistency of cement paste. Cubes 50 × 50 × 50 mm and 100 × 100 × 100 mm were used to evaluate the compressive strength of cement paste and concrete, respectively. Tests were conducted at different test age: 3, 7, and 28 days.

3 Results and discussion

3.1 Properties of the beneficiated ponded fly ash

3.1.1 Particle size distribution and loss-on-ignition

The particle size distribution curves for six samples of ponded fly ash collected after the electrostatic separation from each cell of tanker (1-4) and plate electrodes (A, C) of the experimental setup (Fig. 1) are given in Fig. 2.

The curves of the samples from tanker cells show approximately the same particle size distribution over the range between 0.3 and 300 μm while the curves of the samples from electrodes are over the more narrow range: 0.3-100 μm (anode) and 0.3-200 μm (cathode). Besides, the PSD analysis indicates that the mass median diameter is the smallest for the samples collected from the plate electrodes: 19.89 μm (cathode), 17.93 μm (anode) and decreases from the cells of cathode zone to anode one (Table 2). The highest content of the particles less than 2 μm is observed near the anode zone respectively. On the other hand, the cells 3 and 4, which are near the zone of anode contain the particles with maximum size d_{98} = 151.1 and 142.3 μm, respectively. Obviously, as it was pointed out earlier large particles of fly ash mostly contain unburned mineral material.

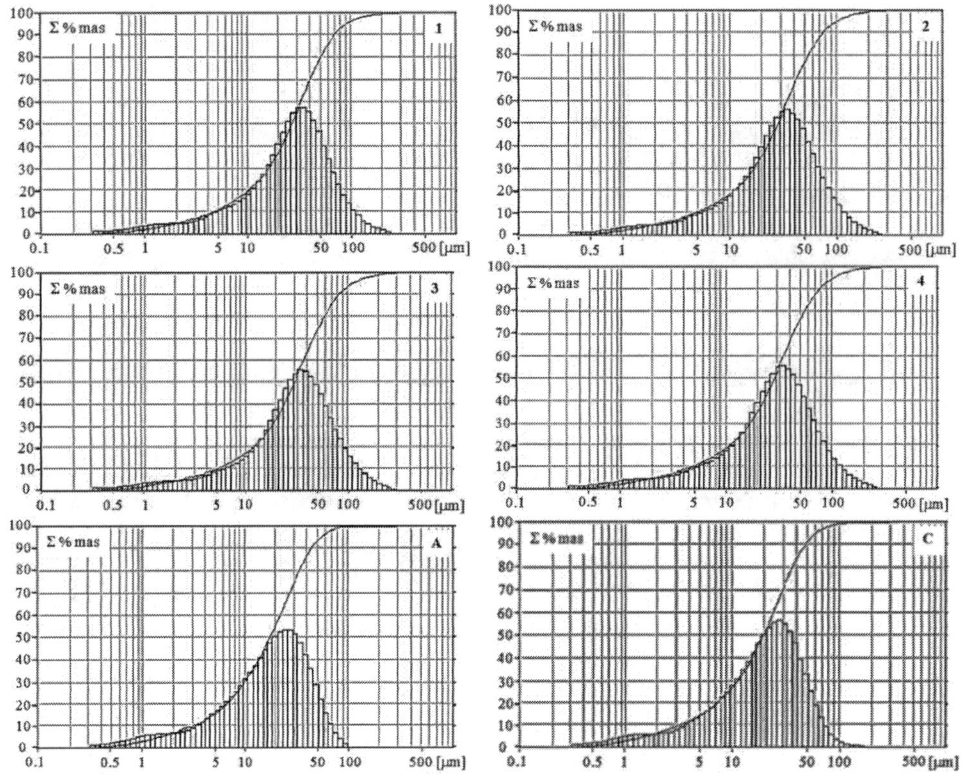

Fig. 2. Particle size composition and cumulative particle size distribution of ponded fly ash.

Table 2. Particle size and loss-on-ignition data for yields obtained after triboelectrostatic separation of PFA.

Sample – zone of separator	Yield		Particles size, µm		Particles less than 2 µm, %	LOI		
	g	%	maximum d_{98}	median d_{50}		g	% yield	%total mass
Cathode (C)	15	1.5	76.461	19.891	7.64	3.3	22.0	4.8
1	45	4.5	123.972	27.302	6.17	21.5	47.8	31.1
2	113	11.3	134.805	28.592	5.89	19.2	17.0	27.7
3	276	27.6	151.125	26.546	5.75	11.4	4.1	16.5
4	482	48.2	142.337	25.859	6.96	12.3	2.6	17.8
Anode (A)	69	6.9	66.576	17.930	8.01	1.5	2.2	2.1
Total	1000	100				69.2		100

Theoretically taking into account the positive electric charge of carbon, its particles should be deflected towards the electrode according to their polarity, i.e. towards cathode. However, the data represented in Table 2 provide evidence of the existence of quite high content of LOI (17.3 % of total yield of PFA after separation) within the range of positive electric field (anode, cells 4-3). This could be explained by the following approach. Electrostatic separation must be carried out on dry material, as fly ash and unburned carbon cannot be charged and separated in wet form due to capillary cohesion forces. Researches, carried out on electrostatic carbon removal of wet-stored (ponded) fly ash after drying, have found that a charge reversal of carbon occurs [4, 18, 28]. It is necessary to mention, that the mechanism that changes the relative work function and thus contact-charging behavior of the material nevertheless has been not yet understood. One suggested mechanism is the redistribution of soluble ions on the surface of carbon particles, possibly further influenced by the pH of the aqueous solution on the ash [4].

3.1.2 Phase analysis

The crystalline components of ponded fly ash have been characterized by using X-ray powder diffraction. The main crystalline phases detected from XRD (Fig. 3) are quartz (SiO_2) (d=0.334 nm, 2θ=26.64°; d=0.228 nm, 2θ=39.49°; d=0.165 nm, 2θ=55.65°) and hematite (Fe_2O_3) (d=0.240 nm, 2θ=37.44°; d=0.184 nm, 2θ=49.49°). The broad "halo", centered at approximately 2θ=26.5° on the diffractograms suggests the presence of a large amount of amorphous material.

The proportions of each of the crystalline phase and the bulk amorphous material determined for the six samples of PFA collected from different zones of the triboelectrostatic separator are reported in Table 3. Evidence from XRD analysis indicates that all samples of PFA are identical in the bulk amorphous material. The differences in the content of the crystalline phases are of a minor nature. In particular, the highest amount of quartz along with the smallest amount of hematite are in the cells located in the operation zone of anode electrode. This could beneficiate the fly ash as pozzolana since ultrafine particles of quartz may exhibit pozzolanic properties as well as play the role of nucleation centers.

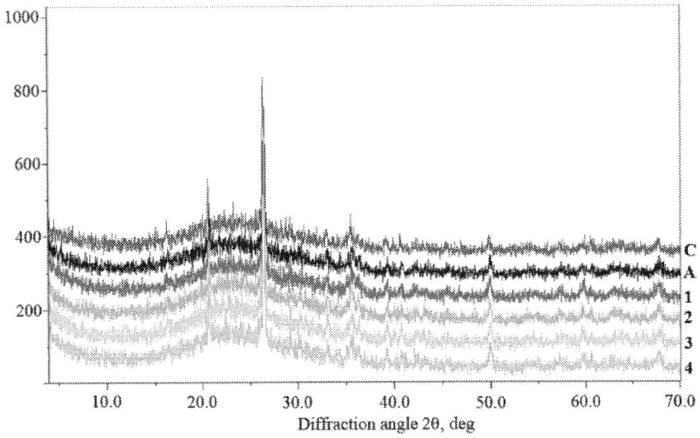

Fig. 3. Diffractograms of ponded fly ash after triboelectrostatic separation.

Table 3. Proportions of crystalline phases and bulk amorphous material presented in PFA.

Sample – zone of separator	Content, %		
	quartz	hematite	amorphous phase
Cathode (C)	19.7	5.3	75.0
1	20.9	4.1	75.0
2	19.0	6.0	75.0
3	20.0	5.0	75.0
4	21.2	3.8	75.0
Anode (A)	21.9	3.1	75.0

3.1.3 Foam index

When a fly ash addition contains a large amount of carbon, the adsorption of the AEA surfactant by the carbon destroys the ability of concrete to hold the required involved air.

This adsorption problem dictates the rejection of large amounts of commercial utility ash [29]. Thus, decreasing content of LOI is one of the key factors to beneficiate low-quality fly ash, in particular ponded fly ash.

Table 4 shows the change in air-entraining admixture requirement during the foam index test. The cement-fly ash proportions were made up in such a way that when the ash content is over the range between 15 and 60 %, the amount of unburned carbon in the samples of ash taken from different zones of the triboelectrostatic separator ranged from 1.32 to 7.17 %.

Table 4. Proportions of cement-fly ash mixtures and foam index.

Sample	OPC		PFA			Water, ml	Solution of AEA, ml
	%	g	%	g	LOI, g		
Anode (A)	40	8	60	12	0.264	50	0.20
4	45	9	55	11	0.286	50	0.25
3	50	10	50	10	0.410	50	0.33
Reference (R)	55	11	45	9	0.621	50	0.55
2	65	13	35	7	1.190	50	0.96
Cathode (C)	70	14	30	6	1.320	50	1.62
1	85	17	15	3	1.434	50	1.82

It follows from the data of Table 4 that a high amount of unburned carbon (LOI) affects the required dosage of air-entraining admixture significantly but not a high content of fly ash. The higher LOI content in the fly ash sample the higher the amount of diluted AEA is needed to produce stable foam.

3.2 Properties of the Portland cement – beneficiated ponded fly ash pastes

3.2.1 Compressive strength

The compressive strength results are given in Table 5. It can be observed that compressive strength of the Portland cement-fly ash pastes of all formulations increases with curing age. Besides, strength increment of samples with partial replacement of Portland cement by ponded fly ash as compared with control formulation (100 % OPC) increases with curing age. Thus, for example, the paste formulation with 15 % replacement of OPC by beneficiated PFA (sample from anode) possesses the following percentage of compressive strength at the age of 3, 7, and 28 days – 87.6, 103.9, and 106.7 %, respectively. Nevertheless, the retardation effect of PFA on the 3-d early compressive strength is clear observed.

Table 5. Compressive strength of cement pastes.

Formulation index		w/b	PFA, %	Compressive strength, MPa (%)		
				3-d	7-d	28-d
1	Control	0.25	0	26.6 (100)	51.9 (100)	66.7 (100)
2	PFA-anode	0.24	15	23.3 (87.6)	53.9 (103.9)	71.2 (106.7)
3	PFA-anode	0.20	45	17.7 (66.5)	35.8 (68.9)	56.8 (85.2)
4	PFA-cathode	0.26	15	19.2 (72.2)	41.6 (80.2)	52.1 (78.1)
5	PFA-cathode	0.28	45	15.5 (58.3)	31.3 (60.3)	41.9 (62.8)
6	PFA-reference	0.29	15	17.2 (64.7)	34.4 (66.3)	44.8 (67.2)
7	PFA-reference	0.32	45	12.2 (45.9)	26.7 (51.4)	39.7 (59.5)

As pozzolanic reaction is a slow process that its contribution to strength occurs only at later ages, the early strength of cement paste will be significantly reduced if a large amount of fly ash is used [8]. It can be seen that at the age of 28 days the formulation with 15 %

beneficiated PFA (sample from anode) as a partial replacement of OPC has the compressive strength value 6.7 % higher as compared with the control sample. When the replacement level is 45 % (sample from anode) the 28-d compressive strength is high enough (85.2 %) too. First and foremost, this is due to the beneficiated characteristics of ponded fly ash after the triboelectrostatic separation process, in particular decreased LOI content, improved granulometric and phase composition that are responsible for the improved rheological properties of fresh pastes (decreased w/b ratio). In this case, the fly-ash particles can act as "ball-bearing" to better the rheological properties of fresh pastes [10].

3.2.2 Influence of the fly ash addition on cement paste hydration

According to the XRD analysis (Fig. 4), it has been found that PFA affects the hydration of cement clinker minerals in particular tricalcium silicate (alite). It can be illustrated by the relative intensity of the most indicative diffraction peaks in the composition of cement pastes with equal level replacement of OPC by PFA. For instance, the relative intensity (as compared with control sample) of the cement pastes with 15 % replacement CP-2, CP-4, and CP-6, respectively: d = 0.278 nm (2θ = 32.17°) – 85.6, 93.4, 87.9 %; d = 0.177 nm (2θ = 51.59°) – 84.4, 94.3, 88.1 %; d = 0.164 nm (2θ = 56.00°) – 91.3, 102.2, 113.9 %.A similar pattern holds for the cement pastes containing 45 % PFA: CP-3, CP-5, and CP-7, respectively: d = 0.278 nm – 81.8, 90.7, 93.1 %; d = 0.177 nm – 96.3, 106.7, 96.2 %; d = 0.164 nm – 107.3, 113.8, and 120.4 %.

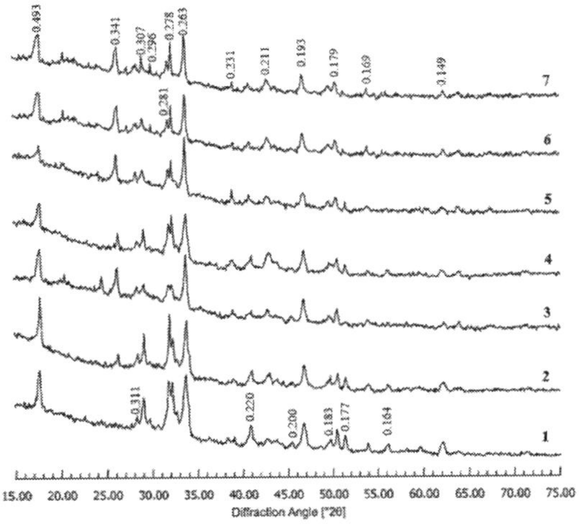

Fig.4. XRD patterns of cement pastes with addition of PFA.

Figure 4 also gives the variations in the relative intensity of the Portlandite (CH) diffraction peaks: d – 0.493 nm (2θ – 17.98°), d – 0.311 nm (2θ – 28.66°), d – 0.263 nm (2θ = 34.05°) as well as calcium hydrosilicate phases (CSH(B) and C_2SH_2): d = 0.307 nm (2θ = 29.06°), d = 0.281 nm (2θ = 31.82°), d = 0.183 nm (2θ = 49.79°); d = 0.220 nm (2θ = 40.99°), d = 0.211 nm (2θ = 42.90°), d = 0.200 nm (2θ = 45.31°).The results indicate that PFA collected from the anode plate is more reactive as compared with the samples from the cathode plate as well as the reference PFA.

3.3 Properties of HVFAC

The formulations and the properties of HVFAC made with ponded fly ash collected from the different zones of the triboelectrostatic separator as well as with the reference ponded fly ash are given in Table 6. The replacement level of OPC by PFA was 45 % and the slump of concrete mixtures got corresponded to the class S2 in accordance with EN 206-1 by means of the w/b ratio of concrete.

Table 6. Mixture proportions and test results.

Sample	Mixture proportions, kg/m^3								S,	f_c, MPa	
	OPC	PFA	CA	FA	A1, l	A2, l	w/b	AC, %	mm	7	28
C	500	0	1080	605	0	0	0.40	1.14	5.5	29.3	48.4
A-4-3	275	225	920	719	5.0	3.5	0.28	5.88	7.3	31.7	53.8
C-2-1	275	225	946	740	5.0	3.5	0.28	4.13	6.4	22.6	45.2
R	275	225	954	747	5.0	3.5	0.28	3.56	5.2	20.5	41.4

Keys: C = control concrete; A-4-3 = concrete with PFA from anode and 4-3 cells of the triboelectrostatic separator tank; C-2-1 = concrete with PFA from cathode and 2-1 cells; R = concrete with reference PFA; A1 = “ART Concrete-K” admixture; A2 = SikaAer®PRO-100 admixture; AC = air content; S = consistence of concrete (slump); f_c = compressive strength of concrete.

It can be seen from the data of Table 6 that concrete mixtures with equal content of the reference ponded fly ash (R) and PFA collected from different zones of the triboelectrostatic separator (A-4-3 and C-2-1) possess distinct properties, in particular air content and slump. Obviously, the A-4-3 sample containing the beneficiated ponded fly ash with LOI of 2.52 % has the highest values. On the other hand, the capacity of the air-entraining admixture decreases with increasing content of LOI – 4.40 % (C-2-1) and 6.92 % (R). Besides, the increased content of LOI adversely affects the slump of concrete mixtures.

The concrete sample containing the beneficiated PFA (A-4-3) and the nanostructured-carbon-based plasticizing admixture exhibited high rate of strength increase: 31.7 MPa (8.2 % increment as compared with control sample) as soon as after 7 days of curing and 53.8 MPa (Δ=11.2 %) after 28 days.

4. Conclusions

From the results of this study, it is possible to conclude that triboelectrostatic beneficiation of ponded fly ash with increased-unburned carbon content (LOI = 6.92 %) can produce low-carbon ash product (LOI = 2.52 %) meeting technical requirements for the high replacing level of cement (45 %) in concretes. The beneficiated ponded fly ash has an improved granulometric and phase composition, a decreased content of unburned carbon that accelerates the hydration process of cement, increases the rheological properties of cement paste and the ability of AEA to hold the required involved air. In a combination with the nanostructured-carbon-based plasticizing admixture the beneficiated PFA exhibits high rate of strength increase when is used in HVFAC.

The authors appreciate the National Research Moscow State University of Civil Engineering (NRU MGSU) for providing the opportunity to carry out this research.

References

1. I. Nawaz, IJIRSET, **2**, 10, pp. 5259-5266 (2013)
2. K. L. Scrivener, V. M. John, E. M. Gartner, Cem. Concr. Res., **114**, pp. 2-26 (2018)
3. R. Cheerarot, C. Jaturapitakkul, Waste Manage.,**24**, 7, pp. 701-709 (2004)
4. L. Baker, A. Gupta, F. Hrach, 2019 World of Coal Ash (WOCA) Conference in St. Louis, Missouri (2019)

5. H. Chung-Ho, L. Shu-Ken, C. Chao-Shun, C. How-Ji, Constr. Build. Mater., **46**, pp. 71-78 (2013)
6. F. Cangialosi, M. Notarnicola, L. Liberti, P. Caramuscio, G. Belz, T.Z. Gurupira, J.M. Stencel, Fuel, **85**, pp. 2286-2293 (2006)
7. S. Donatello, C. Kuenzel, A. Palomo, A. Fernández-Jiméneza, Cem. Concr. Comp., **45**, pp. 234-242 (2014)
8. G. Li, Cem. Concr. Res., **34**, 6, pp. 1043-1049 (2004)
9. R. Siddique, Cem. Concr. Res., **34**, 3, pp. 487-493 (2004)
10. Q. Zeng, K. Li, T. Fen-Chong, P. Dangla, Cem. Concr. Res., **42**, 1, pp. 194-204 (2012)
11. R. T. Chancey, P. Stutzman, M. C. G. Juenger, D. W. Fowler, Cem. Concr. Res., **40**, 1, pp. 146-156 (2010)
12. J.-H. Ji, J. Hwang, G.-N. Bae, Y.-G. Kim, J. Electrostatics, **61**, pp. 57-68 (2004)
13. C. J. Badenhorst, N. J. Wagner, B. R. V. Valentim, K. S. Viljoen, A. C. Santos, A. Guedes, CCGP J., **11**, pp. 89-96 (2019)
14. A. Antoni, V. Satrya, H. Djwantoro, CED, **17**, 1, pp. 38-43 (2015)
15. F. Cangialosi, F. Crapulli, G. Intini, L. Liberti, M. Notarnicola, WIT Transac. Ecol. & Environ., **92** (2006)
16. K. Wierzchowski, B. Białecka, J. Calus Moszko, A. Klupa, Int. J. Environ. Sci. Technol. (2020)
17. Y. Soong, M. R. Schoffstall, T. A. Link, Fuel, **80**, 6, pp. 879-884 (2001)
18. F. Cangialosi, M. Notarnicola, L. Liberti, J. Stencel,J. Hazard Mater., **164**, (2-3), pp. 683-688 (2009)
19. D. P. Bentz, C. F.Ferraris, Cem. Concr. Comp., **32**, 4, pp. 265-270 (2010)
20. I. De la Varga, R. P. Spragg, C. Di Bella, J. Castro, D. P. Bentz, J. Weiss, Cem. Concr. Comp., **45**, pp. 102-110 (2014)
21. Ch. Mounika, V. Srinivasa Reddy, M. V. Seshagiri Rao, S. Shrihari, Int. J. Recent Technol. Eng., **8**, 3, pp. 5289-5293 (2019)
22. D. P. Bentz, T. Sato, I. De la Varga, W. J. Weiss, Cem. Concr. Comp., **34**, 1, pp. 11-17 (2012)
23. A. Durán-Herrera, C. A. Juarez, P. Valdez, D. P. Bentz, Cem. Concr. Comp., **33**, 1, pp. 39-45 (2011)
24. S. A. Zhdanok, E. N. Polonina, S. N. Leonovich, et al., J. Eng. Phys. Thermophy., **92**, 4, pp. 376-382 (2019)
25. S. A. Zhdanok, E. N. Polonina, S. N. Leonovich, et al., J. Eng. Phys. Thermophy., **92**, 4, pp. 12-18 (2019)
26. J. P. Baltrus, R. B. LaCount, Cem. Concr. Res., **31**, 5, pp. 819-824 (2001)
27. *The foam index test: a rapid indicator of relative AEA demand*, Tech Bulletin, GCP Applied Technologies Inc, USA, 2016
28. M. J. McCarthy, M. R. Jones, T. A. Hope, R. A. Carroll, N. Cooke, *Innovative processing of stockpile fly ash* (Working Draft Report, University of Dundee, 2019)
29. I. Külaots, A. Hsu, R. H. Hurt, E. M. Suuberg, Cem. Concr. Res., **33**, 12, pp. 2091-2099 (2003)

MATEC Web of Conferences **315**, 07007 (2020) https://doi.org/10.1051/matecconf/202031507007
ICMSSTE 2020

Development of a technology for producing ceramic refractory material in a composition of montmorillonite clays (bentonite-like) and ferrochrome production wastes

Sarsenbek Montayev [1*], *Bekbulat* Shakeshev[1], *and Sabit* Zharylgapov[1]
[1]Zhangir khan university, 51 Zhangir khan str. Uralsk, Kazakhstan

Annotation. The article presents the results of scientific and experimental studies on the creation of ceramic refractory material. Montmorillonite clay (bentonite-like) of the Pogodaevsky deposit and ferro-dust from self-decaying slags of low-carbon ferrochrome of the Aktobe ferroalloy plant were selected as the most effective types of raw materials in terms of their physico-mechanical properties and chemico-mineralogical composition. The composition of ceramic mass in chamotte - ferro dust - binder and rational technological parameters of molding, drying and firing was developed. The basic physical and mechanical properties of the obtained ceramic samples, which have high strength, average density, low linear thermal expansion coefficient and fire shrinkage, are studied. Ceramic samples based on the proposed raw material composition can withstand temperatures up to 1350 C, which is 200-300 C higher than ceramic samples based on clay raw materials. The high strength indices and refractoriness of samples are explained by the fact that in result of heat treatment of the studied ceramic composition, a densely sintered framework with reinforced refractory minerals such as forsterite Mg2 (SiO4) and spinel (MgAl2O4) is formed. The prospects of this research area are associated with the creation of innovative technologies for the production of popular refractory materials, but also with an environmental issue. Processing of technogenic resources of Kazakhstan will not only reduce the consumption of natural resources, but also reduce the human impact on the environment.

1 Introduction

Currently, Kazakhstan is successfully implementing the state program of industrial and innovative development. In the process of implementing this program, new facilities are being introduced for the production of cement, lime, expanded clay, ceramic bricks, and also metallurgical enterprises are expanding. The development of these industries is inextricably linked with the consumption of refractory materials for lining thermal units, without which it is impossible to produce products.

To meet these needs, the Republic of Kazakhstan imports refractory materials for these enterprises from Russia, Ukraine, China, Austria and other countries.

The high demand for new high-quality refractory materials is also due to the steady increase in their consumption at existing metallurgical plants, as well as in the production of building materials, the technological process of which requires high-temperature firing (1100-14000C).

A preliminary analysis of the types of refractory materials consumption in the Republic of Kazakhstan showed that basically stable demand exists for refractory bricks and tiles, ramming materials, masonry mortars.

In this regard, the development of domestic production of refractories is one of the urgent and strategic tasks of the Republic of Kazakhstan.

The main key factors in the development of domestic production of refractories is the analysis of the availability of natural and technogenic raw materials and their assessment, as well as the development of innovative technologies for their production regarding improving quality, resource and energy conservation.

At the same time, a differentiated approach to each enterprise-consumer is necessary, given their high-temperature production processes and the design of thermal units. According to A.I. Khlystova [1], there is no universal refractory material that meets the requirements of working at high temperatures, with thermal shock, in liquid and gaseous aggressive environments and suitable for all cases of factory practice.

The studies of the authors [2-8] are devoted to solving issues related to the development of effective technologies for refractory materials that can provide future production industries.

* Corresponding author: montaevs@mail.ru

Scientific research has special interest on recycling of battlefield of refractory products formed during repairs of thermal units (furnaces, converters, boilers, etc.) [9].

This is due to lack of refractory factories in Kazakhstan, therefore, enterprises for the production of building materials, the metallurgical industry and the power system located in the country are experiencing a shortage of it. For example, only one plant for the production of lime of JSC "West Kazakhstan Corporation of Building Materials" annually purchases over 50 tons of refractory materials abroad for repair and lining of rotary kilns. In the process, spent refractory materials are thrown into dumps as a waste.

Moreover, the experience of industrialized countries shows that the refractory brick battle is an excellent secondary raw material in the production of piece goods (bricks), as well as unformed refractories (refractory concrete, shotcrete, packed masses and dry mixes, refractory mortars, etc.) The use of battle of refractory products is effective both from an economic and environmental point of view [9].

At the current stage, we are familiar with earlier research works of scientists whose aim was to develop a technology for producing refractory materials using various natural raw materials and industrial waste. Relying on the results of undertaken investigations, unique system-based refractory ceramics were invented. To be specific, the MgO-ZrO2 system were used on the basis of secondary magnesia-carbon brick [10] and high-quality refractory ceramics using oil sludge and bauxite raw materials [11] The given work [12] presents the results of the mentioned investigations on the development of a technology for producing refractory material based on alkali-activated refractory chamotte waste and from the alumina-zirconia-silicon dioxide. The given research work also contains determined possibility of obtaining refractory cellular concrete based on a phosphate binder from waste from the production and processing of aluminium [13].The effectiveness of the use of ferrochromic slag for the production of refractory materials, including high-strength refractory ceramics, has been proved by scientists' researches [14 -18].

An analysis of research by leading scientists shows that further development of the production of ceramic and refractory materials is inextricably linked with the improvement of existing technological processes with the development of new compositions based on technogenic and natural raw materials systems in order to improve energy efficiency and quality of the finished product [19, 20].

Enterprises in Kazakhstan that produce various products under the conditions of operation of thermal units at a temperature of 1000-12500C receive the most dynamic development. (production of ceramic bricks, lime, expanded clay, etc.).

Generally, those enterprises function almost in all areas of the country. That is why, the sintered refectories related to the rough ceramics technology are the most interesting, from the point of view of the rapid scientific developments realization in the existing enterprises.

Sintered refractories are successfully used for the construction of industrial furnaces and apparatuses operating at high temperatures (1000-1800 C). The products of sintered refractories, except refractoriness, provide not only strength at high temperature, but also heat and chemical resistance.

Analysis of refractory products that are delivered to Kazakhstan has showed that chamotte products are in high demand by the market. They make up about 75% of all refractories used in the country. In this case, the main types of refractory materials are refractory bricks, extruded shaped refractory products made by ceramic technology.

2 Goals and objectives of the study

Therefore, the goal of our study is to select raw materials from natural and technogenic resources of Kazakhstan and conduct scientific and experimental work on the development of ceramic compositions to establish the possibility of obtaining refractory ceramic material.

To achieve this goal, it is necessary to solve the following tasks:

- to select of the most effective types of raw materials according to their physical and mechanical properties and chemical and mineralogical composition;

-to develop of the most rational technological solutions for the preliminary preparation of raw materials with the aim of compiling a refractory ceramic composition;

- to conduct scientific and experimental work on the creation of refractory ceramic samples by dosing, mixing, molding, drying and firing.

3 Materials and techniques

High interest consists in creating composite refractory ceramic materials such as bentonite clays, ashes of thermal power plants (TPPs), metallurgical slag, and waste refractory materials which are generated during the repair and dismantling of the lining of high-temperature metallurgical and heat power units.

In this paper, to develop ceramic composition in order to obtain refractory materials, the following natural and technogenic resources of Kazakhstan were evaluated and studied:

1. Montmorillonite (bentonite-like) clay Pogodaevskoye deposit (Western Kazakhstan) was chosen as natural raw material;

2. Ferro-dust from self-decaying slags of low-carbon ferrochrome of Aktobe ferroalloy plant was chosen as technogenic raw materials.

The selected raw materials are located in one region of Kazakhstan, which creates favorable technical and economic conditions for their processing and organization of the construction of a workshop for the production of ceramic refractory material.

At the initial stage of the research, the total reserves of the selected raw materials were specified. A study of the geological exploration revealed that the reserves of Pogodaevsky deposit of montmorillonite (bentonite-like) clays amount to more than 6181 thousand m^3.

According to the statistic data of Aktobe ferrochrome plant, the annual output of ferro-dust is 1.5 thousand tons per year.

The studies on refractoriness of montmorillonite (bentonite-like) clay in Pogodaevskoye deposit have shown that its refractoriness is in the range of (1200 -1250 C).

The chemical and mineralogical composition of studied raw materials was determined using a JSM-6390LV scanning electron microscope with an energy-dispersive microanalysis system, an X'Pert PRO MPD X-ray diffractometer, and an ICP-MS Agilent 7500cx inductively coupled plasma mass spectrometer (JEOL, Japan) Fig. 1.

X-ray phase analysis (XRD) was carried out on a DRON-3 diffractometer with SiKa- radiation in the range of angles 80-640. The sensitivity of method is from 1 to 2%. The powder of the studied raw materials passed through a 0.315 sieve was subjected to X-ray phase analysis.

The chemical composition of montmorillonite clay (bentonite-like) is presented in Table 1.

Table 1. The chemical composition of montmorillonite clay (bentonite-like) Pogodaevsky deposits.

Name of raw material	Oxide content, mass %							
	SiO_2	Al_2O_3	CaO	MgO	Fe_2O_3	SO_3	Na_2O	Lost on ignition
Montmorillonite clay (bentonite-like)	61.51	17.06	2.27	3.21	6.36	1.27	3.57	6.75

4 Results and discussion

According to the results of x-ray phase analysis, it was found that the mineralogical composition of montmorillonite clay (bentonite-like) in Pogodaevsk deposit is represented mainly by montmorillonite (d / n = 5.06; 4.46; 3.79; 3.06; 2.455; 2.28; 2.127; 1.977; 1.817; 1.675 * 10-10 m.)

Ferro dust from self-decaying slags of low-carbon ferrochrome of Aktobe ferroalloy plant is a finely divided gray powder. When sifting on a sieve with a mesh of 0.5 mm, the residue is not more than 0.2%.

The results of the study on their chemical and mineralogical composition and microstructure are presented in Table 2 and in Fig. 1.

Table 2. Chemical composition of dust from self-decaying slag low-carbon ferrochrome of the Aktobe ferroalloy plant.

Name of raw material	Oxide content, mass %								
	Cr_2O_3	CaO	MgO	Al_2O_3	FeO	SiO_2	Na_2O	K_2O	TiO_2	Mn
Ferroalloy	2.001	2.07	29.2	15.91	1.29	18.8	0.51	0.2	0.36	0.10
	2.75	2.84	40.3	21.89	1.77	25.9	0.70	0.3	0.49	0.13

According to the results of chemical-mineralogical and electron-microscopic studies, as well as X-ray diffraction (XRD) analysis, it was found that the studied ferro dust is represented by spinel (MgAl2O4) (d, Å: 4.69-3.73-2.87-2.45-2.02-1.431); forsterite (Mg2 (SiO4)) (d, Å: 5.107-3.882-3.491-2.762-2.349-2.263-2.158-1.747-1.495-1.392); magnetite (Fe3O4) (d, Å: 3.00-2.51-2.11-1.616-1.565.1.476). Thus, it can be argued that the mineral and phase composition is mainly represented by forsterite and spinel, which are the main minerals of refractory materials.

Element	Weight %	Atomic %
O	45.52	59.72
Na	0.41	0.38
Mg	20.45	17.66
Al	10.82	8.42
Si	11.81	8.82
S	0.70	0.46
Cl	0.27	0.15
K	0.43	0.23
Ca	2.83	1.48
Ti	0.48	0.22
Cr	3.39	1.37
Mn	0.19	0.08
Fe	2.70	1.01
Total	100.00	

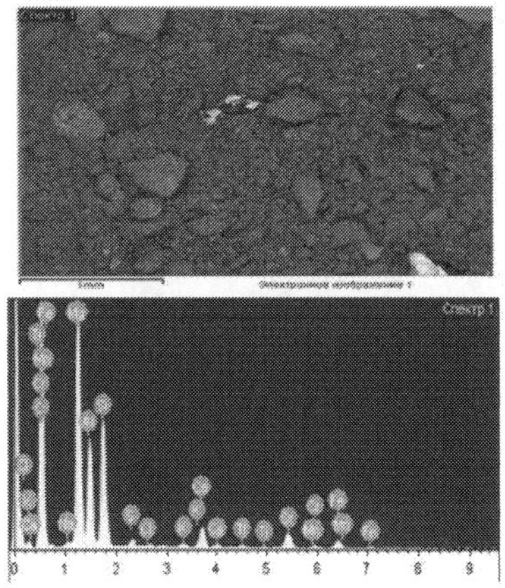

Fig. 1. Elemental composition, microstructure, and spectra of dust from self-decaying slags of low-carbon ferrochrome of the Aktobe ferroalloy plant.

To determine the refractoriness of the ferro dust, the samples were fired in an electric furnace until the onset of deformation. According to the results of experimental work, the refractoriness of ferro dust was established, which is 1300-13500C.

The ratio (CaO + MgO) / (SiO2 + Al2O3) is 1.7–1.9, which makes ferrous dust highly basic. The indicated properties of the dust will increase the heat resistance of refractory materials.

Scientific and experimental work was carried out in the following sequence:

- development of compositions and technology for producing chamotte based on montmorillonite (bentonite-like) clay;
- development of the composition of the ceramic mass in the chamotte - ferro-dust - binder system;
- development of technological parameters of molding, drying and firing of the ceramic composition in order to obtain refractory material.

To achieve the maximum degree of mullitization during the clay firing process, the optimal compositions of the ceramic composition in the system of montmorillonite clay - ferro-dust from self-decaying slags to produce fireclay have been developed. To obtain chamotte based on a optimal composite raw material mixture, the montmorillonite clay was first dried in an oven (ШСП-0,5-70) at a temperature of 100-1100C to a residual moisture content of 5-7%. Then, the dried montmorillonite clay was crushed in a laboratory jaw crusher (DS 80150) to obtain a fraction of ≥ 30 mm. Then, crushed montmorillonite clay and ferro dust from self-decaying slags were dosed in the ratio 1: 2 and loaded into a laboratory ball mill.

The raw material composition was subjected to joint grinding in order to homogenize and to be mechanically activated until they pass through a 0.315 sieve. The thus obtained powders were used for two purposes. One part was used as raw material for the production of chamotte, and the other part as a binder component in the formation of samples.

The first part of the obtained powder was used in order to get chamotte in the following way: a ceramic mass was prepared with a moisture content of 25-27% and granules with a diameter of 10-20 mm were formed. Then they were put to dry in an oven to get residual moisture content of 5-7%.

The dried granules were fired in a laboratory rotary furnace (RSR120 / 1000/13 Germany) at a temperature of 1250 ° C with an exposure, at a maximum temperature, for 2 hours.

The burnt granules were crushed once again in a laboratory jaw crusher and sieved through a set of sieves. Thus, chamotte of various fractions was obtained. Fractions of 1.0-3.0 mm chamotte were selected for scientific and experimental work, because according to the results of preliminary experimental work these fractions showed the most optimal properties at the stage of molding, drying and firing.

In order to form raw samples of refractory material, a raw mixture was prepared in the chamotte-binder system. The second part of the finished powder which was obtained by co-grinding of montmorillonite clay and ferro-dust from self-decaying slags in a laboratory ball mill was used as a binder. It should be noted that this mixture has a plasticity number of 12-15 due to the high amount of plastic montmorillonite clay in it, which ensures reliable adhesion of polyfraction chamotte particles.

The optimal amount of binder accounted for 10-12% of the total weight of chamotte.

The total content of ferrous dust in the composition is 88-90%. The raw material mixture from the ceramic composition in the chamotte-binder system was thoroughly mixed in a spherical bowl with the addition of water in an amount of 8-10% and cylinder samples (50x50x50 mm) were formed by semi-dry pressing. The pressing pressure was 20 MPa. The molded samples were fired without preliminary drying in a muffle electric furnace (SNOL 80/12) according to a specially developed mode. The maximum firing temperature was 1200 C.

The heat-treated samples were cooled in a disconnected furnace to room temperature and were tested to determine the physical and mechanical properties. Physico-mechanical properties of the studied samples are presented in Table.3.

Table 3. Physico-mechanical properties of the samples.

Properties	Unit of measurement	Indicators
Average density	kg/m^3	2200 - 2250
Compressive strength	MPa	15.5 – 16.7
Fire shrinkage	%	0.18 – 0.21
Thermal extension coefficient	$10^{-6}\ C^{-1}$	3.1 – 3.3
Temperature at the initial deformation	^0C	1350

The test results of the physicomechanical properties of the samples show that the samples obtained on the basis of the studied raw material composition have the required compressive strengths. The strength of the calcined samples is 15.5-16.7 MPa. In addition, the samples are characterized by low rates of fire shrinkage of -0.18-0.21%. The samples have a densely sintered structure, which is confirmed by high average density indices - 2200-2250 kg / m3. Ceramic samples based on the proposed raw material composition can withstand temperatures up to 1350 0C, which is 200-300 0C higher than ceramic samples based on clay raw materials. The high strength characteristics and refractoriness of the samples are explained by the fact that in result of heat treatment of the studied ceramic composition, a densely sintered frame with reinforced refractory minerals such as forsterite Mg2 (SiO4) and spinel (MgAl2O4) is formed.

5 Conclusion

- The montmorillonite clay (bentonite-like) of Pogodaevsky deposit and ferro-dust from self-decaying slags of low-carbon ferrochrome of Aktobe ferroalloy plant were selected as the most effective types of raw materials in terms of their physical and mechanical properties and chemical and mineralogical composition in order to obtain ceramic refractory material;

- It was found that the studied raw materials are representing a valuable resource for the production of refractory ceramic materials.

- Compositions and technologies for producing the chamotte were developed based on montmorillonite (bentonite-like) clay in order to increase the content of refractory crystalline phases;

- The composition of the ceramic mass was developed in the chamotte system - ferro-dust - binder and rational technological parameters of molding, drying and firing the ceramic composition in order to obtain a refractory material;

- The basic physical and mechanical properties of the obtained ceramic samples were studied. It was found that those ceramic samples have the high strength, average density and low linear thermal expansion coefficient and fire shrinkage;

- It was found that the obtained ceramic products based on the developed composition may well satisfy the requirements for lining thermal units with a working temperature of up to 13000C.

- The prospects of this research area are associated not only with the creation of domestic technology for the production of refractory materials, but also with the environmental issue of the utilization of industrial waste. Processing of technogenic resources of Kazakhstan will not only reduce the consumption of natural resources, but also reduce the human impact on the environment.

References

1. A.I Khlystov, Improving *Efficiency and Improving the Quality of Refractory Lining Materials* 134 p. (Monograph, Samarsk. state arch. - builds. un-t, 2004)
2. I.V. Gladkih, E.P. Volynkina, *The methodology for assessing the quality of industrial raw materials in the production of refractory and heat-insulating materials for metallurgy* **10**, 42-45 (University proceedings. Ferrous metallurgy, 2011)
3. A. T. Volochko, K. B. Podbolotov, E. M. Dyatlova, *Refractory and refractory ceramic materials* 385 p., ISBN 978-985-08-1640-5 (Minsk: Belarus, Navuka, 2013)
4. V.A. Kononov, Bull. "Ferrous metallurgy" **6**, 33-43(2001)
5. L.B. Khoroshavin, V.A. Perepelitsin, Refract. Ind. Ceram., **4**, 43-47 (2000)
6. Yu .E. Pivinsky, Refract. Ind. Ceram., **1**, Part 1, 11-16 (2000)

7. L. M. Axelrod, et al., Bull. of USTU, **1,** 60-66 (Physicochemistry and technology of oxide-silicate materials, Ekaterinburg. Publishing house of USTU, 2000)
8. A. Nukhuly, B.N. Satbaev, A.K. Svidersky, Bull. of the Innovative Eurasian University, **2**, 158-164 (2007)
9. K.A. Cherepanov, S.M. Abramovich, M.V. Temlyandrov, E.N. Temlyandeva, *Recycling of solid waste in metallurgy,* 212 p. (Moscow: Flint Science, 2004)
10. R. Kusiorowski, Constr. Build. Mater., **231**,117084 (20 January, 2020)
11. N.M. Khalil, Y.Algama. l. Qayid, M. Saleem, Exploitation of petroleum waste sludge with local bauxite raw material for producing high-quality refractory ceramics, Ceram. Int., **44** (15), 18516-18527 (15 October, 2018)
12. B. Coppola, C.Tardivat, S. Richaud, Jean-Marc Tulliani , L. Montanaro, P. Palmero, J. Eur. Ceram. Soc. (February , 2020)
13. V.A. Abyzov, *Procedia Eng.*, **206**, 783-789 (2017)
14. Pattem Hemanth, Kumar Abhinav. Srivastava Vijay, Kumar Vinay, Kumar Singh, J. Asian Ceram. Soc., **2** (4), 371-379 (2014)
15. Yuhan Ren, Qiang Ren, Xiulan Wu, Jinle Zheng, Ou Hai, Recycling of solid wastes ferrochromium slag for preparation of eco-friendly high-strength spinel–corundum ceramics, Mater. Chem. Phys., **239**, 1220601 (2020)
16. Kumar Hemanth, Srivastava Abhinav, Kumar Vijay, Ranjan Manas, Kumar Majhi Vinay, J. Asian Ceram. Soc., **2** (2), 169-175 (2014)
17. Vusumuzi Sibanda, Refilwe Kopong, Sehliselo Ndlovu, Ratombo Zwanga, Int. J. Miner. Process., **157**,46-54 (2016)
18. Rundong Fan, Huizhong Zhao, Han Zhang, Pengda Zhao, Jianwei Chen, Xianghui Wang, Ceram. Int., .**45** (9), 11204-11215 (2019)
19. S.A. Montayev, Zh.T. Suleimenov, *Wall ceramics based on compositions of technogenic and natural raw materials of Kazakhstan* 157 p. (Monograph., Almaty: Gylym, 2006)
20. K.A. Cherepanov, S.M. Abramovich, M.V. Temlyard, E.N. Temlyandeva, *Solid Waste Recycling in Metallurgy* 212 p. (Moscow: Flint – Science, 2004)

On the possibility of using lime dust for the production of building materials of forced carbonate hardening

Aleksandr Bakhtin[1*], *Nikolai* Lyubomirskiy[1], *Sergey* Fedorkin[1], *and Tamara* Bakhtina[1]

[1]V.I. Vernadsky Crimean Federal University, 4 Academician Vernadsky Ave., Simferopol 295007, Russian Federation

Abstract. The paper presents research on the development of building materials with low level of CO_2 emissions based on technogenic recycled materials. The paper addresses the determination of optimal formulation and technological parameters of obtaining materials based on lime dust generated by mechanical deposition in cyclones and bag filters of shaft furnaces, as well as finely dispersed marble limestone with a fraction of up to 5 mm. Studies have shown that it is possible to obtain carbonized material with compressive strength of more than 40 MPa from this recycled material by forced carbonization used during three hours. Moreover, to obtain such numbers, the optimal content of lime dust in raw materials should be in the range of 35-40% wt. when the water content of the mixture is 6-7% wt. The carbonized material obtained with the indicated technological parameters will have an average density of 1.95-2.0 g/cm3 and water absorption by weight of not more than 12%.

1 Introduction

Since the middle of the last century, there has been significant economic growth in the developed countries of the world, associated with the development of industrial production, and, accordingly, with high consumption of natural resources. The growth of industrial production led to certain environmental problems, including problems with climate change caused by significant emissions of greenhouse gases into the atmosphere, especially CO_2. The significance of these problems and, accordingly, the need to respond to them, directs the world economy to the search for a new development vector. It forged the appearance of the concept of a "low-carbon economy", in which the carbon footprint will be the main criterion for evaluating any human activity, especially technological. In particular, in EU programs, including 'A Roadmap for moving to a competitive low carbon economy in 2050' [1], priorities for reducing CO_2 emissions were included in all sectors of industrial production. The leading target is ways to reduce industrial CO_2 emissions as the main industrial gas is causing significant climate change on the planet. Such innovations are fixed at the legislative level in the countries of the EU, the USA, Canada, Australia, and

* Corresponding author: aleserba@gmail.com

China and, first of all, should stimulate the organization and conduct of research work to find ways to reduce carbon dioxide emissions without harming the economy of states. To achieve the identified priorities, the International Energy Agency has identified a list of industrial production, in which it is advisable to organize research work to reduce CO_2 emissions. According to this list, one of the promising industries is the industry for the production of various building materials. The largest CO_2 emission process in this industry is the production of Portland cement and various lime binders, which together account for up to 10% of technogenic CO_2 emissions worldwide (2.9-3.1 Gt per year) [2, 3]. Accordingly, the use of this technogenic CO_2 in the technological cycle as a raw material component without reducing the main production is the most expedient and effective way to reduce the emission of carbon dioxide into the atmosphere. Fundamental and applied scientific research in this direction began at the beginning of the 21st century [4–7] and came down to finding ways to reduce CO_2 emissions in cement production or developing alternative binders to Portland cement. During this process researchers mainly studied the kinetics of the reaction of carbonization of clinker-forming minerals, MgO-containing systems and the change in the structure and properties of the obtained carbonized material. The intensification of research in this direction is observed after the climate conference in Paris and the signing of the relevant agreement by most of the countries of the world. The ratification of the final document of the conference was carried out by 186 countries in a short time, which indicates that the world community understands the seriousness of the consequences if these problems are further ignored. The intensification of scientific research has shown that the building materials industry is a very promising sector of the economy in addressing the issue of reducing carbon dioxide emissions into the atmosphere [8-24].

Conducted researches have led to the formation of the concepts of "environmental construction" and "safe and comfortable living environment" as one of the most relevant global trends in the construction industry, making up the concept of "sustainable development". According to these scientific trends, the most significant distinguishing feature that enhances the environmental aspects of construction is the use of building materials and products with low CO_2 emissions, manufactured using technologies whose principal feature is the absorption and binding of technogenic carbon dioxide into insoluble compounds (sequestration). As a result of these studies, researchers have identified types of raw materials [9,10, 15-21], including secondary [11-14, 22], which have significant potential for CO_2 binding, developed scientific and technological foundations for the implementation of the results in industry [13- 15, 22], as well as released pilot batches of building products hardened in an environment of high concentration of CO_2 [13, 15]. Summarizing the results of the studies, it can be noted that the directions for reducing CO_2 emissions in the building materials industry are identified; their high efficiency is justified, in terms of quantifying CO_2 and obtaining a certain range of high-quality building products that harden due to forced carbonization. However, for the introduction of such technologies, for example, at existing enterprises producing the main product, it is necessary to conduct individual scientific research aimed at exploring the possibility of using recycled materials to organize the production of materials and products using forced carbonization.

In connection with the foregoing, the aim of this work is to study the possibility of using technogenic recycled materials, in particular lime dust and finely divided limestone, to obtain building materials with high physical and mechanical characteristics using forced carbonization.

2 Materials and methods

Lime dust, which is mechanically deposited in cyclones and bag filters of shaft furnaces that burn limestone to produce quicklime, was used as the material for scientific research. Along with it finely dispersed marble-like limestone with a fraction of up to 5 mm was used which is formed during the crushing of the initial rock into a working fraction sent to the shaft furnace for firing. These materials are produced at enterprises producing quicklime and are mainly sent to dumps, since there is practically no demand for this recycled material. The lack of demand is explained by the low quality of these products, in particular, lime dust contains burnt CaO particles, and the amount of active calcium and magnesium oxides, as a rule, does not exceed 25-35%.

Thus, the presence of harmful burnt particles in a complex with a low content of active CaO and MgO causes a lack of demand. Finely dispersed marble-like limestone also does not find wide application in the construction industry due to the high content of dust and clay particles in it. In these studies, various raw materials were prepared from these materials, and cylinder samples with a diameter and a height of 30 mm were prepared by semi-dry pressing, curing of which was carried out in the developed carbonization chamber [25] under conditions of a 35% concentration of CO_2 for 3 hours.

The mineralogical composition was determined using the high-temperature synchronous analyzer STA 8000 from Perkin Elmer (USA) in the temperature range of 30-1000° C at a heating rate of 10° C/min, in a nitrogen atmosphere. The particle size of the investigated raw materials was determined using the HORIBA LA-960 laser analyzer (Japan). The qualitative course of the carbonization reaction was determined by applying a 1% phenolphthalein solution to the cleaved sample. The optimization of the composition of the raw material mixture to obtain prototypes of semi-dry pressing on the basis of the studied materials was carried out using statistical methods of mathematical design of the experiment. Processing of the obtained experimental data was carried out using the StatSoft STATISTICA software package. As a result of processing the experimental data, ES (experimental statistical) models of each investigated parameter are formed, which are second-order equations of the following form:

$$Y = b_0 + \sum_{i=1}^{n} b_i z_i + \sum_{i;l=1}^{n} b_{il} z_i z_l + \sum b_{ii} z_i^2 . \qquad (1)$$

where b_o, b_i, b_{il}, b_{ii} are the correlation coefficients determined as a result of mathematical and statistical processing of experimental data; z_i, z_l are the value of the varied formulation and technological factors. The mechanical characteristics of the prototypes were determined using an automatic test system, based on the MCC8 (Controls) control console. Before testing, the prototypes were dried to zero humidity in an oven at the temperature of 80°C.

3 Results and discussion

Lime dust is a fine powder of light gray color. Laser analysis of particle size (Fig. 1) showed that this recycled material consists of particles ranging in size from 4 to 80 microns. Most of the particle size distribution is made up of particles ranging in size from 15 to 30 microns. The average geometric particle size is 15.5 microns.

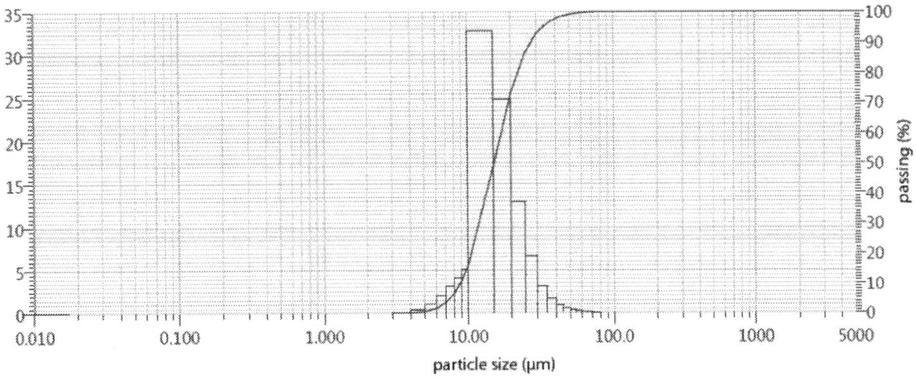

Fig. 1. Particle size distribution and differential curve of particle size distribution of lime dust.

An analysis of the mineralogical composition of calcareous dust was carried out using thermal analysis (see Fig. 2). Calculations of the quantitative content of the corresponding mineral phases were carried out according to equations. The calculation results are presented in table 1.

Table 1. Mineralogical composition of lime dust.

Losses in the temperature range 450-600 °C, %	Content of $Ca(OH)_2/CaO$, %	Losses in the temperature range 600-1000 °C, %	Content of $CaCO_3/CaO$, %	Impurities, %
10.59	43.5/32.9	21.65	49.2/27.6	up to 7.0

Limestone dust in its mineralogical composition has almost 43-44% of $Ca(OH)_2$. The number of particles of calcite is about 49-50% wt. Impurities do not exceed 7% wt. Also this raw material contains 2-3% wt. refractory burnt grains of CaO. Due to the fineness of lime dust grains, the destructive effect of burnt particles on the properties of products made on its basis will be minimal. DTA curve of finely dispersed marble-like limestone with a fraction of up to 5 mm has one endothermic effect in the temperature range of 600-1000°C, corresponding to the decomposition of calcium carbonate. The total weight loss in this range was 41.9% wt. Accordingly, the calcite content is 95.2% wt.

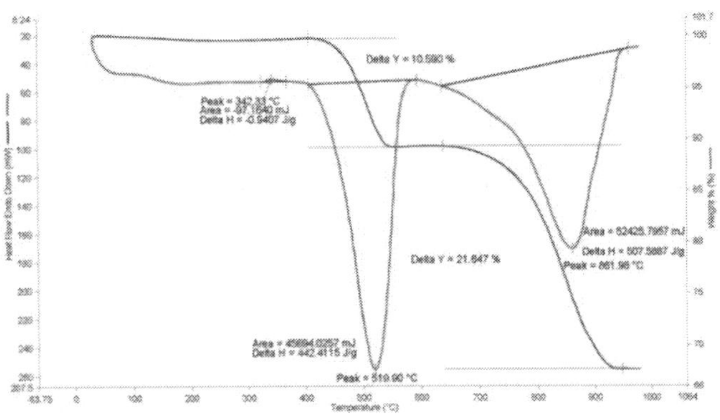

Fig. 2. TG-DTA curves of limestone dust.

To determine the optimal formulation and technological factors for the preparation of semi-dry pressed test samples based on the materials studied, and their subsequent forced carbonization, a rotatable central compositional plan was compiled. The following factors were taken as variable factors: the content of lime dust (Z_1), the water content of the raw material mixture (Z_2), and the pressing pressure of the experimental cylinder samples (Z_3). The levels of variation of the studied factors are presented in table 2.

Table 2. Conditions for the experiment.

Factor	Measurement units	Levels of variation				
		-1.682	-1	0	+1	+1.682
Z_1	% wt.	13.18	20.0	30.0	40.0	46.82
Z_2	% wt.	4.64	6.0	8.0	10.0	11.36
Z_3	MPa	16.6	20.0	25.0	30.0	33.4

Forced carbonization of the samples was carried out under normal conditions (~ 20-30 °C, atmospheric pressure). The concentration of CO_2 in the carbonization chamber was ~ 33-35%. The carbonation time was constant and equaled 3 hours. The main parameters analyzed for carbonized prototypes were: compressive strength (R_c), average density (ρ), water absorption (W_m), water resistance (K_s). Analysis of experimental data (see table 3) showed that the properties of carbonized cylinder samples vary over a wide range depending on the varying technological factors.

Table 3. Planning matrix and experimental data of carbonized prototype cylinders.

№ g	Coded values of factors			Natural values of factors			Analyzed parameters			
	Z_1	Z_2	Z_3	Z_1, (%)	Z_2, (%)	Z_3, (MPa)	R_c, (MPa)	ρ, (g·cm^{-3})	K_s	W_m, (%)
1	-1	-1	-1	20.0	6.0	20.0	21.8	1.99	0.84	11.4
2	+1	-1	-1	40.0	6.0	20.0	30.9	1.82	0.74	15.5
3	-1	+1	-1	20.0	10.0	20.0	20.8	1.98	0.62	11.3
4	+1	+1	-1	40.0	10.0	20.0	32.9	1.98	0.77	13.9
5	-1	-1	+1	20.0	6.0	30.0	31.0	2.06	0.66	10.3
6	+1	-1	+1	40.0	6.0	30.0	43.8	1.90	0.69	13.3
7	-1	+1	+1	20.0	10.0	30.0	18.4	2.02	0.73	10.5
8	+1	+1	+1	40.0	10.0	30.0	28.5	1.91	0.73	13.1
9	-1.682	0	0	13.18	8.0	25.0	14.9	2.09	0.60	9.1
10	+1.682	0	0	46.82	8.0	25.0	41.0	1.80	0.68	15.6
11	0	-1.682	0	30.0	4.64	25.0	30.1	1.92	0.71	13.1
12	0	+1.682	0	30.0	11.36	25.0	12.9	1.92	0.71	12.8
13	0	0	-1.682	30.0	8.0	16.59	30.4	1.88	0.68	14.6
14	0	0	+1.682	30.0	8.0	33.41	39.6	1.98	0.69	11.7
15	0	0	0	30.0	8.0	25.0	36.2	1.94	0.58	12.8
16	0	0	0	30.0	8.0	25.0	36.0	1.94	0.60	12.9
17	0	0	0	30.0	8.0	25.0	36.3	1.95	0.59	12.8
18	0	0	0	30.0	8.0	25.0	35.8	1.93	0.60	12.7

Analysis of the influence of the amount of lime dust in the compositions of raw mixes (see table 3) shows that with an increase in the content of this component from 20 to 40% wt. strength indicators of prototypes increase in average in 1.4-1.7 times. Moreover, due to the increase in the composition of the fine particles, the average density of the material decreases and, accordingly, the water absorption increases. It should be noted that an increase in the content of lime dust, despite an increase in the rate of water absorption, increases the water resistance of the samples. The increase in the initial water content of raw mixes from 6 to 10% wt. with the same other studied factors, reduces the strength of carbonized samples. Since limestone has a denser structure and correspondingly lower water absorption compared to lime dust, most of the water remains in the intergranular space, increases the water-lime ratio and, thereby, slows down the process of carbonation of lime dust. The degree of carbonation of the calcareous component in the experimental samples can be indirectly judged by the change in the size of the carbonized layer (Fig. 3). From the data presented in Fig. 3, it can be seen that the formation of a carbonate stone during the carbonate hardening of a calcareous binder begins from the surface of the samples, gradually moving deeper with a diffusion restriction of carbon dioxide. The degree of lime carbonization depends on all the studied formulation and technological factors. With increasing pressing pressure and the content of lime dust, the thickness of the carbonized layer in the test samples decreases. As the pressing pressure increases, the structure of the material becomes denser, which complicates the movement of gas through the pores of the material.

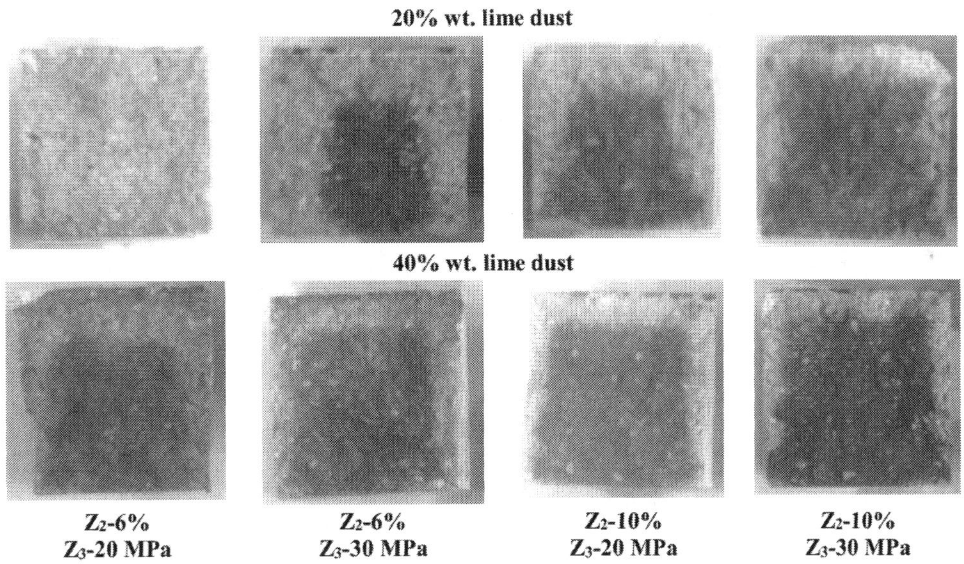

Fig. 3. The change in the thickness of the carbonized layer of the test samples obtained at the water content of the molding mixture (Z_2) 6 and 10% wt. and pressing pressure (Z_3) of 20 and 30 MPa.

Adding of more lime dust, a finely divided component, leads to the formation of smaller pores and also complicates the diffusion of the gaseous medium through the pores in the bulk of the material. It is possible that the complete carbonization of the calcareous component in experimental samples with a content of 40% wt. needs more than 3 hours of forced carbonization, which were insufficiently accepted in the experiment. The analysis of the ES model of the main properties of the test samples showed that the most influential factors on the formation of the properties of carbonized samples in decreasing order of importance are: the content of lime dust (Z_1), the water content of the raw material mixture (Z_2), and the pressing pressure of the experimental cylinder samples (Z_3). The equations obtained by statistical data processing (table. 3) are the following:

$$R_c = 36.52 + 15.80 Z_1 - 11.08 Z_2 + 4.51 Z_3 - 6.74 Z_1^2 - 11.27 Z_2^2 - 1.73 Z_3^2 + 5.10 Z_1 Z_2 + 0.42 Z_1 Z_3 - 7.21 Z_2 Z_3; \quad (2)$$

$$\rho = 1.94 - 0.135 Z_1 + 0.015 Z_2 + 0.043 Z_3 + 0.013 Z_1^2 + 0.012 Z_2^2 + 0.002 Z_3^2 + 0.053 Z_1 Z_2 - 0.024 Z_1 Z_3 - 0.046 Z_2 Z_3; \quad (3)$$

$$K_s = 0.58 + 0.02 Z_1 + 0.05 Z_1^2 + 0.13 Z_2^2 + 0.08 Z_3^2 + 0.04 Z_1 Z_2 + 0.01 Z_1 Z_3 + 0.09 Z_2 Z_3; \quad (4)$$

$$W_m = 12.94 + 3.39 Z_1 - 0.34 Z_2 - 1.45 Z_3 - 0.57 Z_1^2 - 0.14 Z_2^2 + 0.01 Z_3^2 - 0.46 Z_1 Z_2 - 0.26 Z_1 Z_3 + 0.39 Z_2 Z_3. \quad (5)$$

The value of the coefficient b_o confirms the analysis of experimental data. An increase in the content of lime dust in the compositions of the molding sand (coefficient b_1) contributes to an increase in strength characteristics and a decrease in the average density of the material; however, it significantly increases the water absorption by weight and reduces the thickness of the carbonized layer of the samples, reducing their water resistance. To a large extent, the formation of the carbonate structure of the material of the experimental samples, which can be indirectly judged by the thickness of the carbonized layer, depends on the initial water content of the molding mixtures, Z_2 (coefficient b_2). With an increase in water content in the system, the carbonized layer of the samples decreases and the compressive strength decreases, and the decrease in strength is significant (b_2 = -11.08). The growth in pressing pressure, factor Z_3 (coefficient b_3), increases the average density of the samples, their R_c and reduces W_m. Moreover, with an increase in the density of the structure, the thickness of the carbonized layer and the water resistance of the test samples decrease.

4 Conclusions

An analysis of equations 2-5 made it possible to establish that the optimal molding pressure range for pressing, which allows obtaining maximum compressive strength (more than 40 MPa), is in the range from 25 to 30 MPa. Moreover, to obtain such strength indicators, the optimal content of lime dust in raw materials is in the range of 35-40% wt. when the water content of the mixture is 6-7% wt. The carbonized material obtained with the indicated technological parameters will have an average density of 1.95-2.0 $g \cdot cm^{-3}$ and water absorption by weight of not more than 12%. The water resistance of the material by softening coefficient will be 0.65-0.75. It was found that K_s depends on the thickness of the carbonized layer, the value of which increases with increasing thickness of the layer. Accordingly, an increase in carbonation time (this study was taking 3 hours as an example) will growth the water resistance of the carbonized material on the basis of the investigated secondary raw materials. Thus, the data obtained make it possible to consider the possibility of obtaining high-quality building products with physic-mechanical, physicochemical, hydro physical, and other properties sufficient for building materials by means of semi-dry pressing in a short time period based on calcareous dust and fine limestone waste.

The study was conducted with the financial support of the intra-university grant of V.I. Vernadsky Crimean Federal University.

References

1. European Commission. A strategic long-term vision for a prosperous, modern, competitive and climate-neutral EU economy. URL: https://ec.europa.eu/clima/policies/strategies/2050_en
2. K. Scrivener, V. John, E. Gartner Cem. Concr. Res. 114, 2-26 (2018)
3. P. Humbert, J. Castro-Gomes J. Clean. Prod. **208**, 448-457 (2019)
4. E. Gartner Cem. Concr. Res. **34**, 1489-1498 (2004)
5. P. De Silva, L. Bucea, D. Moorehead, V. Sirivivatnanon Cem. Concr. Comp. 28, 613-620 (2006)
6. T.A. Plekhanova, J. Kerieneb, A. Gailiusb, G.I. Yakovlev Constr. and Build. Mater.**21**, 1833-1838 (2007)
7. M. Liska, A. Al-Tabbaa Construct. Build. Mater. **22**, 1789-1797 (2008)
8. D. Kannan Int. J. Prod. Econ. **195**, 391–418 (2018)
9. R. Maddalena, J. Roberts, A. Hamilton J. Clean. Prod. **186**, 933-942 (2018)
10. R. Kajaste, M. Hurme J. Clean. Prod. **112**, 4041-4052 (2016)
11. P. Humbert, J. Castro-Gomes, H. Savastano Construct. Build. Mater. **210**, 413-421 (2019)
12. Z. Ghouleh, R. Guthrie, Y. Shao Construct. Build. Mater. **99**, 175-183 (2015)
13. H. Huang et al. J. Clean. Prod. **241**, 118359 (2019)
14. Z. Ghouleh, R. Guthrie, Y. Shao J. CO_2 Utiliz. **18**, 125-138 (2017).
15. T. Bakhtina, N. Lyubomirskiy, A. Bakhtin Mater. Today: Proceed. **19**, 1998-2004 (2019)
16. T. Bakhtina, N. Lyubomirskiy, A. Bakhtin E3S Web of Conferences **97**, 02024 (2019)
17. N. Lyubomirskiy, T. Bakhtina, A. Bakhtin, S. Fedorkin Mater. Scien. Forum **974**, 187-194 (2020)
18. N. Lyubomirskiy, A. Bakhtin, T. Bakhtina IOP Conf. Series: Mater. Sci. and Engin. **365**, 032032 (2018)
19. C. Unluer, A. Al-Tabbaa, Cem. Concr. Res. **54**, 87-97 (2013)
20. Y. Yi, M. Liska, A. Al-Tabbaa. Adv. Cem. Res. **26**, 114-122 (2014)
21. C. Unluer, A. Al-Tabbaa Constr. Build. Mater. **94**, 629-643 (2015)
22. N. Lyubomirskiy, A. Bakhtin, T. Bakhtina IOP Conf. Series: Mater. Sci. and Engin. **463**, 032065 (2018)
23. V. Gavrish, T. Chayka, G. Baranov, S. Fedorova, O. Gavrish Mater. Today: Proceed. **19**, 1961-1964 (2019)
24. S. Fedorova, L. Nichkova, G. Baranov, Ju. Shagova Mater. Today: Proceed. **19**, 1941-1944 (2019)
25. N. Lyubomirskiy, S. Fedorkin, T. Loktionova, A. Bakhtin, Ukraine Patent 28051 (2006)

Use of glass granules to produce effective gypsum composites

Victoria Petropavlovskaya[1], *Tatiana* Novichenkova[1,*], and *Kirill* Petropavlovskii[2]

[1]TSTU, Production of building materials and structures Department, 170026, Tver, Russia
[2]MGSU, Binders and concretes Department, 129337, Moscow, Russia

Abstract. The possibility of using glass hollow microspheres in self-reinforced gypsum composites in the work is investigated. The need to use microspheres for gypsum products is dictated by modern high requirements for safety and environmental protection to construction sites and the building materials industry. The dispersion characteristics and grain composition of Russian production glass hollow microspheres, as well as the main component, a gypsum binder, were studied. It was established that all compositions contain particles of micro- and nanometer size. It is shown that glass microspheres provide the structure of a composite with a hardened matrix a low density while maintaining the necessary strength without the introduction of expensive and environmentally unsafe chemical additives.

1 Introduction

The development of environmental and environmental aspects of the building materials industry, green building technologies and health-saving building materials has made it possible to make a technical breakthrough in the application of gypsum binders and materials [1–4]. This is due to their compliance with all modern requirements for the safety of production, use and disposal [2].

It is known that the goal of green construction is to reduce energy and material costs from the moment the design of the construction project to its demolition, including the stages of construction, operation and maintenance. At the same time, a high quality of life should be ensured, including a combination of comfortable living conditions with the usefulness and durability of construction projects.

Gypsum does not create a burden on the natural environment at all stages of the life cycle. When gypsum stone is dehydrated during the production of binders, only water vapor is released, which ensures high environmental friendliness of their production [1,2]. Energy costs for obtaining semi-aquatic gypsum are significantly lower than in the case of cement or other mineral binders, which meets the solution to the problem of energy conservation. In the case of the use of finished products based on gypsum, there is also no danger to people, animals and plants, which compares them favorably with modern synthetic materials [3]. It should also be noted that the modern development of materials science

[*]Corresponding author: tanovi.69@mail.ru

allows the repeated recycling of gypsum, which until recently, in the 20th century, seemed impossible [3]. At the same time, issues of resource conservation and environmental protection are successfully resolved [4, 5, 6].

However, today the requirements are increasing for the gypsum binders themselves from the standpoint of their manufacturability, workability and other indicators [3, 4, 5]. Since gypsum materials are most in demand to create a favorable indoor environment with high architectural expressiveness, the weight of the products obtained on the basis of gypsum binders is an important property among the requirements for finishing materials and structures. The combination of high strength and low density of materials can allow creating a new generation of building materials [3, 7, 8].

In order to reduce the weight of gypsum products, various methods are used, such as the plastering of gypsum stone through the use of gas and foaming agents of various nature, the introduction of various porous natural and man-made lightweight fillers, as well as by using a synergistic effect when combining the first and second method [6, 8, 9, 10 – 12].

The average density of mortars using perlite and vermiculite in the form of expanded sand is reduced to 1140 kg/m^3. However, the introduction of hollow glass microspheres (PSMS) has more effective indicators (tens of times) for equally dense mixtures for all the studied properties: ultimate compressive strength, ultimate flexural strength, rheological properties of finished mixtures, etc. For example, the plastic strength of building mixtures with hollow microspheres is 4 ... 8 times higher than the density for compositions with added perlite or vermiculite, which affects the technological properties of solutions. Mixtures with hollow microspheres retain vitality longer. They have a water retention capacity of 90% more. In this case, the mixtures do not lose their shape in the period before crystallization and have high adhesion to the base after it [7].

The introduction of 10% hollow microspheres of the MSO – "0" grade of class B1 of Novgorod production (TU–6–11–156–79) (Fig. 1) into the structure of high-strength gypsum stone makes it possible to obtain materials with a compressive strength of 3.4 MPa and a strength at a bend of 1.78 MPa at the age of 2 hours and an average density in the dry state of 868 kg/m^3. Moreover, the authors note a change in the structure of gypsum stone, which is confirmed by a decrease in the interplanar spacing in dihydrate crystals compared to pure gypsum [8].

Fig. 1. Hollow microspheres MSO – «0» of the Novgorod production.

An important factor in the use of micro fillers is the dispersion of grains [15 – 17]. This is noted by many manufacturers [15]. Domestic production of glass hollow microspheres is significantly superior to foreign analogues in a number of indicators, primarily in terms of strength / density, therefore, in this study, to obtain an effective gypsum composite, we used material from Russian local microspheres.

2 Experimental

Studies to evaluate the effect of the addition of glass micro granules on the properties of a self-reinforced composite were carried out on the basis of a low-grade gypsum binder - G-5 BII gypsum production in the Moscow region. The main properties are listed below in table 1. The properties that are interesting for this study are indicated. First of all, this is the degree of grinding, bulk density and strength.

In this study, the GR microspheres of the Tver Plant were used as filler. Three compositions were studied in the work; the characteristics of their size distributions are shown in Fig. 3-5. Grain formulations are indicated additionally in tables 2 - 7. The best characteristics in the study according to the criterion of ratio and strength were shown by the composition of GR -2, therefore, these results are given in the work - the dependence of strength and density on the content of the additive of glass microspheres GR-2.

It was found that the average diameter of the microspheres in the GR-2 mixture is 16 μm, the distribution mode corresponds to a grain size of 19.73 μm, and the distribution amplitude is:(d90 – d10) / 50 = 1.75.

Table1. GypsumG-5 BII. Properties.

Name of indicator	Value
Degree of grinding, the residue on a sieve with a mesh size of 0.2 mm,%, not more than	14
Bulkdensity, kg / m^3	750-850
The tensile strength of beam samples at the age of 2 hours, MPa, not less than compressivestrength: bendingstrength	 5 2.5

Fig. 2. Hollow microspheres GR-3of the Tver production.

As shown in fig. 4, the composition of GR-2 has the largest distribution in a narrow range from 5 to 30 μm with a small grain content in the range from 0.1 μm to 5 μm and from 30 μm to 50 μm. Other compositions - GR-1 and GR-3 have a smaller number of particles in this range, which, apparently, is reflected in the general particle size distribution of the raw material mixture with a gypsum binder.

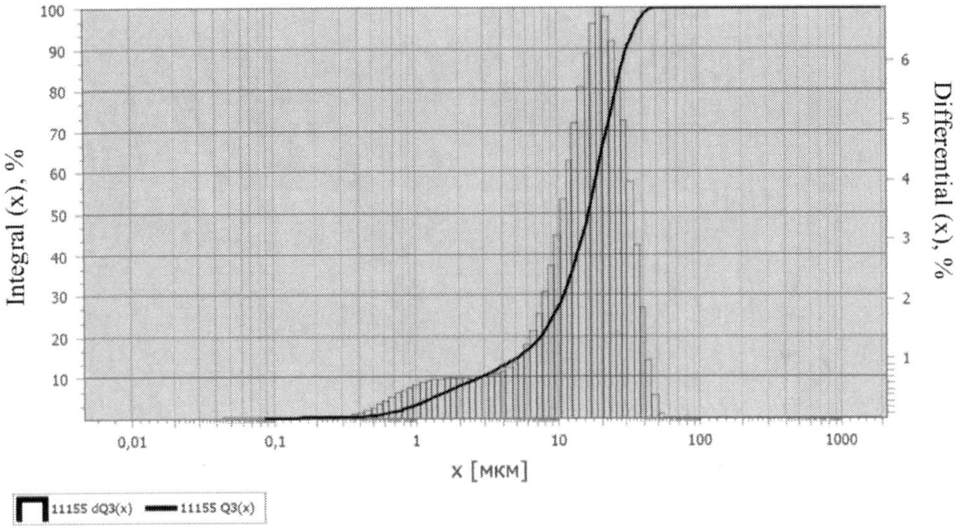

Fig. 3. Integral and differential distribution of glass microspheres GR-1 particles in the mixture.

The composition of the GR-1 mode is slightly higher: 21 μm and the average particle diameter is 17.1 μm. The smallest average diameter corresponds to the composition of GR-3. It is 14.7 μm. The particle distribution mode in the composition of the GR-3 mixture also corresponds to the smaller most probable particle size – 18.49 μm.

Table 2. Grain composition of glass microspheres GR-1.

d (mkm)	Q3(d), %
0.1	0.07
0.5	0.59
1	3.22
5	14.46
10	26.78
15	44.61
20	63.02
25	77.9
30	88.01
35	94.35
40	97.68
50	99.84

Table 3. The weight distribution of glass microspheres GR-1 particles in the mixture.

Q3(d), %	d (mkm)
10	2.81
50	16.43
90	31.2

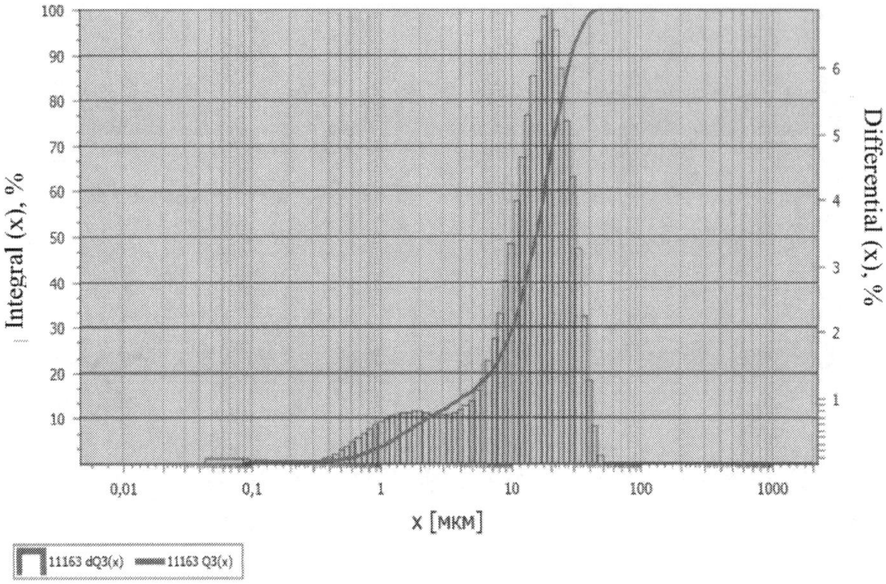

Figure 4. Integral and differential distribution of glass microspheres GR-2 particles in the mixture.

All integral particle distributions in the composition of all mixtures are quite "steep", the distribution shape characteristic, which is called the "shape factor", according to the calculation for the composition of GR-2 is ~ 1.77, for the composition of GR-1 ~ 1.74, and for the composition of GR -3 this characteristic is an intermediate value of ~ 1.75. Systems have a monomodal distribution (fig. 4, 5, 6).

At the same time, gypsum binder is characterized by a large indicator on the characteristic of the distribution form - 3.21. This determines the flatness of the integral distribution curve. Thus, a mixture of gypsum binder with the addition of microspheres has a bimodal distribution of particles, which will affect the packaging of particles in the structure of the composite.

Table 4. Grain composition of glass microspheres GR-2.

d (mkm)	Q3(d), %
0.1	0.14
0.5	0.75
1	3.61
5	16
10	29.16
15	48.34
20	67.32
25	81.79
30	90.95
35	96.23
40	98.71
50	99.97

Table 5. The weight distribution of glass microspheres GR-2 particles in the mixture.

Q3(d), %	d (mkm)
10	2.35
50	15.42
90	29.36

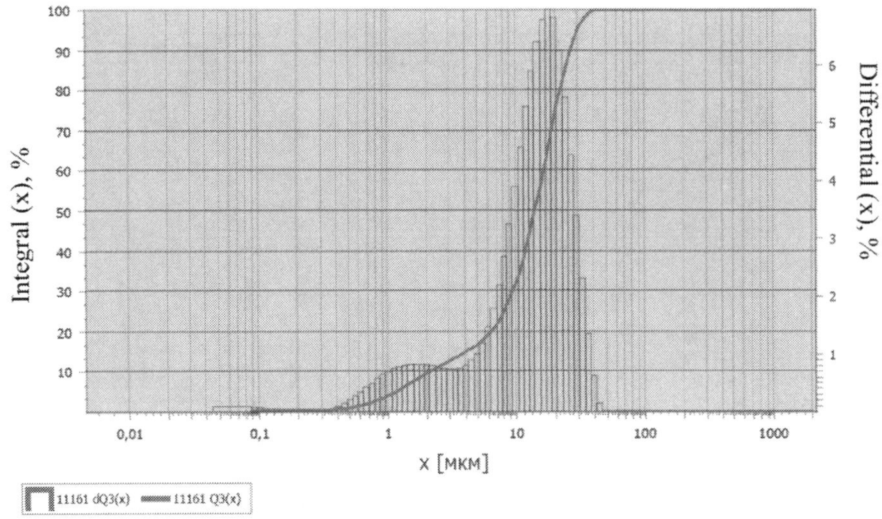

Fig. 5. Integral and differential distribution of glass microspheres GR-3 particles in the mixture.

The test results for the strength and density of the self-reinforced composite are shown in fig. 6, 7. It is shown that the water-solid ratio has little effect on the average density of the composite. This is due to the shape and surface quality of the granules. Micro granules not only increase water demand, but also increase the workability ofthe whole mixture. The effect of W/S is manifested for a mixture without filler or with a small amount.

However, the water-solid ratio affects the strength of the composite. Increased W/T reduces strength. For W/S = 0.6 and W/S = 0.62, the curves are closer in character. The lowest density corresponds to the content of microspheres - 30% and amounts to 858 kg/m^3.

The optimal composition of the mixture according to the criteria of strength and density is characterized by a granule content of 30% atW/S = 0.6.

Table 6. Grain composition of glass microspheres GR-3.

d (mkm)	Q3(d), %
0.1	0.17
0.5	0.84
1	3.91
5	16.7
10	31.83
15	53.36
20	73.01
25	86.66
30	94.33
35	38.14
40	99.56
50	100

Table 7. The weight distribution of glass microspheres GR-1 particles in the mixture.

Q3(d), %	d (mkm)
10	2.14
50	14.23
90	26.83

Table 8. The weight distribution of gypsum.

Q3(d), %	d (mkm)
10	0.71
20	1.4
30	2.27
40	3.25
50	4.5
60	6.24
70	8.67
80	11.71
90	15.65

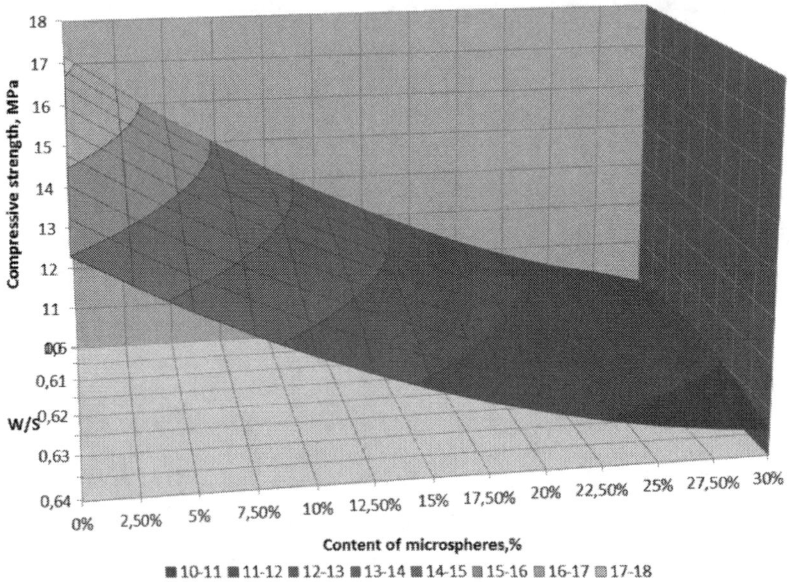

Fig. 6. The effect of the addition of microspheres on the strength of the gypsum self-reinforced composite at various W/T.

The gypsum self-reinforced composite with the addition of GR micro granules has an average strength of ~ 11 MPa at a density of 858 kg/m^3. Without adding expensive chemical modifiers.

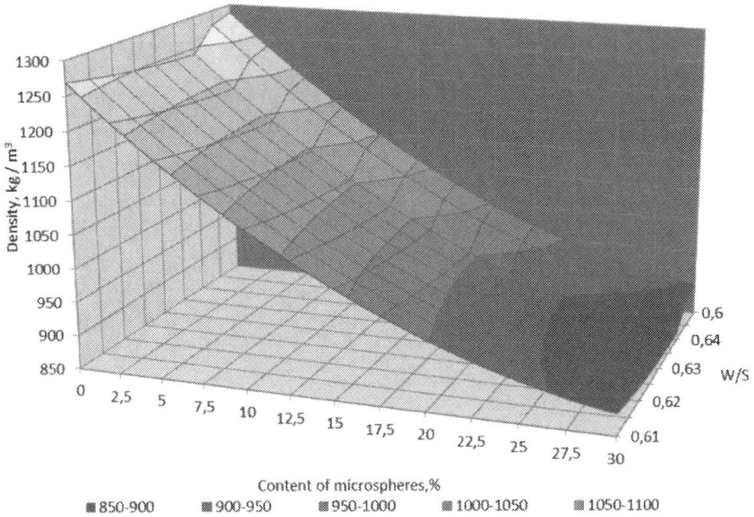

Fig. 7. The effect of the addition of microspheres on the density of the gypsum self-reinforced composite at various W/T.

Thus, the studies confirmed the possibility of obtaining a lightweight gypsum composite based on glass hollow microspheres. The introduction of glass micro filler allows you to use the material to create a wide range of 3 D gypsum products. The granulometric and dispersed composition of the filler does not create technological difficulties for its similar implementation.

References

1. N.A. Sapelin, V.N. Khoglov,Suhiestroitel'nyesmesi. **5**, 30-33 (2012)
2. N.A. Galtseva, *Innovation and modelling in building materials science*(2016)
3. N. Lushnikova, L. Dvorkin,*Sustainability of construction materails.*DOI: 10.1016/B-978-0-08-100370-1.00025-1
4. V.B. Petropavlovskaya, A.F. Buryanov, T.B.Novichenkova, K.S. Petropavlovskii,*The self-reinforced gypsum composites* (De Nova, 2015)
5. V.Petropavlovskaya, A.Buryanov, T. Novichenkova, K. Petropavlovskii,*IOP Conference Series: Materials Science and Engineering,***365**, 032060 (2018)
6. M.S. Lebedev,Tekhnicheskienauki – otteorii k praktike. **7-8 (44)**, 126 – 140 (2015)
7. K.I. Kirillov,*Improvement of thermophysical properties of masonry solutions*, Belgorod: BGTU im. V.G. SHuhova, **4**,pp. 57 (2004)
8. T.E. Haev, E.V. Tkach, D.V. Oshchkin,Stroitel'nyematerialy,**10**, 45–50 (2017)
9. A.A. Pashkevich, E.G. Pervushin, D.V. Oshchkin,*Hollow glass microspheres and cement systems*,M.: NIISF, 134–139 (2006)
10. J.M. Gomez-Soberon,F.G. Cabrera-Covarrubias, J.L. Almaral-Sanchez, M.C. Gomez-Soberon, Advances in Materials Science and Engineering (2018). Article ID 1386946. DOI:.org/10.1155/2018/1386946
11. Yu. Song, David A. Lange, Applied Sciences,**9(5),** 876(2019). DOI: 10.3390/app9050876
12. K.S. Petropavlovskii, A.F. Buryanov, V.B. Petropavlovskaya, T.B. Novichenkova, Stroitel'nyematerialy,**10**, 40–45 (2019)
13. Y.-Sh. Wang, J. L. Provis, J.-G. Dai Role, Cement and Concrete Composites,**93,**186–

195 (2018)
14. V.S. Semenov, T.A. Rozovskaya, Stroitel'nyematerialy,**6**, 16–19 (2015)
15. V.F. Draper, Glass Ind.,v. **62,10**,13-20 (1981)
16. V.B. Petropavlovskaya, T.B. Novichenkova, V.V. Belov, A.F. Buryanov,Stroitel'nyematerialy,**1**, 64–65 (2013)
17. V.B. Petropavlovskaya, T.B.Novichenkova, K.S. Petropavlovskii, A.F. Buryanov,Stroitel'nyematerialy, oborudovanie, tekhnologii XXI veka. **1 (192),** 15-17 (2015)
18. V.V.Budov, L.S. Egorova, Glass and ceramics,**7**, 2–5 (1993)
19. Ahmad Shayan, XuAimin, Cem. Concr. Res.,**36 (3),** 457–468 (2006)
20. N. Schwarz, H. Cam, N. Neithalath,Cem. Concr. Compos. **30 (6),** 486–496 (2008)

Studying the possibility of obtaining heat-insulating building materials based on lime carbonization hardening and rice husk

Vitaliy Nikolaenko[1*], *Nikolai* Lyubomirskiy[1], *Elena* Nikolaenko[1], and *Aleksandr* Bakhtin[1]

[1] V.I. Vernadsky Crimean Federal University, 4 Academician Vernadsky, Simferopol 295007, Russian Federation

Abstract. The paper presents studies of changes in the physicomechanical properties of samples based on lime-limestone compositions with rice husk filler hardening in an environment with a high concentration of carbon dioxide. The samples are semi-dry pressing cylinders made of a mixture of hydrated lime, ground limestone and a filler in the form of rice husk in order to reduce the density and increase the heat-insulation ability of building materials based on this raw material mixture. The studies showed that there is an increase in the strength of the samples due to the transition of portlandite to calcite during forced carbonization. The possibility of using waste products of plant origin by the example of rice husk in the production technology of heat-insulating building materials using artificial forced carbonization which will allow obtain products with the following characteristics: compressive strength 2.2 – 4.0 MPA at a density of 700 - 1200 kg/m³, thermal conductivity – 0.11-0.13 W/m•K, has been proved.

1 Introduction

An integral part of any industrial production is the production of a large number of by-products which are accumulated in the dumps or released into the atmosphere as gaseous waste. These materials often contain undesirable compounds and their deposition is a serious social problem with the impact on the environment and human health.

In view of the current environmental situation, all sorts of decisions to conserve natural resources and reduce the level of air pollution by various industrial wastes are of vital importance. The analysis of the problem state shows that for most industrial and agricultural industries waste generation is characteristic with a low degree of their use which affects the increase in the area of dumps.

Processing plant products is no exception as well. For example, the processing of 1 ton of rice produces about 200 kg of waste in the form of rice husk.

About 600 million tons of rice husk is produced annually in the world as a result of threshing [1]. For the most part, it is burned in furnaces or buried requiring the use of large land. But the most interesting thing is that the husk does not decay in the ground due to the presence of silicon dioxide in it. And when burning husks, substances are released that can adversely affect the nature and human health. Disposal of rice husk is an urgent problem throughout the world, especially in countries where rice is the main cereal product (China, India, Egypt, South Korea, African countries and partially Russia, Uzbekistan).

The resulting husk is widely used and can be disposed in many ways. However, most methods of utilization of rice husk are associated with pyrolysis (to obtain ash) and the release of a large number of caustic gaseous products that pollute the atmosphere and harm human health [2-6].

The problem of disposal of the above wastes can be solved through their use in the production of heat-insulating building materials reminiscent of arbolite in structure. The use of eco-materials from agricultural waste is an increasingly common solution in the field of construction. This necessitates the development of research in the field of obtaining economical and environmentally friendly building materials and the introduction of developed resource-saving technologies in production.

In addition, climate change on the planet due to the annual increase of the amount of CO_2 emitted by industrial enterprises requires new technological solutions for the disposal of gaseous waste [7].

In this regard, it is relevant to study the use of CO_2 in the technology of building materials based on technogenic lime binder and limestone waste with a filler of plant origin in order to utilize by-products of production and reduce carbon dioxide emissions into the atmosphere.

As a result of the transition of hydrated lime to calcium carbonate, the material is hardened, its water resistance and frost resistance increase. The durability of calcareous materials is proven by time because a large number of ancient architectural monuments were built using precisely lime mortars and limestone stones (palaces of the ancient city of Knossos of XVII - XV centuries BC on the island of Crete, St. Sophia Cathedral in Kiev of XI century AD, ancient defensive buildings and structures on the coast of Lisbon, the Great Wall of China, etc. [8]). However, this process is

* Corresponding author: author@email.org

lengthy and can take years or even decades. It can be intensified by forced carbonization of materials based on lime due to the disposal of specially prepared flue gases from lime kilns [9-12].

The authors of the article obtained the optimal parameters for producing carbonized calcareous-carbonate building wall materials formed by means of semi-dry pressing [13-16]. A new direction in the study of materials based on lime carbonization hardening is the production of heat-insulating materials due to the introduction of fillers that increase porosity and lower the density of finished products.

In order to give the materials greater heat-insulation potential, it was decided to introduce a rice husk filler into the composition of the lime-limestone compositions. In addition, according to modern views, the structure of the material does not remain unchanged but continuously undergoes changes in space and time under the influence of both external and internal factors [17,18] which makes it possible to predict an increase in the strength of products over time.

The aim of this work is to study the changes in the physicomechanical properties of forced carbonized samples based on lime, limestone filler and filler of plant origin in the form of rice husk.

2 Materials and methods

The study of the properties of the proposed material was carried out on samples-cylinders with a diameter of 125 mm pressed onto a screw mechanical press. The raw mix was a composition of slaked lime, sieved through a sieve with a mesh size of 0.63 mm, ground limestone with a specific surface of 1700 cm²/g and rice husk crushed in a special device to a specific surface of 2400 cm²/g. Dry raw materials were mixed with water. In the preparation of samples, the percentage of raw materials and the moisture content of the raw mix were varied. The carbonization of the pressed-in products was carried out in a tooling by supplying carbon dioxide of 100% concentration to the screw press housing for 5 minutes.

To study the effect of organic compounds contained in rice husks on the course of the forced carbonization process, cylinder samples with a diameter of 30 mm were formed by the method of semi-dry pressing (pressing pressure – 30 MPa). In this case, the raw material mixture was a composition of lime sifted through a sieve with a mesh size of 0.63 mm, a marble-like limestone of fractionated composition and a special solution obtained by digesting plant waste (rice husks, wood shavings, and straw). Extracts from plant waste were obtained by digesting the waste in boiling water and further filtering the resulting broth. To assess the effect of the presence of organic compounds in the grinder of the raw material mixture, samples mixed with plain water were also formed. Forced carbonization of these samples was carried out in a special carbonization chamber [19] for 60 minutes.

Studies were carried out on changes in the phase composition of secondary calcite obtained by forced carbonization of samples from a raw mixture mixed with decoctions of plant waste using a thermogravimetric derivatographic analysis on STA 8000 synchronous analyzer (TGA/DSC/DTA) (PerkinElmer (USA). Chemical analysis of rice husk was carried out using X-ray fluorescence analysis on Epsilon 3XLE ED spectrometer (PANalytical).

The physicomechanical characteristics of the test samples were measured using standard methods for determining the compressive strength and coefficient of thermal conductivity.

3 The main section with the results and their analyses

Studies on the qualitative effect of organic additives contained in plant waste (rice husk, sawdust, and straw) showed that mixing the original composition of slaked lime and fractionated limestone with the resulting broths does not negatively affect the course of the forced carbonization process. As can be seen from the obtained thermograms, at the end of the forced carbonization process, practically no free portlandite remains in the cylinder samples (Fig. 1).

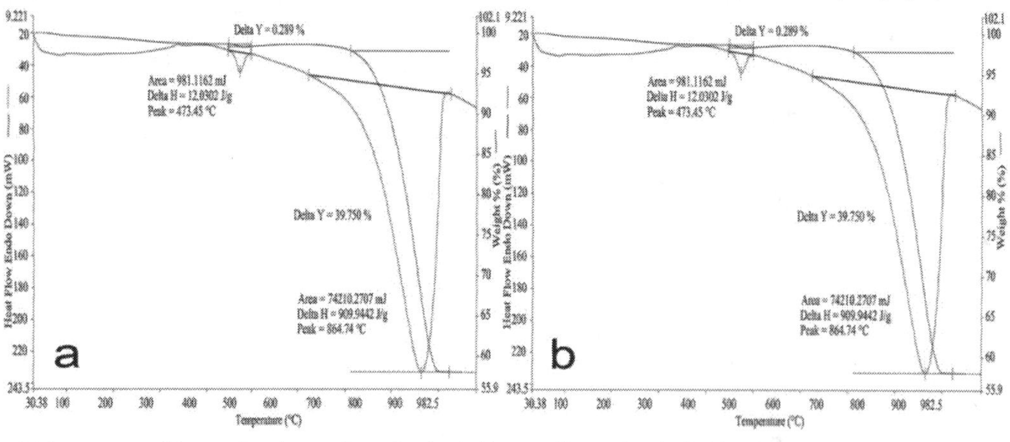

Fig. 1. Thermograms of the samples: a) control sample mixed with water; b) sample mixed with a decoction of rice husk.

Physicomechanical characteristics also indicate the absence of a negative effect on the process of forced carbonization (Table 1).

Table 1. Physicomechanical characteristics of the samples obtained by mixing with water and extracts from waste of plant origin (rice husk, sawdust, and straw).

Raw mixture (as a percentage of dry weight)	Type of fluid	Physicomechanical characteristics		
		R_{pr}, MPa	ρ, kg/m³	Δm, g
Lime (10%) + Limestone (90%)	Water	38.5	1913	2.037
	Extract of rice husk	40.3	1910	2.041
	Extract of sawdust	39.6	1917	2.032
	Extract of straw	39.1	1913	2.021

In the future, we decided to use rice husk for research for several reasons.

Firstly, rice husk contains starch particles (10-15%) left over from rice grains which, in turn, contains amylopectin – a polysaccharide that promotes the growth of smaller particles of secondary calcite in comparison with natural lime carbonization in the presence of water. Studies on the effect of introducing rice broth into lime mortar during the construction of the Great Wall of China indicate an improvement in the physicomechanical properties of the mortar [20,21].

Secondly, in the case of burning rice husk, carbon dioxide is formed which is a necessary component for the formation of secondary calcite during forced carbonization, and the resulting ash can act as a filler in the structure of the material.

Changes in the physicomechanical properties of the samples obtained from the composition of lime, limestone and rice husk are shown in Table 2.

Table 2. Physicomechanical properties of the samples.

Composition of the dry raw mix, %				Humidity, %	Δm, g	ρ, kg/m³	R_{pr}, MPa
$Ca(OH)_2$	$CaCO_3$	Rice husk	Proportion				
15	75	10	1:5	20	87.6	1267	3.3
18	72		1:4		90.8	1230	4.2
22.5	67.5		1:3		106.2	1211	4.6
13.3	66.7	20	1:5	20	51.8	1095	3.1
16	64		1:4		68.4	1051	3.6
20	60		1:3		101.2	1000	4.9
11.7	58.3	30	1:5	22	23.6	863	2.6
14	56		1:4		46.5	830	2.7
17.5	52.5		1:3		62.3	801	3.2
10	50	40	1:5	24	29.5	754	1.8
12	48		1:4		38.0	731	2.0
15	45		1:3		80.1	716	2.4

As it can be seen from the data in Table 2, there is a regular increase in weight due to the transition of portlandite to calcite, i.e. the carbonization reaction takes place and after opening the press-tool, the sample does not crumble which indicates the successful operation of lime carbonization hardening as a binder matrix with a filler. The strength of the samples decreases with an increase in the amount of rice husk filler introduced, while the density, respectively, with the introduction of a larger amount of rice husk decreases which favorably affects the heat-insulation properties of the obtained material – the heat conductivity of a material with 40% rice husk content lies in within the range of 0.11 – 0.13 W/m•K, which, along with a strength of 2.5 MPa, makes it possible to classify it as a structural and heat-insulating material.

It should be noted that the introduction of 30% and 40% of the rice husk into the samples during testing of the samples leads to non-fragile destruction (the samples are deformed before destruction). In this regard, the compressive strength of the samples at given fractions of rice husks was determined when the 10% deformation of the samples was achieved in height.

Figure 2 shows a destroyed sample with the shown qualitative reaction of calcium hydroxide to phenolphthalein. As it can be seen from the photo, the sample still contains a significant amount of portlandite particles that have not reacted with CO2. This fact allows us to assert that the experimental samples have a certain potential for further carbonate hardening and, consequently, to increase their strength characteristics. It is known [13-16] that the strength characteristics of carbonate hardening materials largely depend on the degree of carbonization of the calcareous component. We can say that five minutes of forced carbonization is not enough for these samples, i.e. the processing time of samples based on a lime-limestone composition and rice husk should be increased or more intensive methods of their forced carbonization should be applied. However, even with such a treatment time with carbon dioxide, a heat-insulating material is obtained which complies with the requirements of regulatory documents in terms of strength and heat-transfer properties.

Fig. 2. General view of the destroyed sample stained with phenolphthalein (a). A series of samples prepared for testing (b).

4 Conclusion

As a result of the studies, the possibility of obtaining heat-insulating building materials based on compositions of lime, limestone and rice husk treated with carbon dioxide gas of 100% concentration was established. The results obtained show the possibility of obtaining products with a strength of 2.2 to 4.0 MPa at a density of 700 – 1200 kg/m^3 with a thermal conductivity coefficient in the range of 0.11 – 0.13 W/m•K.

The proposed type of material has been little studied and has many directions for research on optimizing the raw material composition and materials used: the use of lime-containing industrial wastes such as lime dust [17,22,23] or carbide lime; studying the effect of starch compounds of rice husks on the structure formation of the material; studying changes in the strength of products over time and etc.

The production of this material is one of the possible ways to solve the problems of pollution of the atmosphere and the environment by involving various industrial waste in the process of creating building materials.

This work was partially supported by the V.I. Vernadsky Crimean Federal University Development Program for 2015–2024.

References

1. Information from Food and agriculture organization of the United Nations website http://www.fao.org
2. M. Chabannes, J.-C. Bénézet, L. Clerc, E. Garcia-Diaz, *Use of raw rice husk as natural aggregate in a lightweight insulating concrete: an innovative application,* Construction and Building Materials **70**, pp. 428–438 (2014)
3. M.M. Younes, H.A. Abdel-Rahman, Magdy M. Khattab, *Utilization of rice husk ash and waste glass in the production of ternary blended cement mortar composites,* Journal of Building Engineering **20**, pp. 42-50 (2018)
4. E. Chabi, A. Lecomte, E.C. Adjovi, A. Dieye, A. Merlin, *Mix design method for plant aggregates concrete: example of the rice husk,* Construction and Building Materials **174** , pp. 233–243 (2018)
5. R.L. Reaño, *Assessment of environmental impact and energy performance of rice hus utilization in various biohydrogen production pathways*, Bioresource Technology **299** (2019)
6. T.-H. Liou, P.-Y. Wang, *Utilization of rice husk wastes in synthesis of graphene oxide-based carbonaceous nanocomposites*, Waste Management **108**, pp. 51-61 (2020)
7. Information from the European Commission's of climate strategies and targets website https://ec.europa.eu/clima/policies/strategies/
8. C. Borges, A. Santos Silva, R. Veiga, *Durability of ancient lime mortars in humid environment*, Construction and Building Materials **66**, pp. 606–620 (2014)
9. H. Pöllmann, *Mineralogical Strategies to reduce CO2 in the fabrication of alternative cements* Proc., IBAUSIL, **1**, pp. 110-129 (2015)
10. D.R. Moorehead, *Cementation by the carbonation of hydrated lime*, Cement and Concrete research, **16**, pp. 700–708 (1986)
11. P.De Silva, L. Bucea, D.R. Moorehead, V. Sirivivatnanon, *Carbonate binders: Reaction kinetics, strength and microstructure*, Cement & Concrete Composites **28**, pp. 613–620 (2006)
12. K. Van Balen, D. Van Gemert, *Modelling lime mortar carbonation*, Materials and Structures **27**, pp. 393-398 (1994)

13. N.V. Lyubomirskiy, S.I. Fedorkin, A.S. Bakhtin, T.A. Bakhtina, T.V. Lyubomirskaya, *Research in influence of regimes of forced carbonate hardening on properties of materials on the basis of lime-limestone compositions of semidry pressing*, Stroitel'nye materialy **8** , pp. 7–12 (2017)
14. S.I. Fedorkin, N.V. Lyubomirskiy, M.A. Lukyanchenko, *System based on lime of carbonizing hardening* Constr. Mater. **11**, pp. 45-47 (in Russian)
15. N.V. Lyubomirskiy, S.I. Fedorkin, *The influence of carbon dioxide pressure on the kinetics of forced carbonation of limestone semi-dry pressing and the formation of its strength*, Construction and technogenic safety **3** (2016), pp. 28-38 (in Russian)
16. N.V. Lyubomirskiy, A.S. Bakhtin, T.A. Bakhtina, E. Yu. Nikolaenko, V.V. Nikolaenko, *Influence of calcium bicarbonate on the structure and properties of materials based on lime carbonation hardening*, International research journal **11-4**, pp. 86-93 (2016) (in Russian)
17. V. Nikolaenko, N.V. Lyubomirskiy, E. Nikolaenko, *Changes in the physicomechanical properties of materials of carbonate hardening on the basis of lime and lime stone systems with the course of time*, Materials Today: Proceedings, **5-19**, pp. 1917-1921 (2019)
18. N.V. Lyubomirskiy, A.S. Bakhtin, T.A. Bakhtina, *Change of physical and mechanical properties of limecarbonate calcium materials of forced carbonate hardening in time*, Construction and technogenic safety **8**, pp. 67-73 (2017).
19. S. I. Fedorkin, N. V. Lyubomirskiy, V. G. Nosatov, T.A. Loktionova, *Automatic installation and technique of studying of process of carbonization of lime*, Construction and technogenic safety **19-20**, pp. 74-78 (2007)
20. Y. Fuwei, Z. Bingjian, M. Qinglin, *Study of Sticky Rice-Lime Mortar Technology for the Restoration of Historical Masonry Construction*, Accounts of Chemical Research 6-43, pp. 936-944 (2010).
21. W. Shanwei, W. Sheliang, L. Zhe, Y. Wenjuan, L. Bo, L. Binbin, *Characterization of sticky-rice lime binders from old masonry relics in north China: The primary contribution for conservation*, Construction and Building Materials, **250** (2020)
22. M.A. Latifa, S. Naganathanb, H.A. Razakc, K. N. Mustaphab, *Performance of lime kiln dust as cementitious material*, Procedia Engineering **125**, pp. 780 – 787 (2015)
23. A. Arulrajah, A. Mohammadinia, A. D'Amico, S. Horpibulsuk, *Effect of lime kiln dust as an alternative binder in the stabilization of construction and demolition materials*, Construction and Building Materials **152**, pp. 999–1007 (2017)

The effect of biomineral composition on the biochemical parameters of animal blood

Nurgul Montayeva [1*]
[1] Zhangir khan university, 51 Zhangir khan str. Uralsk, Kazakhstan

Abstract. The article presents the results of scientific and experimental studies on the creation of a biomineral composition intended for the prevention of diseases and improve animal health. The possibility of creating biomineral compositions based on wormwood of bitter and siliceous rocks - flasks of Western Kazakhstan, taking into account their chemical, mineralogical and biological properties. Based on experimental studies, the choice of siliceous rock, flasks of the West Kazakhstan region, is scientifically justified by the criterion of a high content of amorphous and finely divided silica, which gives the following advantages: - highly dispersed silica, guarantees effective sorption of mycotoxins, salts of heavy metals, chemical toxins, radionuclides, gases and other metabolic products in the animal's body due to the huge sorbing surface. Wormwood as a biological active component enhances immunity and acts as an antifungal, antimicrobial, anti-inflammatory agent. As a result of scientific and experimental work, it was found that the proposed biomineral composition has high biological activity and significant pharmacological action, which is manifested by an improvement in the morphological and biochemical parameters of the blood of cows.

1 Introduction

The Republic of Kazakhstan has adopted the State Program for the Development of the Agro-Industrial Complex of the Republic of Kazakhstan for 2017 - 2021, where one of the main priority areas is the development of animal husbandry. However, there are objective threats associated with animal diseases for the successful development of animal husbandry. Infectious, non-infectious, invasive (parasitic) animal diseases cause great economic harm to the farms of the Republic of Kazakhstan. Among the common diseases of animals is mycotoxin poisoning as a result of eating contaminated herbs, plants, feeding animals rotted, moldy feed. These factors provoke digestive disorders, poisoning of varying severity, metabolic failures. Often during poisoning, refusal of feed is noted, which leads to the death of animals. Among the main factors contributing to the mass disease of animals is a poor-quality, unbalanced, meager diet, weakening of resistance and immunity.

An important direction in improving the technology of preventive measures is the introduction into production of new tools and methods for the prevention and treatment of animal diseases, including the use of drugs with bioactive properties that can have a

* Corresponding author: montayeva-n@mail.ru

regulatory effect on the intensity of metabolic processes, enhance functional activity organs and systems of the body, increase the level of natural resistance of animals [1-3].Особенное место в этом направлении занимает обширный класс биоминеральных веществ, выполняющие важную роль в построении структурных частей и тканей животного организма [4-6].

It is well known that the inorganic part of bone tissue consists of calcium phosphate and magnesium, silicon, calcium carbonate, potassium and sodium, potassium, magnesium and sodium chlorides and other compounds. Other elements are part of similar organic compounds that perform various functions in physiological and biochemical transformations. Among the inorganic natural raw materials is an extensive group of aluminosilicate minerals that have valuable specific properties - sorption, ion exchange, binders, thixotropic. This group includes siliceous rocks and bentonite (montmorillonite) clays [7–9].

It should be noted that research in this area is actively being carried out abroad. The authors argue that bioactive supplements usually play a central role in animal diets in order to stimulate performance and improve well-being and health, especially during stressful periods of life [10].

The authors of the studies carried out detailed studies of clay minerals as sorbents for mycotoxins in the diets of lactating goats. The effects of bentonite and montmorillonite on consumption, digestibility, blood chemistry, rumen fermentation, milk yield and composition, as well as milk aflatoxin were revealed [11].

According to many scientists, montmorillonite clays contribute to the inactivation of mycotoxins, which may be present in dry animal feed [12-16].

They are especially important when using ruminant synthetic nitrogen-containing substances in feeding against the background of diets insufficiently supplied with sugar in the winter-stall period, and when feeding green protein-rich feeds in the summer in the summer [17].

There is a positive experience in the use of natural minerals in agriculture as bioactive additives that contribute to the normalization of metabolic processes, better absorption of nutrients, trace elements and vitamins. The properties of minerals, such as high adsorption, catalytic, and ion-exchange activity, are especially appreciated. Due to these properties, it adsorbs alkaloids, paints, heavy metals, toxins of microorganisms, nitrates and nitrites, and other harmful substances [18-21].

One of the promising raw materials of Kazakhstan for using them as a mineral component in the creation of biomineral active substances is carbonate-siliceous rocks with high sorption properties and useful minerals for animals. These siliceous rocks are sedimentary deposits and are distinguished by the presence of well-preserved remains of fauna and flora, consisting of biogenic silica (radiolarians and diatoms) [22-23].The author of the work has established a close relationship between mineral, protein, carbohydrate, lipid and vitamin exchanges. In this paper, it is noted that ensuring the productivity of cows with a sufficient number of macro- and microelements helps to increase their productivity, improve reproductive ability and maintain the health of animals. This is possible only through the additional use of mineral dietary supplements. It should be noted at present that the main feeds cannot satisfy the increased demand of highly productive animals for inorganic substances [24].

2. The main idea and hypothesis of research

Creation of domestic active biomineral substances for animals based on the composition of natural mineral raw materials and local plant bioactive components of the Republic of Kazakhstan.

Carrying out scientific and experimental work in this direction is also dictated by the fact that at present, due to the lack of domestic producers of effective bioactive feed additives, the Republic of Kazakhstan is completely dependent on imported manufacturers.

For the successful implementation of this idea and hypothesis, the unique biological and mineralogical properties of wormwood and siliceous rocks of Kazakhstan are of particular scientific and practical interest.

However, in order to use the siliceous rock of Kazakhstan of a particular deposit and wormwood, it is necessary to carry out scientific and experimental work taking into account the biological and chemical-mineralogical characteristics of each type of investigated raw materials.

3. The purpose and objectives of research

Development of biomineral composition compositions based on natural, safe and readily available raw materials of Western Kazakhstan and assessment of their impact on the morphological and biochemical parameters of cattle blood.

To achieve this goal, it was necessary to solve the following tasks:

- selection of raw materials for the development of biomineral compositions;
- to study the basic biological and chemical-mineralogical characteristics of the selected raw materials;
- development of biomineral composition compositions;
- conducting studies on their effect on morphological and biochemical parameters of animal blood.

4. Research methods and materials

The chemical and mineralogical composition of the studied raw materials was determined using a JSM-6390LV scanning electron microscope with an energy-dispersive microanalysis system, an X'Pert PRO MPD X-ray diffractometer, and an ICP-MS Agilent 7500cx inductively coupled plasma mass spectrometer (JEOL, Japan).

X-ray phase analysis (XRD) was carried out on a DRON-3 diffractometer with radiation in the range of angles 80-640. The sensitivity of the method is from 1 to 2%. Silica rock powders — flasks passing through a 0.315 sieve — were subjected to X-ray analysis.

The sorption activity of the flask was carried out with respect to methylene blue by the photocolorimetric method. The method was based on the ability of the flask to adsorb various substances in aqueous solutions. As a test indicator from dyes, a 1% solution of methylene blue in purified water was selected and used.

The sorption capacity was judged by the change in the optical density of the solution after its contact with the flask.

Blood biochemical parameters were determined using the CHEM WELL analyzer, which has the following features: it performs the whole range of biochemical and immunochemical analyzes, is fully automatic, Russified, and equipped with a quality control program.

A hemotological analysis of blood was performed on a Mindray BC-2800 Vet veterinary automatic hematology analyzer designed to determine 18 parameters with differentiation of leukocytes into 3 subpopulations and plotting distribution histograms. The analyzer uses two independent measurement principles: the Coulter method and the colorimetric method. During each measurement cycle, the sample is automatically aspirated, diluted and mixed before each parameter is determined. These research methods were carried out on the basis of M. Auezov South Kazakhstan State University and Zhangir Khan West Kazakhstan Agrarian Technical University.

As a mineral raw material, siliceous rock was chosen - a flask of the Taskalinskoye deposit of the West Kazakhstan region.

As a result of the studies, it was found that the flasks of the Taskalinskoye deposit are light, dense microporous rocks composed mainly of the smallest (<0.005 mm) particles of opal - cristobalite silica. Table 1 shows the mineralogical composition of the investigated raw materials.

Table 1. Mineralogical composition of siliceous rock - flasks of the Taskala deposit of the West Kazakhstan region, %

Name of species	Opal	Clay minerals	Calcite	Quartz	Mica	Glauconite	Organo-gene residues
Siliceous rock - flask	54-78	15-22	until 6	4-7	2-4	2-3	until 12

In addition, it was found that the flask of the Taskalinskoye field has a relatively low average density in the range of 1200 - 1350 kg / m3, and is characterized by good moisture capacity, high hydraulic and adsorption activity.

As a plant biological stimulant, wormwood (Artemísia absínthium) was chosen.

The choice of this plant is justified by the fact that wormwood is included in the list of medicinal herbs and is growing everywhere in almost all regions of Kazakhstan.

For further studies, bitter wormwood collected after full maturation, i.e. with a completed growing season.

The complete completed vegetative period of wormwood (mature wormwood) occurs after the first frost.

After the first frosts, wormwood dries up in natural conditions. At the same time, it was proved that useful substances in wormwood with a full vegetative period are higher than 1.5 - 2 times than in wormwood with an incomplete vegetative period.

In addition, the use of dried bitter wormwood excludes the process of special drying in the process of preparing a biomineral composition.

Essential oils are the main biological substance in wormwood, and contain organic acids, fatty acids, tannins as accompanying substances.

The herb and leaves of wormwood contain 0.5-2% of essential oil, the components of which are cineol (1), thujol alcohol, thujone (14.15) (about 10%), tuyol (5) (10-25%) cadinene (16), fellandren (17.18), pinene (19.20), b-caryophyllene (21), y-sepinen, bisabolen (22),, chamazulenogen.

As the complex chemical and biological composition of wormwood is shown, they can be classified as biologically active substances that increase appetite, improve digestion, and as antifungal, antimicrobial, anticonvulsant, diuretic, anti-inflammatory, wound healing, carminative, sedative.

5. Results and discussions

To prepare a biomineral composition, the collected stalks of wormwood are crushed using a grinder to form a loose fraction with sizes of 0.5 - 2.5 cm. In parallel, siliceous rock - flask is prepared by preliminary high-speed drying in a rotary kiln at a temperature of 100 - 110 ° C to a residual humidity 7 - 8%. This heat treatment provides thermal sanitation of the used mineral raw materials. The dried siliceous flask rock is ground in a porcelain ball mill before passing through a 0.14 mm sieve. Finished powders of wormwood and siliceous rock-flask breed are weighed on an electronic balance and the mixture is prepared in the following ratios wt%: wormwood 60–70, siliceous rock — flask 30–40. Then, loose mixtures of wormwood are mixed with ground flask using a mixer. The thoroughly mixed mixture is further co-milled using grinding equipment. Thus obtained biomineral composition has a particle size of 0.1 to 3.0 mm The studied compositions of the biomineral composition are presented in table 2.

Table 2. The investigated composition of the bio-mineral composition

Composition №	Wormwood, wt. %	Siliceous rock - flask, wt. %
1	60	40
2	65	35
3	70	30

As the optimal composition, composition No. 1 was chosen, since it is in these ratios that the adsorption properties of the siliceous flask are maximally manifested. According to the results of preliminary studies, we found that a decrease in the content of siliceous rock - flask due to an increase in the content of wormwood with a slide significantly reduces the adsorption capacity of the biomineral composition. Moreover, the content of wormwood in the amount of 60% by weight of dry matter is quite sufficient for the manifestation of biological activity. In the composition, it dominates in mass and volume due to the low average density in a loose state.

Scientific and experimental studies on the influence of the developed compositions of the biomineral composition on the cattle in the conditions of the peasant farm "Kazyna" in the Shieli district of the Kyzylorda region (Kazakhstan).

The formation of groups of animals for the experiments was carried out according to the principle of analogues, which took into account age, breed, body weight, physiological state, productivity, health status. The number of animals in the groups was determined by the appropriateness of an objective assessment of the results obtained and their statistical reliability.

The control groups of animals received the main diet, envisaged by the feeding technology and designed to obtain high productivity.

The animals used in the experiment were divided into two groups of 10 animals each, taking into account body weight, age and productivity. All experimental cows were clinically healthy.

Cows of the control group received feed of the main diet, the experimental group of animals was additionally fed with a biomineral composition at the rate of 3% of the dry matter of the diet.

For three months, the physiological state of the animals was evaluated by morphological and biochemical studies of animal blood. dynamics was monitored Changes in the metabolic processes of the cows were monitored by the following indicators: the number of red blood cells and the level of hemoglobin, total protein and protein fractions, glucose, urea, the level of aminotransferases (AST, ALT), total calcium and inorganic phosphorus, as well as some trace elements (Mg, Zn, Fe, Cu) and carotene.

According to the results of the data obtained using the analyzer, a significant ($P < 0.05$) increase in erythrocytes and hemoglobin is observed. At the same time, the hemoglobin level in the first month of the experiment increased by 6.2% relative to background values, and after 60 and 90 days its concentration increased to 8.3% and 12.2%, respectively. The indices of these increases were 104.1 ± 41.2 g / l and 105.6 ± 38.5 g / l against 93.1 ± 29.7 g / l of background indicators.

The increase in the number of red blood cells in experimental animals over the entire period of research varied in the range of 5.3–8.4%.

In the control group, an increase in the level of hemoglobin is not observed while maintaining the initial (background) number of red blood cells (6.21 ± 0.9x10).

The results of studies on the effect of biomineral composition on biochemical blood parameters of cows are presented in table 3.

Table 3. The effect of biomineral composition on biochemical blood parameters of cows

Indicators	Background	Control		Experimental	
		after 60 days.	after 90 days.	after 60 days.	after 90 days.
Total protein, g / l	87.2±10.4	85.3±11.4	89.2±10.7	88.9±12.9	95.2±16.2
Protein fractions:% albumin	40.1±6.7	41.2±4.3	39.1±5.1.0	42.9±4.8	45,6±5.9
a-globulins	13.4±0.9	12.9+1.1	11.4+0.7	13.9+2.6	16.4±2.4
Glucose, mmol / L	1.41±0.3	1.44+0.50	1.72±0.4	2.17±0.5	2.3±0.8
Urea, mmol / L	4.7±0.9	6.4±0.7	5.2±0.3	4.9±0.66	5.5±0.5
AST, U / ml	26.5±3.2	36.1±3.6	37.1±5.7	35.5±2.9	38.7±3.6
ALT, U / ml	15.7±2.0	14.6±3.4	22.3±5.1	15.6±3.8	23.1±2.7
Total calcium, mmoy/l	1.57±0.2	1.58±0.3	1.8±0.32	1.76±0.54	1.98±0.6
Phosphorus inorganic-nich., mmol / l	2.4±0.25	2.5±0.3	2.6±0.28	2.7 ±0.42	2.84±0.45
Iron, mmol / L	20.58±3.1	21.3±3.4	22.0±1.9	26.5±3.8	27.3±3.3
Zinc, micromol / l	13.9±1.8	12.9±0.8	14.4±2.0	15.2±2.3	15.1±2.9
Copper, mmol / L	13.4±2.2	13.5±3.4	13.7±4.5	13.9±3.4	14.1±2.4
Manganese, mmol	1.06±0.5	1.07±0.32	1.08 ±0.24	1.08±0.8	1.09±0.2
Carotene, mM / L	0.47±0.02	0.49±0.03	0.48±0.03	0.49±0.05	0.5±0.01

Note: * - confidence level P <0.05

As the results of the study show, the use of the biomineral composition had a positive effect on the biochemical homeostasis of experimental animals.

The total protein content in blood serum when prescribing biomineral increased and amounted to 95.2 ± 16.2 against 87.2 ± 10.4 g / l of background indicators and 89.2 ± 10.7 indicators of control animals by the end of the experiment . Moreover, the largest difference in protein content by groups was noted from the second month of research.The amount of albumin in serum before the experiment, on average in groups, was 40.1%. With the introduction of the biomineral composition, an increase in the level of this fraction in experimental cows was noted after 60 days — by 4.1%, and after 90 days — by 16.6%. Moreover, the reliability of the results on the level of albumin was already evident from the second month of application of the proposed biomineral composition (P <0.05).

The main indicator of carbohydrate metabolism is glucose, in the experimental group it increased during the research period by 1.3 times relative to control animals by 1.6 times relative to background studies.

The results of the experiment showed that the use of biomineral composition in the diet of cows influenced the absorption of mineral elements. Prior to the experiment, the phosphorus level in the blood serum of cows was at the upper limits of the physiological norm (2.4 ± 0.25 mmol / L).

However, after 60 days of applying the biomineral composition, the calcium content in the blood of experimental cows increased from 1.58 ± 0.3 to 1.76 ± 0.54, and by the end of the experiment to 1.98 ± 0.6, reaching physiological limits.

Analysis of the obtained data on the effect of the proposed biomineral composition on the morphological and biochemical parameters of the blood of cows showed positive dynamics in terms of hemoglobin, red blood cells, carbohydrate metabolism, protein content, albumin, phosphorus and calcium.

Thus, one can note the activation of the processes of biological protein synthesis and, first of all, its albumin fraction under conditions of using a biomineral composition, the metabolic activity of which is stipulated by the content of a wide range of macro- and microelements, physicochemical and biological structural features. This determines the further stage of protein biosynthesis. In this case, Mg, Zn, and Cu ions can be included. The concentration of these elements within small limits is quite sufficient for the manifestation of the biological activity of siliceous rocks - flasks.

6. Practical recommendations

The proposed biomineral composition is recommended to be used as a feed additive for therapeutic and prophylactic action in case of gastrointestinal diseases of non-infectious etiology, anemia, correction of mineral metabolism, mixed mycotoxicosis, normalization and optimization of immuno-biochemical status, intensification of metabolic processes, stress loads, and as a means of increasing animal productivity.

7 Conclusion

1. The possibility of creating Biomineral compositions based on wormwood and siliceous rock - flask of Western Kazakhstan with regard to their chemical-mineralogical and biological properties.

2. When choosing wormwood, its chemical and biological composition was taken into account as a criterion of biological activity that contribute to increased immunity, appetite, improves digestion, and as an antifungal, antimicrobial, anti-inflammatory, wound healing, carminative, sedative.

3. Based on experimental studies, the selection of siliceous rock - flask of the West Kazakhstan region on the criterion of a high content of amorphous and finely divided silica is scientifically substantiated, which gives the following advantages:

- highly dispersed silica, guarantees the effective sorption of mycotoxins, salts of heavy metals, chemical toxins, radionuclides, gases and other metabolic products in the animal body due to the huge sorbing surface;
- the presence of amorphous silicon in the composition of the siliceous rock - flask contributes to the growth and hardening of tissues during the development and formation of the skeleton and is involved in bone mineralization.

4. Developed the most optimal compositions and method for preparing compositions in the system biomineral wormwood - siliceous rocks - flask differing therapeutic and prophylactic action;

5. As a result of scientific and experimental work, it was found that the proposed biomineral composition has high biological activity and significant pharmacological action, which is manifested by the improvement of morphological and biochemical parameters of blood of cows.

References

1. A.A. Akopov, A.V. Abraghyan, *Prevention of metabolic disorders in cows with the help of microelements* **5**, 54 (Moscow: Veterinary Medicine, 1986)
2. T.I. Bokova, Feeding of farm animals and feed production, **9**, 9-10 (2008)
3. M.P. Kirillov, Feeding of farm animals and feed production, **3**, 34-37 (2006)
4. I.L. Zinchenko, I.E. Pogorelova, *Mineral and vitamin nutrition of cows*, 89-167 (Moscow: Kolos, 1980)
5. N.I. Kleimenov, M.Sh. Magomedov, A.M. Venediktov, *Mineral nutrition of livestock on complexes and farms* 190 p. (Moscow: Rosselkhozizdat, 1987)
6. N.I. Kleimenov, A.P. Yaroshkevich, Agricultural Biology **4**, 34-89 (1994)
7. G.S. Avakyan, *Bentonites* 104 p. (Moscow: Science, 1982)
8. Л.A. Matyushevsky, *Results and prospects of scientific research on the problems of animal pathology and the development of means and methods of therapy and prevention* 89-94 (Voronezh, 1995)
9. G.A. Machabeli *Silicon accumulation and bentonites Origin and practical use of siliceous rocks* 176-185 (Moscow: Nauka, 1987)
10. P.J. van der Aar, F. Molist, J. D. van der Klis, The central role of intestinal health on the effect of feed additives on feed intake in swine and poultry. Anim. Feed Sci. Technol., **233**, 647 (2017)
11. G.A. Gouda, H.M. Khattab, M.A. Abdel-Wahhab, S.A. Abo El-Nor, H.M. El-Sayed, S. M. Kholif, Small Ruminant Res. **175**, 15-22 (2019)
12. Maria Alonso-Debolt, *Use of thermally treated clays in animal feeds* (U.S. Patent 5,935,623, 10 Aug. 1999)
13. M. Trckova et al., *Kaolin, bentonite, and zeolites as feed supplements for animals: health advantages and risks* (Veterinarni Medicina-UZPI, Czech Republic, 2004)
14. P.W. Kemp, T.H. Nougher, *Animal feed containing molasses bentonite and zeolite* (US Pat. 5908634, United States, 1999)
15. EFSA Panel on Additives and Products or Substances used in Animal Feed (FEEDAP). Scientific Opinion on the safety and efficacy of bentonite as a technological feed additive for all species, EFSA J., **10** (7), 2787 (2012)
16. Dennis R. Taylor, *Animal feed containing selected montmorillonite clay as additive and method for selecting the clay* (U.S. Patent 5,192,547, 9 Mar. 1993)
17. F.R. Arakelyan, *Scientists of the Yerevan Veterinary Institute* 17-18 (Yerevan: Sat. scientific Tr. 1986)
18. F.G. Akhmedov, A.V. Ivanov, M.Ya. Tremasov, Vet. Med., **2,** 47-50 (2001)
19. G.S. Pulatov, A.D. Ignatiev, V.P. Nelyubin, *Biological properties of zeolites* **35**, 30-33 (Tr. UzNIVI, 1983)
20. L. Oustercoat, Agriculture Abroad, **11,** 15 (1970)
21. I.V. Khvorova, *Siliceous rocks* 163-176 (Moscow: Nauka, 1983)
22. S.G. Lumbunov, K.V. Luzbaev, E.A. Alexandrova, Netradits, natures, resources, innovations technologies and products, **6,** 322-342 (2002)
23. D.G. Kozmanishvili, V.N. Nanobashvili, *Preservation of biologically active substances in premixes based on zeolites* **21,** 81-83 (Tbilisi: Tr. VNIIKP, 1982)
24. E.V. Gromov, Ecological Bulletin of the North Caucasus, **2**, 80-94 (2005)

MATEC Web of Conferences 315, 09001 (2020) https://doi.org/10.1051/matecconf/202031509001
ICMSSTE 2020

The prospects of using the silver nanoparticles composition in sodium alginate matrix

Anatoly Kubyshkin[1], *Olga* Pisareva[1,*], *Yevgenia* Bessalova[1], and *Irina* Fomochkina[1]

[1]V. I. Vernadsky Crimean Federal University, Medical Academy Named after S.I. Georgievsky, 5/7 Lenina boul, Simferopol, 295051, Russia

Abstract. A wide range of anti-inflammatory properties of the solution of nanosilver has found its application in medical practice. In this work, nano-silver was used in the treatment of burn wounds in combination with a proteinase inhibitor. An experimental model of a second-degree burn with the use of a nano-silver solution in combination therapy is accompanied by a decrease in the inhibitory potential at the local level and stimulation of regeneration processes, which indicates the anti-inflammatory effects of nano-silver particles. There is also a decrease in the morphological signs of inflammation in the burned areas of the skin of the therapeutic use of a solution of nano-silver. The results indicate that nano-silver solution for prophylactic and therapeutic use has the ability to block a cascade of inflammatory reactions in the focus of damage, as it is evidenced by a decrease in the degree of alteration and intensification of reparative processes in damaged tissues.

1. Introduction

Since ancient times, the unique properties of silver, which were widely used in medicine, has been known. The study of the biological properties of metal nanoparticles in the context of nanopharmacology, as well as from the standpoint of nanotoxicology is currently the leading priority task of modern areas of science and technology [1, 2]. Modern advances in nanotechnology, including methods for producing nanomaterials, evaluating efficacy and safety, and using them for therapeutic and diagnostic purposes, are becoming one of the priority goals of modern materials science and medicine [3]. In connection with the mentioned above, the development and study of the mechanisms of action of silver nanoparticles compounds for the correction of various human body pathological conditions, including difficulties to treat by standard methods, is of particular interest [4].

According to a number of scientific studies, there is evidence that silver in the nanoscale range may have a pronounced bactericidal, virucidal effect, including having a broad antiseptic effect on pathogenic biota [3, 5]. Due to the high specific surface of silver nanoparticles, the area of contact of silver nanoparticles with the bacterial cell wall increases, due to which an expressed bactericidal effect is manifested [6]. The intensive development of nano-biotechnology allows us to influence the structure of silver, changing it to the nanoscale range

* Corresponding author: lolya.anat@gmail.com

(with particle sizes up to 100 nm), potentiating the biological effects of nanocompositions using silver nanoparticles.

At the same time, one of the urgent problems of modern medicine is timely and effective therapy for burn injuries of the skin; thermal burns occupy one of the leading places among all types of injuries: first of all, the dermis directly suffers, which plays an important role in the protective and immune reactions of the body [7]. An important role in the treatment tactics of burn injuries of the skin from thermal effects is taken by drug therapy, aimed primarily at accelerating reparative processes in damaged tissues [8, 9]. To date, the list of drug supply with antioxidant, anti-inflammatory and reparative properties is quite large [10, 11]. However, none of the standard treatment regimens is universal, and in certain cases there is a high percentage of complications in this category of patients associated both with infectious agents and the scarring, contractures etc.

Drugs in the form of ointments have rather effectively proven themselves in the complex therapy of burn injuries. A promising direction of nanopharmacology is the enhancement of the properties of drugs with the additional introduction of silver nanoparticles into these compositions [12, 13]. Their combined use is relevant in view of the increased antibiotic resistance of microflora due to irrational antibiotic therapy and the recently increased multiresistant strains.

The most common theory of the physicochemical properties of silver nanoparticles regarding to their effect on a bacterial cell is the theory of "adsorption". It is based on the action of electrostatic forces between the cell wall of a microorganism, on which surface there is a negative charge, as well as silver ions, which have a positive charge when they are adsorbed between each other [14, 15]. However, the possibility of using silver nanoparticles requires further preclinical, as well as clinical trials due to insufficiently studied their possible toxic effects on cells, tissues and organs, and body systems [16].

In this regard, the validation of the nanosilver solution developed at Taurida National University named after V.I. Vernadsky with the participation of employees of the Institute of Biology of the Southern Seas is relevant and timely. The composition of the solution includes: silver nanoparticles with a size of 10–20 nm (0.1%), sodium alginate (0.6%) and distilled water (99.3%). Currently, a number of studies have been carried out on the antimicrobial activity of silver nanoparticles, but their stabilization from the standpoint of toxicological safety, as well as the achievement of a satisfactory therapeutic effect with sodium alginate solution, has not been studied sufficiently [17]. In this regard, the **purpose of the study** is to investigate the effects of transdermal application of a solution of nano-silver on the burn skin surface of experimental animals.

2. Materials and methods

A series of experiments was carried out on white Wistar male rats weighing 180–210 g, in accordance with the principles of the European Convention for the Protection of Vertebrate Animals used for experimental or for other scientific purposes (Strasbourg, 1986). Experimental studies were carried out in aseptic conditions of the operating vivarium, the laboratory of the department of general and clinical pathophysiology, the department of pathological anatomy with a sectional course, and the histological laboratory with immunohistochemistry and electron microscopy of the Central Research Laboratory of the Medical Academy named after S.I. Georgievsky (structural unit) V. I. Vernadsky Crimean Federal University. The experiment was approved by the ethics committee (protocol No. 2 of September 11, 2015).

A preliminary series of experimental modeling of a burn allowed us to identify the most appropriate method for obtaining a burn in size and depth corresponding to the II degree. Simulation of a II degree burn was performed after intravenous administration of sodium thiopental, at the rate of 20 mg / kg weight. The burn was reproduced by application to the previously depilated skin area in the interscapular region, bounded by a wooden stencil 1.5 × 1.0 cm in size, by a thin-walled glass container with water at a temperature of 90 ° C for 10 sec [18]. Throughout the experiment, the damaged area of the skin remained open.

Experimental groups of animals were distributed as follows:

Group 1 (K) – intact animals (n=10).

Group 2 (G) – II degree burn model, observation for 7 days (n=10).

Group 3 (Y) – II degree burn model, treatment with antibiotic erythromycin for 7 days (n=10). 0.5 g erythromycin was crushed in a porcelain mortar to a fine powder and mixed with 10 ml of sodium alginate (0.6%). Then 1 ml obtained 10% solution was used for the local applications of a burn wound.

Group 4 (W) – II degree burn model, treatment with combination of nano-silver solution with the protease inhibitor aprotinin for 7 days (n=10).

For the experiment, 3 ml of Gordox (aprotinin) was mixed with 2 ml of the original composition of nano-silver stabilized with sodium alginate (0.1% nano-silver), and then mixed with 5 ml of nano-silver in the form of a gel (0.1% nano-silver and 1.6% high molecular alginate sodium). Thus, the composition contained 30% of the aprotinin, 0.07% nano-silver, and 0.8% of a mixture of low and high molecular weight sodium alginate.

On the 7th day, a set of research methods was used: studies of the activity of nonspecific proteinases and their inhibitors in serum and supernatants of skin homogenates by enzymatic methods [19], macroscopic description, histological examination with hematoxylin and eosin staining, and picrofuchsin by van Gieson, transmission electron microscopy (TEM).

The animals were removed from the experiment under thiopental anesthesia, then euthanasia was performed and samples of the damaged skin area 1 cm x 1 cm x 1 cm were taken, followed by fixation in a 10% solution of neutral formalin to evaluate histological changes. To assess the histological changes at the tissue level, a skin fragment was fixed in 10% neutral formalin solution, followed by standard wiring and paraffin infiltration in a LOGOS processor (Italy) according to the manufacturer's recommendation. The use of a Leica DM2000 microscope (Germany) and a camera made it possible to obtain digital photographs of histological specimens. For the ultrastructural study, the pieces were fixed in a solution of glutaraldehyde, the sample preparation was performed manually, strictly following the protocol, and the PEM-125 transmission electron microscope (Sumy, Ukraine) was used for visualization.

3. The results of the study and their discussion

Modeling a thermal II degree skin burn demonstrates disorders in state of proteinase inhibitory system, both local and systemic level. Among the mediators of biochemical inflammation, non-specific proteinases occupy the main positions due to their destructive influence on the skin structure. Thus, an increase in the level of activity of trypsin-like proteinases in the supernatants of skin homogenates was established at all stages of the study. The dynamics of changes in nonspecific proteinases and their inhibitors looks like a progressive proteolysis on the background of a severe exhaustion of the inhibitory potential, which leads to the development of the inflammatory process. Subsequently, a cascade of inflammatory reactions is initiated, resulting in proteolytic destruction of collagen fibers, histohematogenous barriers, edema and hemorheological changes at the level of microcirculation, including acute decrease and damage of reparative properties in alterized tissues. It is important to monitor the potential of pharmacological correction, as well as the dynamics of changes of protease-inhibitory system at the local level. Functional indicators of proteinase inhibitory system can be considered as a criterion for the effectiveness of the proposed selected nanofarmacotherapy.

Considering the results of histological examination of damaged skin areas, it was found that the main morphological changes are exudative-necrotic foci and impaired hemodynamics. On the 7th day, the burn area is covered with a scab, which begins to partly reject, and at its border with the underlying tissues there is a rather wide zone of demarcation inflammation with a prevalence of neutrophilic leukocytes in the exudate. Inflammatory infiltration extends to the deep layers of the dermis, in some areas penetrating into the subcutaneous fat. Plots of granulation tissue begin to appear at the bottom and edges of the burn wound. Hemodynamic disorders are presented in the form of pronounced plethora, edema, small hemorrhages. There is a slight increase in the manifestations of edema in the deeper parts of the dermis (Fig. 1A).

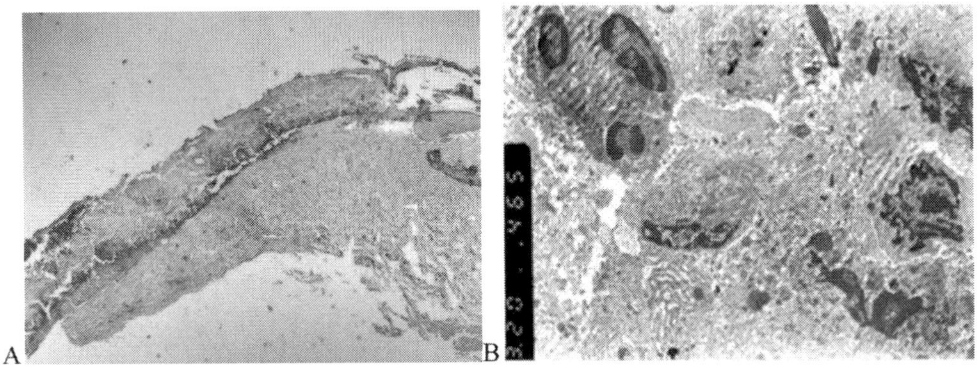

Fig. 1. A.Thermal burn. Paraffin sections, hematoxylin and eosin stain. 7th day, 40x. B. Ultrathin cuts. Transmission electron 7th day. 5800x.

According to electron microscopy, edema, leukocyte-macrophage infiltration and activation of cells of the fibroblastic differentia are the most demonstrative (Fig. 1B). On the 7th day, the damaged areas of the skin are infiltrated with lymphocytes, actively phagocytic macrophages and degranulating neutrophilic granulocytes. Multiple invaginations in lymphocyte nuclei, central arrangement of the nucleolus and marginal heterochromatin are signs of their active state. Fibroblasts are few, have an elongated shape, small processes, and revealed a large number of free ribosomes in their cytoplasm. Intracellular edema and granular dystrophy are characteristic of most cells in the dermis. Expansions of the granular endoplasmic reticulum take the form of bag-like elongated cavities separated by thin sections of the cytoplasm with fine-grained contents. Mitochondria of irregular polygonal shape, densely filled with laminar cristae, have a dark matrix. In the extracellular space, a small amount of collagen fibrils is noted, amorphous substance prevails.

In groups with drug correction, the following pathological changes occurred: on the 7th day the scab looks uneven, there are areas of partial scab rejection with the formation of blisters containing exudate. Granulation tissue with multiple fibrils and active fibroblasts significantly developed. Demarcation inflammation is represented by diffuse focal lymphatic macrophage infiltration spreading to the underlying tissues. On the periphery of the burn zone and in the tissues of the dermis there are signs of spasm, edema, plethora, a small amount of hemorrhages. The epithelium at the edges of the wound is hyperplastic, with symptoms of acanthosis, hyper- and parakeratosis. The growth of granulation tissue and the formation of young connective tissue are noted (Fig. 2A).

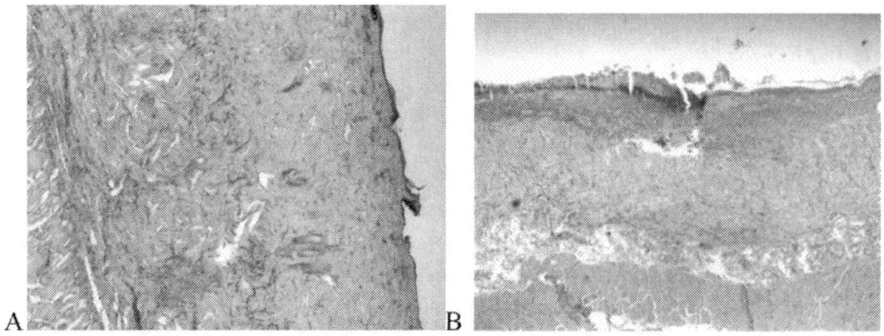

Fig. 2. Fragment of the damaged skin area with the use of an antibiotic 40x. B.Fragment of the damaged skin area using nanofarmacocorrection 100x. Paraffin cut, stained with hematoxylin and eosin.

In the group receiving the combined nanofarmacocorrection (nano-silver + aprotinin), on the 7th day there is a thinning and rejection of the scab over large areas, epithelialization of the wound surface is pronounced to a large extent and spreads almost to the center of the damaged skin area (Fig. 2B). A large amount of granulation tissue with fine connective tissue fibers was detected in the visual fields. Instead of a random arrangement of fibers, their predominantly horizontal direction is noted, parallel to the surface of the epidermis. The layer of granulation tissue in the center is thicker, only as it approaches the edges it becomes thinner. Hemodynamic changes are reduced, as well as inflammatory infiltration in the damaged area.

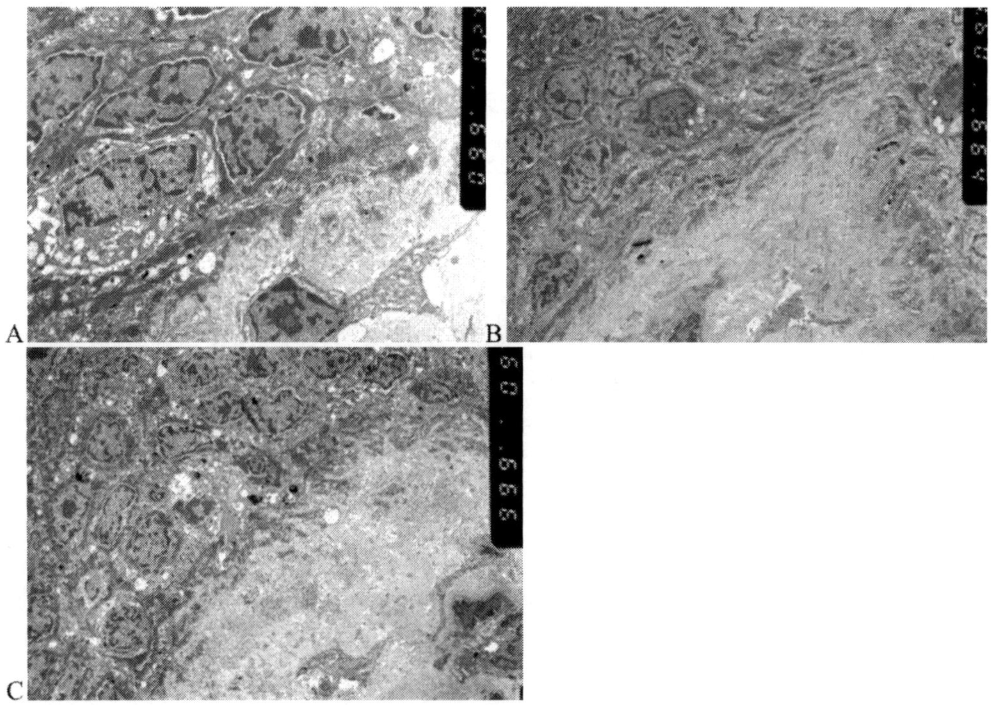

Fig. 3. A. Fragments of the epidermis and dermis of the skin in case of burn correction with erythromycin solution for 7 days. TEM. 5000x. Fragments of the epidermis and dermis (A) and the area of pronounced proliferation of the epidermis (B) of the skin during the drug correction of the burn by the combined method (nano-silver + aprotinin) for 7 days. TEM. 4000x.

When conducting electron microscopy, it revealed a similar dynamics of changes at the ultrastructural level. With the correction of burn with erythromycin, there is a good preservation of all skin structures (Fig. 3A). Kernels of keratinocytes have shallow invaginations of the karyolemma. The perinuclear space of most cells is slightly expanded and continues into the cisterns of the granular endoplasmic reticulum (GER) surrounding the nucleus. Ribosomes, polysomes, and also fragments of GER cisterns are located in the perinuclear zone of the cytoplasm. A large number of mitochondria, small, most of them contain a dark matrix and a large number of cristae. In some fields of view there are cells that have a blade nucleus, small and medium-sized vacuoles, formed as a result of the expansion of the GER cistern and the Golgi complex (Fig. 3A). It was established that the revealed structure of keratinocytes reflects the active synthetic activity of cells. The capillaries of the dermis have a thickened basement membrane, endotheliocytes are intact. A thicker layer of connective tissue is formed by circular fibers forms around the vessels (Fig. 3A). Some cells have a normal ultrastructure of the nucleus and cytoplasmic organelles. In another part of the fibroblasts, hydropic changes were noted with the formation of vacuoles as a result of the swelling of mitochondria, the expansion of the GER cistern and the Golgi complex.

In the experimental group with the use of combined treatment using nanosilver, the best preservation of the skin structures was found: keratinocytes, fibroblasts, dermis vessels. Intercellular contacts and complexes of tonofilaments with desmosomes are restored (Fig. 6 B). The degree of vacuolization is reduced, but retained (Fig. 6 C). In keratinocytes, the ultrastructure is not changed (Fig. 6 B). Kernels contain numerous invaginations. Euchromatin is predominant, and heterochromatin in the same amount is represented by small-size karyosomes and marginal heterochromatin. In the hyaloplasm, a significant number of polisomes are found. The cytoplasm of cells contains a huge number of mitochondria with a preserved structure, a dark matrix, a large number of cristae. In the growth layers of the epidermis, there are single young cells with a large lobar nucleus, moderate depletion of organelles (Fig. 6C). The appearance of such keratinocytes can be considered as a result of the activation of reparative processes. In the papillary layer of the dermis there are groups of differentiated fibroblasts with signs of the functional activity of synthetic processes.

Thus, the conducted studies allowed demonstrating a number of positive properties of nanofarmacotherapy in the treatment of burns of the skin, in more detail the properties of silver nanoparticles are shown in the diagram (Fig. 4).

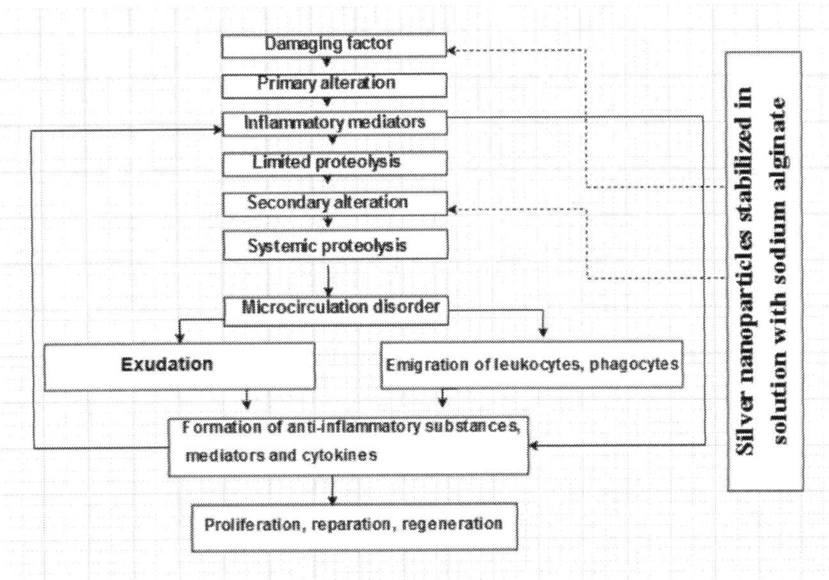

Fig. 4. Pathogenetic scheme of pharmacological use of silver nanoparticles in the development of the inflammatory process.

4. Conclusion

In this work, it is proved that the imbalance in the system of nonspecific proteases and their inhibitors at the local level is characterized by activation of proteases against the background of depletion of the inhibitory potential, which indicates the pathogenetic expediency of using drugs of the protease inhibitor group as a local treatment of thermal burn skin damage. Along with the use of the proteinase inhibitor Aprotinin, we should not forget about the important pathogenetic significance of the nano-silver particle solution, since it has a number of properties necessary for treating thermal skin burns: antimicrobial activity (reduction of infectious complications) does not reduce the processes of local immune reactivity, accelerates tissue regeneration processes. That is why the most effective combination for experimental correction is a combination in the form of applications with a solution of nano-silver and aprotinin.

Based on the above experimental data on animal models (white rats-males of the Wistar line), the efficacy and feasibility of using combined drugs based on silver nanoparticles in the presence of a wide spectrum of bactericidal, bacteriostatic, virulocidal, fungicidal, antiseptic, including the number of activators of reparative processes of damaged areas of the skin after thermal effects of extreme temperatures. Thus, thanks to the achievements of nanotechnology, in particular nanopharmacology and nanotoxicology, one of the priority directions of modern medicine and nanocomposite production technology, one of the leading directions at present is the development, study of the effectiveness and safety of the application of the above-mentioned advanced treatment methods. This therapy will reduce and adjust the qualitative and quantitative composition of treatment regimens, optimize, reduce the risk of the likelihood of side effects, prevent and prevent the development of antibiotic resistance and the emergence of new multiresistant microorganism strains, reduce the number of complications and adverse outcomes in patients with this pathology.

References

1. E. M. Egorova, A. A. Kubatiev, V. I. Shvec, The science, .**6** (74), p.350 (2014)
2. M. V. Taran, N. F. Starodub, A. M. Katsev et al., Biophotonics, **9032**, 1-7 (2013)
3. I. S. Chekman, B. A. Movchan, M. I. Zagorodnyj, The art of healing, **51(5)**, 32-34 (2008)
4. T. M. Bojchuk, N. J. Andrijchuk, L. I. Vlasyk, Clinical and experimental pathology, **4(42)**, 151-157, (2012)
5. M. Amin, F. Anwar, M. Ramzan Saeed Ashraf Janjua et al., Int. J. Mol. Sci., **13**, 9923-9941 (2012)
6. A. I. Gusev, Fizmatlit, **2**, 414 (2007)
7. A. N. Guillory, R. P. Clayton, D. N. Herndon, Int. J. of Molecular Sciences, **17**, 53-53 (2016)
8. A. J. Singer, R. A. Clark, The New England Journal of Medicine, **3(41),** 738-746 (1999)
9. L. K. Branski, A. Al-Mousawi, H. Rivero, M. G. Jeschke, A. P. Sanford, D. N. Herndon, Surgical Infections(Larchmt), **10(5),** 389-397 (2009)
10. A. A. Alekseev, A. E. Bobrovnikov, S. B. Bogdanov, L. I. Budkevich, M. G. Krutikov, YU.I. Tyurnikov, All-Russian public organization "Association of Combustiologists". "A world without burns", 1-12 (2015)
11. V. Singh, L. Devgan, S. Bhat, S. M. Milner, Annals of Plastic Surgery, **59**, 109-115 (2007)
12. T. V. Surakova, N. Yu. Zhidomorov, T. R.Grishina, A. A. Kodin, Russian medical Journal, **20(11)**, 575-581 (2012)
13. I. N. Yurkova, D. A. Panov, V. I. Ryabushko, Scientific notes of Taurida National University. V. I. Vernadsky, Series "Biology, Chemistry", **22 (61)**, 203-207 (2009)

14. E. G. Akopova, A. H. Kade, E. F. Kurnosenkova, Kuban Scientific Medical Bulletin, **1**, 8-11 (2007)
15. E. K. Baranova, A. A Revina, L. I. Vojno, Materials of the 1st Russian Scientific and Methodological Seminar, **1**, 53-60 (2003)
16. V. F. Moskalenko, L. G. Rozenfel'd, B. O. Movchan, Human and medicine. Materials of the I National Congress, **1,** 167-168 (2008)
17. G. V. Odegova, V. A. Burmistrov, P. P. Rodionov, The use of silver drugs in medicine, **5**, 53 – 60 (2004)
18. A. A. Mironov, Ya.Yu. Komissarchik, V.A. Mironov, The science J., 990, (1994)
19. A. V. Kubyshkin, V. Z. Kharchenko, P. F. Semenets L. L. Aliev, I. I. Fomochkina, L. V. Anisimova, *Methods of determining the activity of non-specific proteinases and their inhibitors in blood and biological liquids* (Methodical recommendation, Kiev, 2010)

Influence of pH liquid for mixing gypsum binders on the strength of gypsum composites

Victoria Petropavlovskaya[1], *Maria* Zavadko[1], and *Tatiana* Novichenkova[1,*]

[1]TSTU, Production of building materials and structures Department, 170026, Tver, Russia

Abstract. The influence of fluid acidity on the processes of hardness, mechanism, kinetics, properties of gypsum stone are of great interest.Studies of these processes for modified composites based on gypsum and heavy high density and basalt additives can improve the modifier's effectiveness. The influence of the pH of a liquid on the processes of structure formation and the properties of composites is determined in this work. It is shown that the pH should be shared by other structure-forming factors. This is the chemical composition of basalt waste, its percentage, the interaction of its oxides with pore fluid and natural minerals are the products of their reaction. The regulation of the material composition of composites can enhance the properties and also lead to degradation of structures.

1 Introduction

From the point of view of the effectiveness of the application, composite materials are significantly superior to conventional materials; therefore, it is precisely their development that is currently being given special attention. Modification of the gypsum binder and the creation of gypsum composite materials allowed at the time to expand its scope and range of manufactured materials and products based on it to impressive volumes: plaster, putty, gypsum board, blocks, plates, screeds. Plaster composite materials have a number of advantages for the consumer, regardless of their purpose and nomenclature. This is fire resistance, chemical resistance, the ability to regulate the humidity in the room, the ability to provide the necessary level of noise and heat insulation. Moreover, such materials are unique not only in terms of environmental safety and creating the optimal microclimate in the premises, but also in terms of their production [1, 2]. Their use in the decoration of residential premises also helps to increase the level of comfort and the ability to implement almost any design decisions.

Together with a gypsum binder, various types of other mineral binders, fillers, chemical additives affecting the setting time, ductility, and strength can be used to obtain composite materials with enhanced performance characteristics. It should be noted that special interest is currently shown in the use of waste as fillers, as well as chemical additives. The prospects for composites thus obtained are also due to the saving of expensive raw materials by replacing parts of it, recycling into actual production waste that is currently

*Corresponding author: tanovi.69@mail.ru

disposed at landfills [3-5]. So, the authors of [6] studied the possibility of obtaining a composite material - a modified gypsum stone with increased water resistance due to the introduction of various industrial wastes into its composition and proved the effectiveness of this approach.

Improving the characteristics of composite gypsum materials still remains an urgent topic, due to the low strength and water resistance of the gypsum binder. Such work is primarily associated with work on the structure of the resulting composite. The authors of [7–9] note that controlling the process of structuring composite gypsum materials is one of the tasks of modern science. The control of this process is the control of a whole complex of factors: the structure of the gypsum and filler crystals [10], their shape, packing order and size. Optimization of the internal structure of the composite at all levels makes it possible to obtain the maximum number of particle contacts, the densest packing, and, consequently, an increase in strength characteristics [11]. The author of [12] also notes the same, according to the results of studies of the influence of fillers of various structures on gypsum composites.

As noted above, the design of composite gypsum materials takes into account a number of parameters [13)], which are often regulated by the requirements and rules for each type of material obtained [14]. However, not all parameters of composite gypsum materials are regulated, since they do not directly affect the properties of materials, but indirectly, changing other parameters. Among these is the acidity of the mixing solutions, which directly affects the structure formation processes of gypsum stone (crystal distribution, their morphology) and, as a result, the strength indicators, their growth rate, and the quality of the materials obtained [15]. It should be noted that the effect of the pH of the mixing liquid on composites containing anhydrite, α and β modifications of hemihydrates is not the same, due to the different mechanism of their hydration. As a rule, the pH region with minimal activity of calcium ions corresponds to high solubility and increased strength of binders [16-17].

At the moment, the effect of the pH of the mixing fluid on the hardening process, its mechanism and speed, and the strength of the resulting gypsum composite matrix is still not sufficiently studied for applied applications and is of great interest [18–20].

In the framework of this work, the effect of pulverized waste from the production of basalt fibers and lime on the strength of the obtained modified gypsum stone was evaluated in terms of the pH of the mixing fluid of calcium sulfate hemihydrate. Taking into account the fact that the pH value of gypsum solutions is not constant, it constantly changes due to the processes that occur during hydration and structure formation to identify the effect of pH on the hardening processes of gypsum stone. Initially, aqueous solutions of basalt waste and lime were obtained, which were then used to mix gypsum binder. In addition, the pH of pure gypsum mortars is associated with the presence of impurities in gypsum, so conducting research using the described methodology is the most correct.

2 Materials and methods

As a component in the work, we used gypsum binder β-modifications of the G-5 brand with a water demand of 60-70%, setting time - not earlier than 7-8 minutes, setting point - not later than 13 minutes, compressive strength - 5 MPa, bending - 2.0-2.5 MPa. Also, the components used were waste products of basalt fiber production (dust from dust filters) whose chemical composition is presented in Table 1. Slaked lime was characterized by the content of active CaO + MgO - 69%, CaO - 9%.

The content in% by weight of waste from the production of basalt fibers and slaked lime was taken as variable parameters. The waste content varied from 0 to 15% in increments of

7.5, the lime content from 0 to 5% in increments of 2.5. Consistencies were accepted as standard and amounted to 70%.

To determine the normal density of the gypsum dough, a clean container was filled with the required amount of distilled water, a weighed portion of the dry mixture in the amount of 300 g was poured into it. The dough was mixed with a hand mixer until it became homogeneous and poured into a Suttard viscometer. After 15 seconds from the moment of filling the test, the viscometer rose vertically. If the spreading diameter of the formed gypsum pellet was 180 ± 5 mm, the value of the water-solid ratio was taken as normal density, otherwise the tests were repeated.

Studies of the basic physical and mechanical properties of gypsum composites were carried out in accordance with GOST 23789 79, GOST 8462 - 85, GOST 7025 - 91. To prepare gypsum dough and a mixture based on it, the required amount of a portion of the waste from the production of basalt fibers and / or lime was measured, mixed into for 3 seconds, closed with distilled water, measured pH and the same solution closed the dry mixture of pre-weighed gypsum binder. Samples of 20x20x20 mm in size were prepared from the solution prepared in this way. Samples were injection molded.

Table 1. The chemical composition of the waste production of basalt fibers.

Element(Oxide)	**%**
SiO_2	41.13
MgO	14.66
CaO	13.81
Fe_2O_3	7.34
Na_2O	6.59
Cl	4.29
K_2O	4.14
Al_2O_2	3.99
S	1.57
ZnO	1.11
MnO	0.428
F	0.320
TiO_2	0.262
P	0.191
CuO	0.0904
SrO	0.0323
Cr_2O_3	0.0207
Co_3O_4	0.0117
NiO	0.0106

3 Results and discussion

The studies found that the hydrogen indicator of distilled water from 7 rises to 9 when managing waste production of basalt fibers in an amount of 7.5% and up to 10 with an amount of 15%. With the introduction of lime (2.5%), together with the waste (7.5%), the pH increases to 13 (Fig. 1).

On the 7th day of hardening of the samples, samples containing 2.5% of lime in the composition gained maximum compressive strength, the pH being 12. The strength of samples containing 2.5% of lime and 7.5% of the waste from basalt fiber production exceeded the strength of control samples by 10% (Fig. 2,3), the pH was 13.

Thus, the optimum pH of the gypsum binder mixing fluid is between 12-13. Apparently, an increase in pH to these parameters leads to a sharp increase in the activity of Ca2 + ions, an increase in solubility, and an increase in the strength of the composite gypsum stone.

However, it is not correct to consider the processes that take place during hardening of a composite gypsum binder system containing lime as an alkaline component, only from the influence of pH, because lime in clinker-free composite gypsum binders in the long-term hardening causes the formation of ettringite. It was his formation that caused the decrease in strength on the 28th day of hardening of the resulting stone due to the occurrence of high internal deformations (Fig. 2, 4).

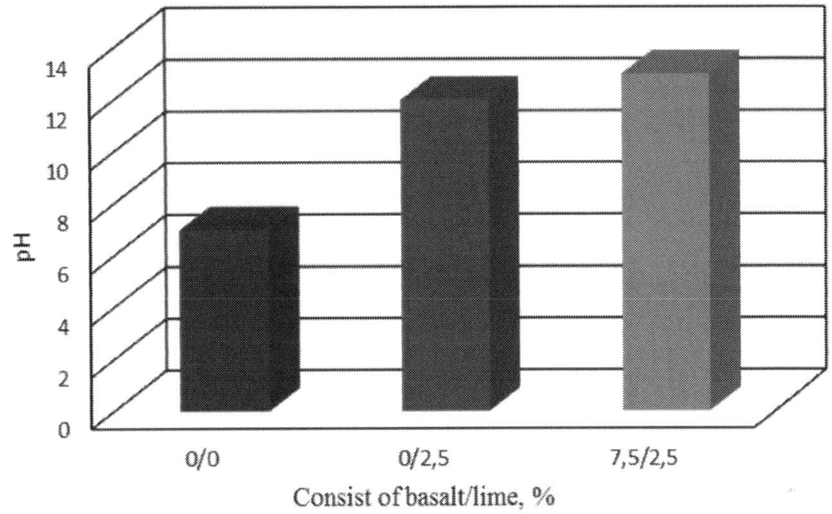

Fig. 1. Effect of lime content and waste of basalt fiber production on pH of solutions.

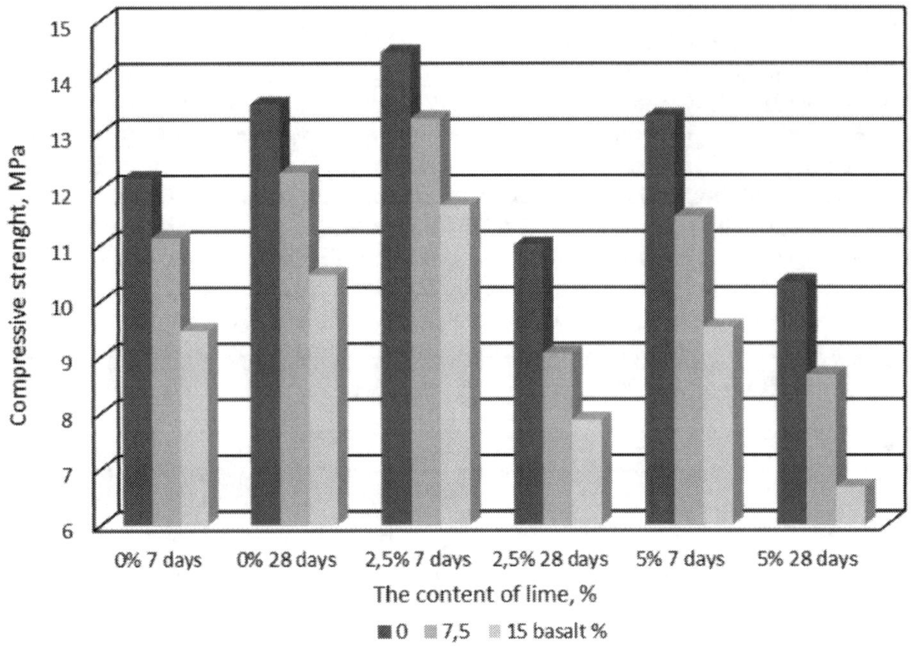

Fig. 2. Effect of lime content and waste of basalt fiber production on the strength of gypsum composites on days 7 and 28.

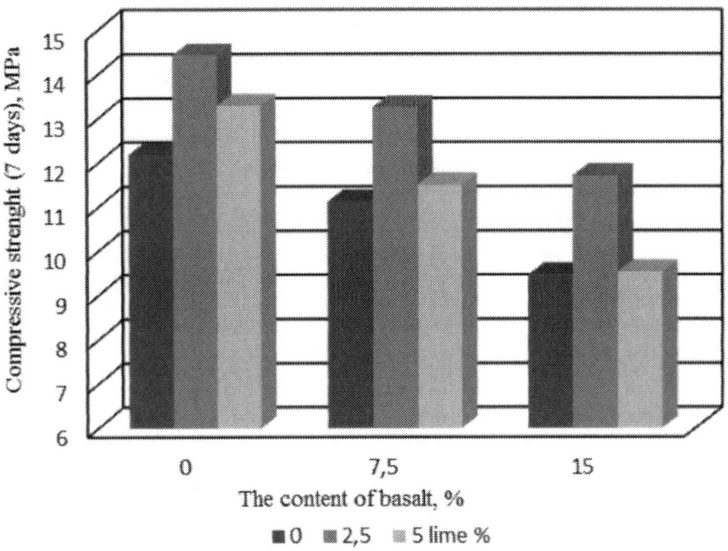

Fig.3.Effect of lime content and waste of basalt fiber production on the strength of gypsum composites on the 7th day of hardening.

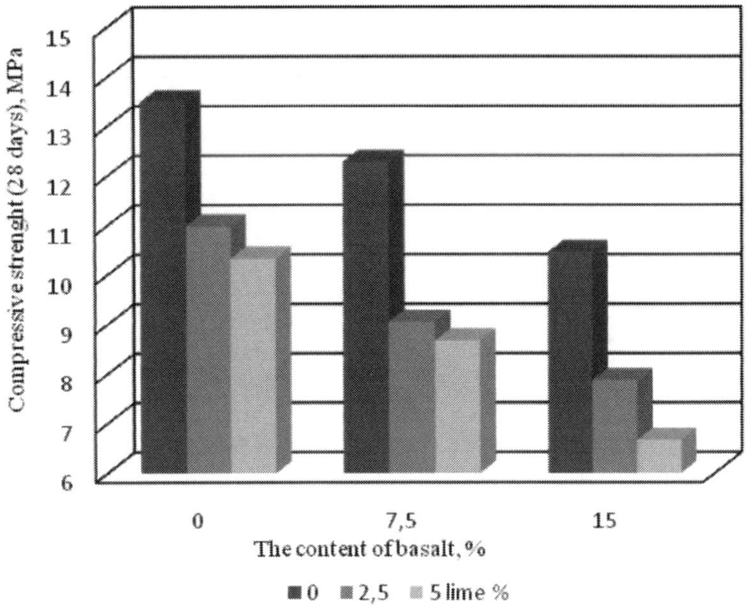

Fig. 4. Effect of lime content and waste of basalt fiber production on the strength of gypsum composites on day 28 of hardening.

4 Conclusion

Thus, the pH of the gypsum mixing fluids affects the strength properties of composite gypsum. Increasing the pH to 12-13 leads to an increase in solubility and an increase in the strength of gypsum stone. However, the use of lime as an alkaline component in gypsum composite systems is impractical due to the formation of ettrigite, which was observed on day 28 of hardening.

References

1. Hassan Z. Harraz, Conference: Geology Department, Faculty of Science, TantaUniversity (2019). DOI: 10.13140/RG.2.2.34749.41446
2. A.S. Edamenko, L.Y. Matveeva, A.V. Yastrebinskaya, Solid State Phenomena, **299**, 1086-1090 (2020). DOI:10.4028/www.scientific.net/SSP.299.1086
3. P.V. Sáez, M. del Río Merino, M. Sorrentino, C. P. Amores, J.S. Cruz Astorqui, C. V. Arrebola, Materials, **13 (1)**, 193 (2020). DOI: 10.3390/ma13010193
4. A.San-Antonio-Gonzalez, M. del Río Merino, C. Viñas, P. VilloriaSáez, International Conference on Construction Research Housing: Past, Present and Future (2013)
5. Q.Wang, R. Jia, Construction and Building Materials, **226**, 11-20 (2019). DOI: 10.1016/j.conbuildmat.2019.07.289
6. A.A. Khalil, A. Tawfik, A. Hegazy, Construction and Building Materials, **167**, 55-64 (2018)
7. V.G. Klimenko, Journal of Physics: Conf. Series, **1118** (2018). DOI: 10.1088/1742-6596/1118/1/012019

8. V.B. Petropavlovskaya, N.P. Bardov, V.V. Matveychuk, High-tech technologies and innovations, pp. 325–330 (2019). DOI: 10.12737/conferencearticle_5cecedc3ad41d4.31814792
9. A.F. Buryanov, V.B. Petropavlovskaya, T.B. Novichenkova K.S. Petropavlovskii, Journal of Physics: Conf. Series, **1425**, (2019) DOI: 10.1088/1742-6596/1425/1/012194
10. J. Dewanckele, W. De Boever, T. De Kock, M.A. Boone, M.N. Boone, L. Brabant, G. Fronteau, L.VanHoorebeke, V. Cnudde, P. Jacobs, Conference: workshop crystallization in porous media, At CrysPoM III (2012)
11. S.T. Sargut, P. Sayan, B. Kiran, Chemical Engineering & Technology, **33(5)**, 804-811 (2010). DOI: 10.1002/ceat.200900567
12. M. Dolezelova, J. Krejsova, A. Vimmrova, International Multidisciplinary Scientific GeoConference-SGEM, **19**, 363-370 (2019). DOI: 10.5593/sgem2019V/6.3/S09.047
13. K. Liu, Liwu Mo, Min Deng, Advances in Cement Research, **2 (8)**, 487–493 (2015). DOI: 10.1680/adcr.14.00076
14. J. de Brito, I. Flores-Colen, Materials for Construction and Civil Engineering, pp.123-184 (2015). DOI: 10.1007/978-3-319-08236-3_3
15. V.B. Petropavlovskaya, T.B. Novichenkova K.S. Petropavlovskii, A.F. Buryanov, MATEC Web of Conferences, **298**, (2019). DOI: 10.1051/matecconf/201929800129
16. J. Qingxu, Lakesha N. Perry, Jeffrey W. Bullard, Cement and Concrete Research, **129** (2020). DOI: 10.1016/j.cemconres.2019.105969
17. N. Singh, B. Middendorf, Progress in Crystal Growth and Characterization of Materials, **53 (1)**, (2007). DOI: 10.1016/j.pcrysgrow.2007.01.002
18. K. Kayabali, T. Beyaz, E. Kolay, Bulletin of Engineering Geology and the Environment, **65**, 65-71(2006). DOI: 10.1007/s10064-005-0027-9
19. M–B. Kodzoev, S. Isachenko, S. Kosarev, A. Basova, A. Skvortzov, M. Asamatdinov, A. Zhukov, MATEC Web of Conferences, **170** (2018). DOI: 10.1051/matecconf/201817003022
20. A.Grishina, E.V. Korolev, E3S Web of Conferences **91**, 02016 (2019). DOI: 10.1051/e3sconf/20199102016

Analysis of radial-shear rolling process parameters of aluminum alloys based on FEM modeling

Yury Gamin[1*], *Alexander* Koshmin[1] and *Dinh Xuan* Ta[1]

[1]Department of Metal Forming, National University of Science and Technology "MISIS", 119049, 4 Leninskiy prospect, Moscow, Russia

Abstract. This article presents the features of temperature change at different deformation-velocity parameters of radial-shear rolling process. Based on modeling the maps of temperature and strain rates distribution and zone of intensive heating are obtained. It is shown that with the elongation ratio $\mu \geq 1.5$, the deformation heating increases more intensively at lower heating temperatures of the workpiece. The rods are obtained from 1050A alloy (Al99.5 %) with mechanical properties (UTS ~ 111 MPa, YS ~ 109 MPa, δ ~ 15%), reflecting the partially recrystallized structure.

1 Introduction

Aluminum alloys are still one of the main constructional materials in aircraft industry, electrical engineering and other industries due to their combination of properties [1-3]. Modern trends of researchers' activities are connected, first of all, with the creation of new materials by change in chemical composition [4-6], improvement or search for new methods of processing materials [7-10]. In the field of materials science and metal forming in the last decades much attention is given to methods of severe plastic deformation (SPD) which allow significantly to increase the mechanical properties of the material due to ultra-fine grained (UFG) or nanostructures [11, 12]. At the same time, the main problem in application of SPD methods in industry is the inability to obtain semi-finished or large products. Thus, the urgent issue is development and studying of processing material methods which allow obtaining the product of acceptable sizes with high level of mechanical properties that are in demand in modern industry.

Radial-shear rolling (RSR) is one of methods for producing of long rods from different steels and alloys. [13-15]. Due to macro-shear deformation, the deep working of material structure takes place in deformation zone. [16-18]. Due to special calibration and large feed angle ($\beta \geq 18°$) the radial deformation (reduction) of workpiece takes place and the cross-sectional nonuniformity of deformation is reduced [19].

Formation of properties during deformation treatment depends on many factors, such as chemical composition of initial workpiece, temperature, strain rate and degree of

* Corresponding author: gamin910@gmail.com

deformation, deformation scheme, etc. Moreover, the influence degree of these parameters depending on alloy can vary greatly.

Based on the aforesaid the urgent scientific task is to study the characteristics of SRS process as method for obtaining semi-finished products from aluminum alloys To solve this task the combination of experimental study and FEM modeling is proposed [20-23], which makes possible to reduce the volume of laboratory experiments and study in detail the main parameters affected on formation of properties.

2 Experimental materials and Methods

Depending on deformation temperature it is possible to obtain the different combination of microstructures that will define mechanical properties of material. As it is known, many aluminum alloys have a tendency to intensive deformation heating. Heat effect at screw rolling is quite significant and depends on number of deformation cycle and value of reduction by each roll, for this reason in the study the analysis of temperature change is carried out at different elongation ratio and strain rates.

2.1 Experimental rolling

The cylindrical ingot with diameter of 60 mm from commercial-purity aluminum 1050A (Al99.5 %) was used as initial workpiece. The workpieces were rolled in 5 passes to a diameter of 14 mm with the total elongation ratio μ=18.4. Regimes of deformation are presented in table 1.

Table 1. Regimes of deformation.

Pass Number, i	Diameter of workpiece D_{i-1}, mm	Diameter of the resulting rod D_i, mm	Elongation ratio μ_i
1	60	42	2.04
2	42	31	1.84
3	31	24	1.67
4	24	17	1.99
5	17	14	1.47

The RSR mini-mill with work rolls with diameter of 90 mm which are turned by the feed angle of β=20° and toe angle δ=10° was used for rolling (Fig. 1).

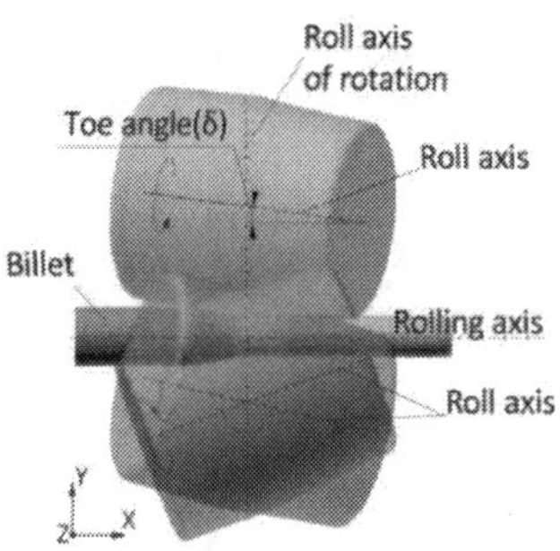

(a)

(b)

Fig. 1. Principal scheme of RSR (a) and RSR mini-mill (b).

Before rolling, the workpieces were preheated to a temperature of T_0=200 °C and 250 °C. Rotary velocity of work rolls was set through the controlling system of main drive to 30 rpm.

As a result, two batches of rods at different temperatures were obtained. Room-temperature tension tests were conducted for as-processed bar specimens with a universal testing machine, model Zwick Z250 (the loading rate was 10 mm/min). According to test results the ultimate tensile stress (UTS), yield stress (YS) and elongation (δ) were obtained.

2.2 FEM simulation

Modeling of RSR process was carried out using software package QFORM 3D. To determine the dependences of temperature changes in the deformation zone during the RSR process, the workpieces were deformed at different elongation ratio μ=1.05; 1.1; 1.2; 1.6; 2.0; 2.4 (T_0=200 °C; n=30 rpm) and rotation velocity of work rolls n=10; 30; 60; 90 rpm (T_0=200 °C; μ=2.0). The main model parameters required for modeling are given in Table 2.

Table 2. Model parameters required for modeling of RSR process.

Parameter	Value
Toe angle δ (°)	10
Feed angle β (°)	20
Material of the roll	40Cr
Temperature of the roll (°C)	25
Ambient temperature (°C)	25
Coefficient of heat transfer between the material and the tool α (W/K·m^2)	30000
Coefficient of heat transfer between the material and the air α_{air} (W/K·m^2)	30

Temperature change of work tool and workpiece during deformation was calculated over total volume of contacted bodies in accordance with equation of transient heat conduction with account for heat transfer in environment on free surface:

$$q_n = b\alpha(T_1 - T_2), \quad (1)$$

where q_n - heat-transfer rate through heat transfer surface, W/m^2;
α – the heat transfer coefficient (W/K·m^2), which takes into account the complex of coefficients of heat transfer between workpiece and lubricant and between lubricant and tool;
T_1 - the workpiece temperature, K;
T_2 - the tool temperature, K;
$b=0.05$ – the pause coefficient which shows how many times it is necessary to reduce a heat transfer rate in the absence of tight contact between the workpiece and tool (without workpiece deformation).

3 Results and discussion

Based on modeling results the temperature distribution maps in the gorge in depending on different elongation ratio μ and velocity rotation of work rolls n were constructed (Fig. 2).

At small value of elongation ratio (μ=1.05-1.1) there is not deformation heating. In zones of contact with the rolls, the decrease in temperature is caused by a lower tool temperature. Beginning with μ=1.2 the surface heating of the rod is observed, which increases with growing of radial reduction of rod.

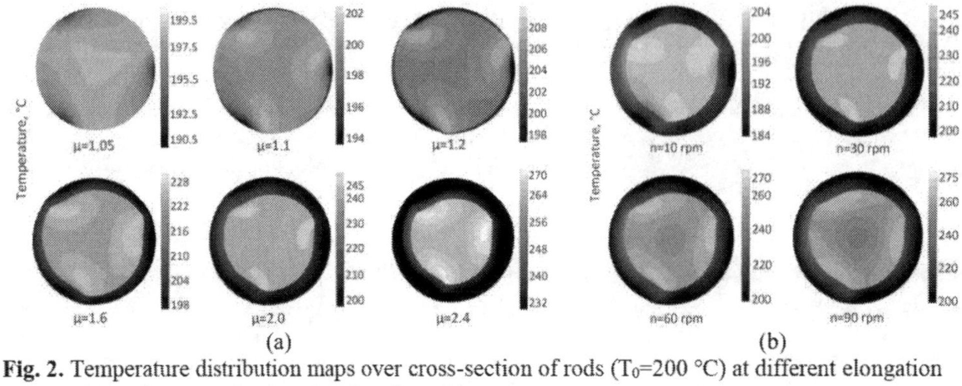

Fig. 2. Temperature distribution maps over cross-section of rods (T_0=200 °C) at different elongation ration μ (a) and rotary velocity of work rolls n (b).

The most heating during RSR process can be observed in the near-surface layers of rod in contact zones with tool where deformation is located. This is also confirmed by the strain rate distribution maps over cross-section of the rod in the deformation zone (Fig. 3).

The change of temperature fields in depending on rotation velocity of work rolls is shown in Fig. 2 (b). With increasing in the rotation velocity of work rolls the temperature gradient also increases, which is due to more intensive heating of near-surface layers. At rotation velocity of 10 rpm the minimal temperature of rod is observed on surface of contact with tool, and total difference of temperatures is no more than 20 °C. At rotation velocity n=60 и 90 rpm the increasing in temperature of surface layer is grown to 70…75 °C, while temperature of center does not increase by more than 40 °C.

Determination of deformation zones and distribution nonuniformity of their influence can be observed by using values of strain rate (Fig. 3). In the deformation zone with increasing in reduction, the strain rate increases on the contact surfaces of rod with workpiece. Moreover, the values of strain rates in the central zone of workpiece do not practically change.

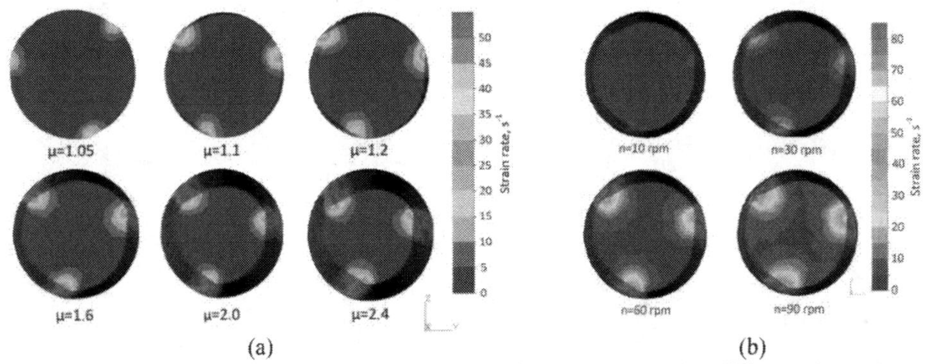

Fig. 3. Strain rate distribution maps over cross-section of rods (T_0=200 °C) at different elongation ration μ (a) and rotary velocity of work rolls n (b).

The change in rotary velocity has significant effect on distribution of strain rates over cross-section of rod (Fig. 3, b). It can be observed that at rolling with rotary velocities n=60 and 90 rpm the strain rate in peripheral layer is an order of magnitude greater than in the center.

The formation processes of structure and properties by workhardening and dynamic recrystallization occurring during the deformation process are, first of all, associated with the deformation scheme, temperature and strain rate. The distribution data of these parameters which are obtained based on modeling, allow understanding the features of the zonal heterogeneity of the structure. For opportunity to evaluate obtained temperature distribution maps and to determine connection with formation of properties the graph of temperature changing ΔT at the exit of deformation zone as a function of reduction per pass was plotted (Fig. 4).

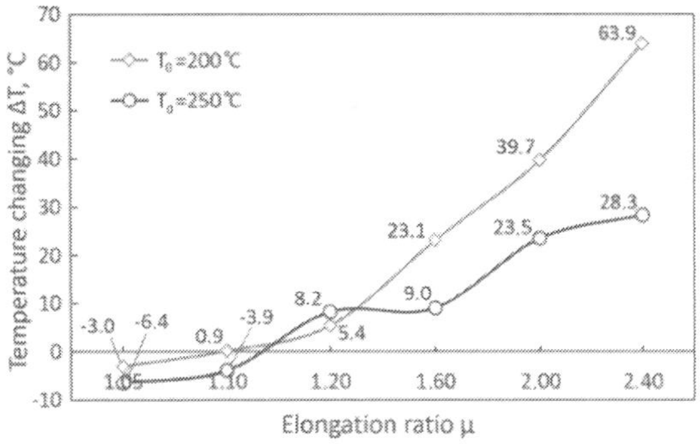

Fig. 4. The dependence of temperature changing ΔT on elongation ratio μ.

The temperature changing ΔT was calculated as difference between T_0 and average value of rod temperature at three points: in the center, middle of the radius, and near the surface. At small value of elongation ratio ($\mu \leq 1.2$) the difference in temperature heating for rods heated to 200 and 250 °C is insignificant. The heat effect is almost completely compensated by heat exchange with work rolls. The increasing in elongation ratio of more than 1.2 leads to temperature growth after deformation, moreover the heating of workpiece at T_0=200 °C is greater than at T_0=250 °C. Since the recrystallization temperature for commercial-purity aluminum is 250 °C according to obtained data on temperature changing and strain rates after RSR process it is possible to assume that in near-surface layers where temperature heating exceed this value the recrystallized structure will be formed, and in the central zone the deformed ones will be developed. This assumption is corresponded with the results of measuring the mechanical properties of the obtained rods (Table 3).

Table 3. Mechanical properties of A1050 alloy after RSR.

Heating temperature T_0 [°C]	YS [MPa]	UTS [MPa]	δ [%]
200	110	115	1,0
250	109	111	15

In accordance with the selected rolling regimes (Table 1), the maximum elongation ratio per pass does not exceed 2. For workpieces heated to T_0=200 °C, the temperature increase is ~40 °C, which does not exceed the temperature of the onset of recrystallization. These rods have high strength and low ductility that corresponding to properties of work-hardened metal. Upon heating at 250 °C the deformation heating in the surface layers is higher than recrystallization temperature which is expressed in a significant increase in ductility (δ~15%). It should be noted that with an increase in ductility, there is practically no decrease in UTS and YS.

4 Conclusion

In the article the parameters analysis of radial-shear rolling of 1050A aluminum alloy is carried out. Based on the FEM modeling, it is shown that nonhomogeneous dynamic change in the temperature-deformation parameters in the deformation zone occurs during the RSR process.

During rolling of aluminum alloy the significant heat effect is observed, which caused by local zones of deformation and friction. The maps of temperature and strain rates distribution at different elongation ratio and rotation velocity were obtained. The comparison of the data on change of temperature and mechanical properties of rods rolled at different temperatures shows that with a certain elongation ratio, temperature and rates, it is possible to obtain a structure where the peripheral part of rod will be completely recrystallized and fine-grained while the axial zone will be deformed which determines the mechanical properties of the material.

The data on distribution of temperature-deformation parameters obtained on basis of FEM modeling make possible to understand the features of zonal heterogeneity of structure and to predict the properties of rods after rolling.

The research was supported by Russian Science Foundation (project No. 19-79-00054).

References

1. G.E. Totten, D.S. MacKenzie. *Handbook of aluminium. Vol. 1. Physical metallurgy and processes.* (N.Y., Marcel Dekker Inc., 2003)
2. L.F. Mondolfo. *Aluminum Alloys: Structure and Properties* (Butterworths: London, UK, 1976)
3. E.A. Starke Jr., J.T. Staley. Progress in Aerospace Sciences, **32**, pp. 131-172 (1996)
4. N. Belov, N. Korotkova, T. Akopyan, K. Tsydenov. Metals, **9**(12) (2019)
5. Akopyan T. K., Belov N. A., Naumova E. A., Letyagin N. V. Materials Letters, **245**, pp. 110-113 (2019)
6. C. Watanabe, R. Monzen, K. Tazaki. J. Mater, Sci. **43**: 813 (2008)
7. B. Romantsev, A. Goncharuk, A. Aleshchenko et al. Int. J. Adv. Manuf. Technol. **97**: 3223 (2018)
8. B.A. Romantsev, Y.V. Gamin, A.V. Goncharuk et al. Metallurgist, **61**: 217 (2017)
9. V.A. Sheremet'ev, A.A. Kudryashova, X.T. Dinh et al. Metallurgist, **63**: 51 (2019)
10. A.V. Zinov'ev, A.N. Koshmin, A.Y. Chasnikov. Metallurgist, **63**: 422 (2019)
11. I. Sabirov, N.A. Enikeev, M.Yu. Murashkin, R.Z. Valiev. Bulk Nanostructured Materials with Multifunctional Properties. (Springer, 2015)
12. R.Z. Valiev, A.P. Zhilyaev, T.G. Langdon. Bulk Nanostructured Materials: Fundamentals and Applications. (John Wiley & Sons, 2013)
13. S.P. Galkin, B.A. Romantsev, D.X. Ta, Y.V. Gamin. Chernye Metally, **4**, 20-27 (2018)
14. T.K. Akopyan, N.A. Belov, A.S. Aleshchenko et al. Mater. Sci. Eng. A, **746**, 134-144 (2019)
15. S. Dobatkin, S. Galkin, Y. Estrin et al. J. Alloys Compd. **774**, 969-979 (2019)
16. T.K. Akopyan, A.S. Aleshchenko, N.A. Belov et al. Phys. Metals Metallogr. **119**: 241 (2018)

17. B.V. Karpov, P.V. Patrin, S.P. Galkin et al. Metallurgist, **61**: 884 (2018)
18. A. Stefanik, A. Morel, S. Mroz, P. Szota. Arch. Metal. Mater., **60,** 2, 809–813 (2015).
19. A.V. Fomin, A.S. Aleshchenko, I.M. Maslenniko et al. Metallurgist, **63**: 477 (2019)
20. G.Y. Deng, Q. Zhu, K. Tieu et al. J. Mater. Process. Technol., **240**, 200–208 (2017)
21. Gontarz, A., Tomczak, J., Pater, Z., Bulzak, T. Materials, **12**, 2917 (2019).
22. Çakırcalı, M., Kılıçaslan, C., Güden, M. et al. Int. J. Adv. Manuf. Technol. **65**: 1273 (2013)

Modeling of the stamping process of box-type forging

*Vagid*Kadymov[1], *Evgeny*Sosenushkin[2*], and *Elena* Yanovskaya[2]

[1]Moscow State University of Humanities and Economics, Russia
[2]Moscow State Technological University "STANKIN", Russia

Abstract. A new mathematical model for manufacturing rectangular of box-type forgings is proposed. Along the outer contour of the forging there are walls with thickness comparable to the thickness of the bottom part of the forging. To calculate the power parameters of the technological process under study, the authors use the theory of flow in a relatively thin plastic layer as described by A. A. Ilyushin. A thickness-averaged two-dimensional mathematical model of the plastic layer spreading process is described by nonlinear partial differential equations of the first order with respect to the contact pressure and the flow velocity components. For a metal that is in a flat deformed state, the condition of complete plasticity is assumed, under which the tangent stresses in the layer are small in comparison with normal stresses and they can be ignored, and two normal stresses in the cross section can be considered equal in the first approximation. The third normal stress is equal to the contact pressure of the layer on the tool. In this case, it is sufficient to set a single boundary condition on the border of the regions.

1 Introduction

We consider box forging of rectangular shape, the walls of which have rounded corners. When stamping, the metal is deformed in accordance with the specified movement of the working parts of the stamp, while the latter are assumed to be rigid non-deformable. The thickness of the bottom part of the forging $h(t)$ is much smaller than its other dimensions: 2L-in the direction of the x-axis and 2B-in the direction of the y-axis.

The flanged area represents a plastic layer of constant thickness $h = h(t)$ and occupies an area bounded by the external contours:

$$G1:\ f_1(x, y) = 0. \tag{1}$$

For a deformable metal, the condition of complete plasticity is assumed [1, 3], according to which the tangent stress in the layer is small in comparison with normal

*Corresponding author: evgeny.sosenushkin@yandex.ru

stresses and it can be ignored, and two normal stresses in the cross- section (vertical) in the initial approximation can be considered equal:

$$\sigma_x \approx \sigma_y = -q; \ \tau_{xy} \approx 0. \tag{2}$$

The third normal stress, constant in the thickness of the layer , is equal to the contact pressure of the layer on the tool:

$$-\sigma_z = p, \tag{3}$$

and is determined from the plasticity condition [1, 6]:

$$\sigma_s = p + q. \tag{4}$$

To calculate the energy-power parameters of the process under study, it is proposed to use the theory of flow averaged over the thickness of the layer in a relatively thin plastic layer as described by A. A. Ilyushin [1]. We set a boundary value problem with respect to the contact pressure $p = p(x, y)$ and two components of the flow velocity $u = u(x, y)$ and $v = v(x, y)$ [1, 3, 6], which includes two quasi-static equilibrium equations and the incompressibility condition:

$$\frac{\partial p}{\partial x} = -\frac{2\tau_s}{h}\frac{u}{\sqrt{u^2 + v^2}}, \tag{5}$$

$$\frac{\partial p}{\partial y} = -\frac{2\tau_s}{h}\frac{v}{\sqrt{u^2 + v^2}}, \tag{6}$$

$$\frac{\partial u}{\partial x} + \frac{\partial v}{\partial y} - \frac{d\lambda}{dt} = 0, \tag{7}$$

where $\lambda(t) = \ln\frac{h_0}{h(t)}$ is the degree of deformation according to A. A. Ilyushin (similarto the measure of accumulated deformation in a particle).

2 Equations and mathematics

In the forging under consideration, we select an element of symmetry $S_i = \{\|(x, y)|a_{i0} \le x \le a_{i2}; b_{i0} \le y \le b_{i1}\}$ with an external border (G1) in the form of walls into which the metal flows. In this case, the lines $x = a_{i0}$ and $y = b_{i0}$ serve as symmetry lines of the flow area (G2) and coincide with the axes of the selected coordinate system (Fig.1). In this case, the boundary conditions take the following form:

$$G1: f_1(x, y) = 0, p = 2\tau_s \tag{8}$$

$$G2: v_n = 0 \tag{9}$$

In this case, the external contour, where the metal flows freely, coincides with the level line (constant pressure). On the other hand, it is known [1, 3, 6] that the current lines are straight, orthogonal to the level lines. Consequently, the quasi-static equation of equilibrium $\frac{\partial p}{\partial \xi} = \frac{2\tau_s}{h}$ along the known flow line $\xi = \xi(x, y)$, is easy to integrate:

$$p(\xi) = \frac{2\tau_s}{h}\xi + 2\tau_s. \tag{10}$$

Thus, the contact pressure field in the considered flow stage is determined, which means that we can find the required force for the extrusion process. The solution of the problem of flow of a plastic layer of constant thickness $h = h(t)$ in the region bounded by the contour G1 is considered below. The selected area S_i is further divided into subdomains S_{in} for convenience (see Fig. 1).

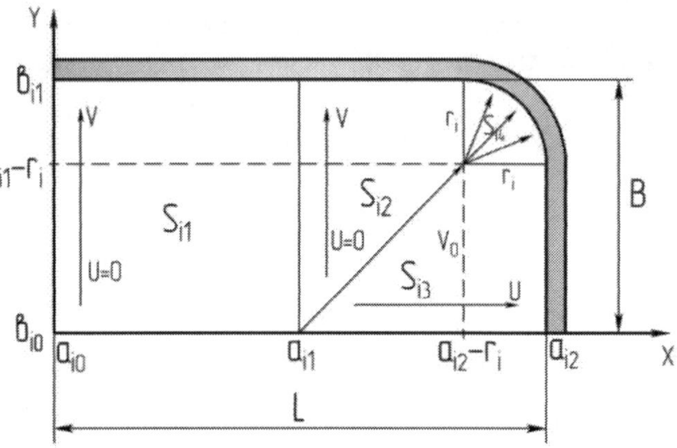

Fig. 1. Design scheme

The equations of the boundary value problem (5) – (7) for the specified subdomain $S_{i1} = \left\{(x,y)\middle| \begin{array}{l} a_{i0} \le x \le a_{i1} \\ b_{i0} \le y \le b_{i1} \end{array}\right\}$ are written:

$$\frac{\partial p}{\partial x} = 0; \tag{11}$$

$$\frac{\partial p}{\partial y} = -\frac{2\tau_s}{h_i}; \tag{12}$$

$$\frac{\partial v}{\partial y} = -\frac{1}{h_i}\frac{dh_i}{dt}. \tag{13}$$

The boundary conditions: $y = b_{i0} = 0 : v = 0$; $y = b_{i1} : p = 2\tau_s$.

The pressure in the area is found by integrating the equation (12):

$$p_{i1} = 2\tau_s - \frac{2\tau_s}{h_i}(y - b_{i1}). \tag{14}$$

The component of the flow velocity after integrating the incompressibility equation (13) takes the form:

$$v_{i1} = -\frac{1}{h_i}\frac{dh_i}{dt}y \cdot \tag{15}$$

Stamping force in the area S_{i1}:

$$P_{S_{i1}} = \iint\limits_{S_{i1}} p_{i1}dxdy = (a_{i1} - a_{i0})\left[2\tau_s(b_{i1} - b_{i0}) - \frac{\tau_s}{h_i}(b_{i1} - b_{i0})^2\right] \tag{16}$$

Let's consider the plastic area S_{i2}: $S_{i2} = \left\{(x,y)\middle| \begin{matrix} a_{i1} \le x \le a_{i2} - r_i, \\ x - (a_{i0} - b_{i0}) \le y \le b_{i1} \end{matrix}\right\}$

The equations of the boundary value problem are given in the form of the system (11) – (13). Boundary conditions in the plastic area $S_{i2}: y = x - (a_{i0} - b_{i0}): v = 0;$ $y = b_{i1}:$ $p = 2\tau_S \cdot$

By analogy with the area S_{i1}, the pressure is determined from equation (14):

$$p_{i2} = 2\tau_s - \frac{2\tau_s}{h_i}(y - b_{i1}) \cdot \tag{17}$$

The flow rate is found by integrating the incompressibility equation (13):

$$v_{i2} = -\frac{1}{h_i}\frac{dh_i}{dt}\left[y - x + (a_{i0} - b_{i0})\right] \cdot \tag{18}$$

Let's go to the definition of the stamping force in the area S_{i2}:

$$P_{S_{i2}} = \iint\limits_{S_{i2}} p_{i2}dxdy = -\tau_s\left\{\left[(a_{i2} - r_i - b_{i1}) - (a_{i0} - b_{i0})\right]^2 - \left[(a_{i1} - b_{i1}) - (a_{i0} - b_{i0})\right]^2\right\} +$$

$$+\frac{\tau_s}{3h_i}\left\{\left[(a_{i2} - r_i - b_{i1}) - (a_{i0} - b_{i0})\right]^3 - \left[(a_{i1} - b_{i1}) - (a_{i0} - b_{i0})\right]^3\right\}, \tag{19}$$

The next area of plastic flow S_{i3}: $S_{i3} = \left\{(x,y)\middle| \begin{matrix} y - (a_{i0} - b_{i0}) \le x \le a_{i2}, \\ b_{i0} \le y \le b_{i1} - r_i \end{matrix}\right\}$

The equations of the boundary value problem (5) – (7) in the considered area have the form:

$$\frac{\partial p}{\partial x} = -\frac{2\tau_S}{h_i}; \tag{20}$$

$$\frac{\partial p}{\partial y} = 0; \tag{21}$$

$$\frac{\partial u}{\partial x} = -\frac{1}{h_i}\frac{dh_i}{dt}. \tag{22}$$

The boundary conditions in the area $S_{i3}: x = y - (a_{i0} - b_{i0}): u = 0;$ $x = a_{i2}: p = 2\tau_s$.

The pressure in the area S_{i3} is found by integrating the equation (20):

$$p_{i3} = 2\tau_s - \frac{2\tau_s}{h_i}(x - a_{i2}) \cdot \qquad (23)$$

The flow velocity is determined by integrating the incompressibility equation (22):

$$u = -\frac{1}{h_i}\frac{dh_i}{dt}(x - a_{i2}) \cdot \qquad (24)$$

The stamping force in the area S_{i3}:

$$P_{S_{i3}} = \iint_{S_{i3}} p_{i3}\,dxdy = -\tau_s\left\{\left[(b_{i2} - r_i - a_{i1}) - (a_{i2} - b_{i0})\right]^2 - (a_{i2} - a_{i0})^2\right\} +$$
$$+\frac{\tau_s}{3h_i}\left\{\left[(b_{i1} - r_i - a_{i2}) - (a_{i0} - b_{i0})\right]^3 - (a_{i0} - a_{i2})^3\right\} \qquad (25)$$

Let's consider the following plastic area S_{i4}:

$$S_{i4} = \left\{(x,y)\left|\begin{array}{c} a_i - r_i \le x \le a_{i2}, \\ b_{i1} - r_i \le y \le (b_{i1} - r_i) + \sqrt{r_i^2 - \left[x - [a_i - r_i]\right]^2} \end{array}\right.\right\}$$

For convenience, when performing calculations, we will proceed to the cylindrical coordinate system [3, 6], in which the limits of changing the geometric parameters are set:

$$0 \le \rho \le r_i; \quad 0 \le \varphi \le \frac{\pi}{2} \cdot \qquad (26)$$

The boundary conditions in the area S_{i4}:

$$\rho = r_i : p_{i4} = 2\tau_s \,. \qquad (27)$$

The current lines form a bundle of straight lines centered at a point with coordinates $(a_{i2} - r_i; b_{i1} - r_i)$.

The equations of the boundary value problem in the domain S_{i4} in cylindrical coordinates take the form:

$$\frac{\partial p}{\partial \rho} = -\frac{2\tau_s}{h_i}; \qquad (28)$$

$$\frac{\partial p}{\partial \varphi} = 0; \qquad (29)$$

$$\frac{\partial u}{\partial \rho} = -\frac{u}{\rho} - \frac{1}{h_i}\frac{dh_i}{dt} \cdot \qquad (30)$$

The contact pressure is determined from the integration of equation (28):

$$p_{i4} = 2\tau_s - \frac{2\tau_s}{h_i}(\rho - r_i) \qquad (31)$$

The deformation force for the considered plastic area S_{i4} in the cylindrical coordinate system takes the form:

$$P_{S_{i4}} = \iint_{S_{i4}} p_{i4} \rho d\rho d\varphi = \frac{\pi}{2}\left(\tau_s r_i^2 + \frac{1}{3}\frac{r_i^3}{h_i}\right) \tag{32}$$

3 Results

The paper presents an approach to calculating the power and kinematic parameters of the process using the example of box forging technology. A characteristic flow area is identified, for which the boundary value problem of plastic flow is formulated, based on the flow theory of A. A. Ilyushin. To solve the problem, the entire region $S_{i,j}(x,y)$ of the plastic flow is divided into the sum of subdomains with a characteristic metal flow, which allowed us to obtain a complete analytical solution of the problem. Using the presented results, it is possible to construct areas of more complex configuration and calculate the results using original numerical methods and software packages based on the obtained result.

4 Conclusion

The presented mathematical model allows us to determine the nature of the change and estimate the pressure values at the contact of the deformable metal with the tool, as well as the values and intensity of changes in the force and deformation parameters of the technological process of extrusion of box forgings.

References

1. A.A.Ilyushin,*Plasticity* (Fizmatlit, Moscow, 2004)
2. L.Prandtl, ZAMM, J**3**(6)(1923)
3. I.A.Kiyko. *Theory of plastic flow*(Mosc. Un-t, 1978)
4. D.V.Georgiyevsky, Mech of SolidsJ**1**. (2006)
5. V.A.Kadymov, Math Methods in the Appl SciJ**38**,16(2015)
6. V.A.Kadymov *Math modeling of contact problems of plastic flow*(Palmarium Ac. Publishing, 2016)
7. E.N. Sosenushkin, V.A.Kadymov, E.A. Yanovskaya, T.V. Gureeva,*Extrusion mechanics of an aluminium alloy forging with longitudinal ribs*, Tsvetnye Metally. **3(915)**,69-75 (2019)

Residual life assessment of welded joints in the girth weld root

Leonid Shron[1,*], *Vladimir* Bogutsky[1] , *Elmar* Yagyaev[2] , and *Illa* Tabolin[3]

[1]Sevastopol State University, 299053, str. University, 33, Sevastopol, Russia
[2]Crimean Engineering and Pedagogical University named after Fevzi Yakubov, per.Uchebnyj, 8, Simferopol, 295015, Russia
[3]Closed joint-stock company «Chelyabinsk interactive cable networks", st. Ordzhonikidze, 54-b, Chelyabinsk, Russia

Abstract. The article analyzes the influence of the structural crack-like defect in the girth weld root welding the bottom to the collector body on the crack growth resistance. A mathematical model has been developed to estimate the residual life resource of welded joints with an internal crack-like defect in the root of the weld under creep. Results of metallographic studies of bench tests results for the welded joints long-term strength formed the basis of the calculation model. In all tested models the cracks start at the top of the narrow structural gap in the weld root and develop along almost the entire weld axis, perpendicular to the wall of the pipe and have a discrete growth pattern. As a result, discrete rupture of material occurs on a length of rc, and the initial crack is incremented by this length. After that, a "plastic core" of rc length is formed in the newly formed crack tip, additional damage is accumulated, and the crack development process continues until the stress level in the section weakened by the crack reaches a critical value. The obtained dependence characterizes the crack growth time from the initial crack-like defect to its critical length.

1 Introduction

The experience of operating steam boilers indicates cases of sudden detachment of the bottom from the end of the collector along the weld metal, on the inside of which there was a linear crack-like defect in the form of lack of fusion in the weld's root. This is explained by the fact that the bottom wall, loaded with internal pressure, imitates the conditions of tight sealing and leads to the appearance of a concentrated bending moment that additionally affects the value of axial tensile stresses near the tip of the crack-like defect in the root of the weld. This problem is a pressing challenge and requires a solution.

Numerous experimental and theoretical studies are devoted to the study of crack growth under creep conditions [1-6]. Obviously, a reliable interpretation of the experimental data is impossible without determining the stress condition not only near the crack tip, but also on the path of its propagation. In this work, when determining the stress condition in the vicinity of the tip of a stationary crack loaded under steady creep conditions, we used the approach described in [7]. According to the works [7] and [8], we assume that the stress

* Corresponding author: shronlb@mail.ru

diagram σ_y on the extension of the crack consists of three areas: nonlinear deformations and the implementation of the fracture mechanism (r_c length), singular (r_s length), regular (b-l-r_s length) and has the nature of the dependence presented in Fig 1.

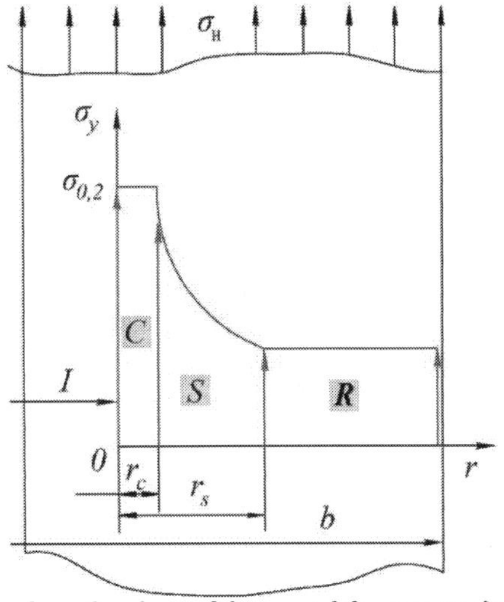

Fig. 1. Distribution of σ_y stress along the plane of the normal fracture crack: C is an area of nonlinear deformations, S is an area of singular stresses, R is an area of regular stresses.

On the $[0r_c]$ segment, the stress $\sigma_y = \sigma_{0,2}^c$, where $\sigma_{0,2}^c$ is the yield strength of the material at creep temperature. Such a redistribution of the asymptotic σ_y stresses on the small-scale yield segment is the well-known Irwin approximation [9] with the only difference that in our case the r_c and r_s segments are subject to determination, as well as the $f_I(\lambda)$ correction function. Here, the relative (dimensionless) crack length is still represented as $\lambda = l/b$.

2 Materials and Methods

According to the theory of plasticity. power law hardening elastoplastic bodies have the components of the stress tensor near the tip of a stationary crack determined by the Hutchinson-Rice-Rosengren singular solution (the so-called HRR asymptotics, [10, 11]). For steady creep (for n> 1), the valid asymptotic behavior is:

$$\sigma_{ij} \sim r^{-\frac{1}{n+1}} . \tag{1}$$

Preserving the stress dimension in (1) and assuming the structure of the stress intensity factor (SIF) in the form:

$$K_I = \sigma_н \cdot l^{\frac{1}{n+1}} \cdot f_I(\lambda),$$

we write the approximate expression for the σ_y stress at the crack extension in the form:

$$\sigma_y \approx \sigma_н \cdot \left(\frac{r}{l}\right)^{-\frac{1}{n+1}} \cdot f_I(\lambda) . \tag{2}$$

Multiplying and dividing expression (2) by $(2\pi)^{\frac{1}{n+1}}$, assuming n=1, we obtain, up to a constant factor $\sqrt{2}$, the well-known expression of σ_y for a normal fracture crack in linear fracture mechanics:

$$\sigma_y = \frac{K_I}{\sqrt{2\pi \cdot r}} \cdot \varphi_y(\theta), \tag{3}$$

where $K_I = \sigma_н \sqrt{\pi \cdot l} \cdot f_I(\lambda)$, $\varphi_y(\theta)$ is a limited function of the polar angle, assuming a value of 1 at $\theta = 0°$. Of course, the formal approach to constructing asymptotic expression (2) demonstrated above is a rough approximation, but at the same time, for engineering applications, it not only qualitatively reflects the stress distribution σ_y over the entire crack cross section propagation, but also allows one to estimate the residual life of specific structural elements operating under creep conditions .

Using the approach developed in [7], as applied to the problem posed, a nonlinear algebraic equation is obtained for the desired quantity r_s with three variable parameters λ, μ, n in the form:

$$Y = Y(r_s, \lambda, \mu, n) \tag{4}$$

where $\mu = \sigma_н / \sigma_{0,2}^c$, $\sigma_н$ is nominal tensile stress in the undamaged section of the pipe wall (Fig. 1). The numerical solution of (4) with respect to r_s allows us to find r_c and $F_I(\lambda)$ by the following formulas:

$$r_c(\lambda) = r_s(\lambda) \cdot \left[\frac{\mu \cdot (1-\lambda)}{1-\lambda-\lambda \cdot r_s(\lambda)}\right]^{1+n}, \tag{5}$$

$$F_I(\lambda) = \left[\frac{1-\lambda}{1-\lambda-\lambda \cdot r_s(\lambda)}\right] \cdot [2 \cdot r_s(\lambda)]^N, \tag{6}$$

$$N = \frac{1}{1+n}.$$

Figure 2a shows the results of the calculation of the correction function $F_I(\lambda)$ in a graph. The calculation uses the data obtained during bench tests for the long-term strength of the welded joint of the bottom and pipe wall in Fig 2b. The initial relative length of the crack-like defect is λ = 0.315; the parameters n and μ are 2 and 0.051, respectively.

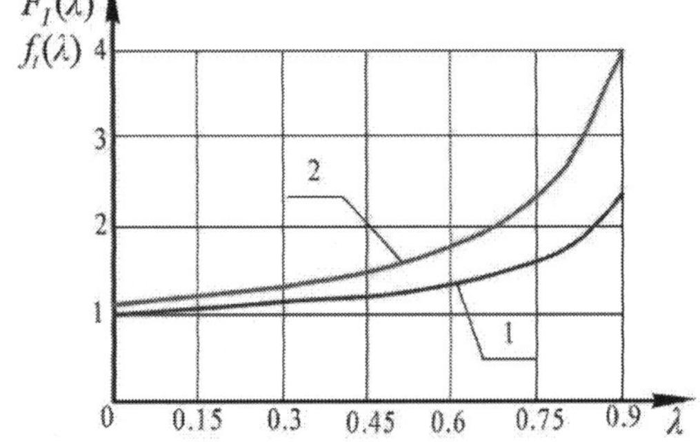

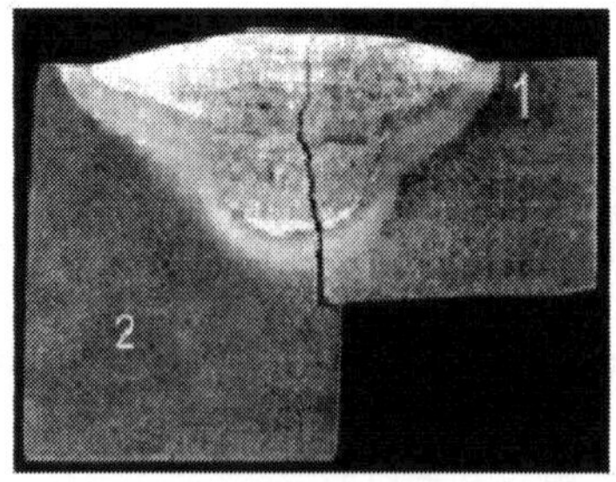

Fig. 2. a) Graphs of correction functions: curve 1 - $f_I(\lambda)$, linear asymptotics, n=1, μ=0, curve 2 - $F_I(\lambda)$, non-linear asymptotics, n=2, μ=0.0513; b) The weld template of the pipe joint (2) with the bottom (1) (model material is 12X1MF steel, temperature is 600C, test time is 3000 hours, increase by 100).

We verify the validity of the obtained expressions (4), (5) and (6) with the boundary value λ = 0. In this case, the external contour (configuration) of the loaded element does not affect the intensity of the stresses near the crack tip (the so-called case of a body of infinite size with a normal fracture crack).

When λ=0, expression (4) admits a transformation to the form:

$$r_s(\mu, n) = \frac{n}{1 + n - \mu^n} \tag{7}$$

and, as a consequence of this,

$$F_I(\mu, n) = [2 \cdot r_s(\mu, n)]^N \tag{8}$$

From (7) and (8) it is easy to see that the value of the correction function is $F_I(\mu=0, n=1)=1$ and coincides with the value $f_I(\lambda=0)=1$ in the framework of linear fracture mechanics, which is quite acceptable for engineering applications taking into account the assumptions made above.

Figure 3 shows the family of curves $F_I(\mu,n)$ for various discrete ratios of μ and n parameters.

For engineering applications, it can be assumed that curves 1 and 4 limit the range of values $F_I(\mu,n)$ for acceptable ratios of the parameters of μ and n. We also note that as the parameter n increases, the family of curves $F_I(\mu,n)$ asymptotically approaches the lower curve ($\mu = 0$).

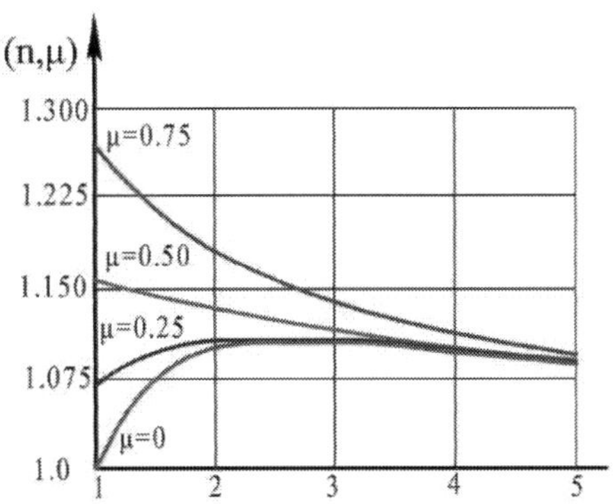

Fig. 3. Family of correction functions $F_I(\mu,n)$ at λ =0.

3 Result and Discussion

The results obtained allow us to proceed to the development of an engineering method for estimating the resource of welded joints of the bottoms of boiler collectors with a crack-like defect in the root of the girth weld.

A significant number of works [12, 13] are devoted to the construction of calculated phenomenological models of fracture of materials under creep conditions. The main task of such calculation models is to determine the time to failure of a loaded structural element. The calculation models are focused on the description of brittle fracture during creep proceeding by nucleation of pores along the grain boundaries orthogonal to the direction of maximum tensile stresses σ_{max}, with subsequent formation of pore chains, their merging first into separate microcracks, and then combining into one macrocrack, the development of which leads to complete destruction of the structural element. In the work [12], when describing the fracture mechanism, a scalar parameter ω was introduced; it characterizes the damage to the material and obeys a kinetic equation of the form:

$$\frac{d\omega}{dt} = A \cdot \left[\frac{\sigma_{max}}{1 - \omega}\right]^n, \quad (9)$$

where A and n are material constants, determined from the long-term strength diagram in the coordinates "$Lg(\sigma)$~$Lg(t)$"; the range of variation of the parameter ω lies in the interval $0 \leq \omega < 1$.

Equation (9) is easily integrated (the parameter ω depends only on time t), which allows us to obtain an expression for finding the time to destruction t_c depending on the level of maximum tensile stresses σ_{max}:

$$t_c = \left[A \cdot (1 + n) \cdot \sigma_{max}\right]^{-1}. \quad (10)$$

It should be noted that subsequently several modifications of the kinetic equation (9) were proposed, with few exceptions, they concerned the modification of the right-hand side by combination with the parameter ω and the introduction of additional constants of material, without affecting the basic structure of the equation. This circumstance confirms the high efficiency of the structure of the kinetic equation (9). A fairly complete review of modern scientific literature on the issues of long-term destruction is given in [14-20].

The modification of equation (9) is based on the following considerations. We use the concept of the scalar damage function ψ, fully preserving the structure of the kinetic equation (9):

$$\frac{d\psi}{dt} = A\left[\frac{\sigma_{max}}{1 - \psi}\right]^n. \quad (11)$$

We assume that the scalar function ψ is directly proportional to the modulus of a radius vector $\vec{r}$, that is

$$\psi = \sqrt{\vec{r} \cdot \vec{r}}. \quad (12)$$

and write the full differential of the expression (12):

$$d\psi = \frac{\vec{r} \cdot d\vec{r}}{\sqrt{\vec{r} \cdot \vec{r}}} \quad (13)$$

There is the sign of scalar multiplication between vectors $\vec{r}$ in the last two expressions. Next, we represent the radius vector in the form of a linear combination of vectors with components ω and λ:

$$\vec{r} = \omega \cdot \vec{i} + \lambda \cdot \vec{j}, \quad (14)$$

where $\vec{i}$ and $\vec{j}$ are unit vectors. Then (14) gives the expressions for $d\vec{r}$:

$$d\vec{r} = \vec{i} \cdot d\omega + \vec{j} \cdot d\lambda, \quad (15)$$

and scalar product of vectors $\vec{r} \cdot \vec{r}$:

$$\vec{r} \cdot \vec{r} = \omega^2 + \lambda^2 \quad (16)$$

In view of (15) and (16), we rewrite expression (13):

$$d\psi = \frac{\omega \cdot d\omega + \lambda \cdot d\lambda}{\sqrt{\omega^2 + \lambda^2}} \quad (17)$$

Dividing both sides of equality (17) by dt, we obtain in final form a modified expression for the kinetic damage equation:

$$\omega \cdot \frac{d\omega}{dt} + \lambda \cdot \frac{d\lambda}{dt} = \frac{A \cdot \sigma_{\max}^{n} \sqrt{\omega^2 + \lambda^2}}{\left[1 - \sqrt{\omega^2 + \lambda^2}\right]^n} \tag{18}$$

As a result of the proposed modification, the structure of the kinetic damage equation became "two-speed" (two-parameter), where $d\omega / dt$ is the growth rate of ω and $d\lambda / dt$ is the growth rate of the relative crack length λ. It is easy to see that at $\lambda = 0$, equation (18) goes over into the original evolutionary damage equation (9). At $\omega = 0$, equation (18) is transformed into an expression that can be interpreted as the kinetic equation of damage during crack growth under conditions of high-temperature corrosion.

To describe the fracture mechanism based on equation (18), we make the necessary explanations regarding the domain of definition of the function ψ. As follows from equation (18), the modulus of the scalar function ψ, like the parameters ω and λ, takes values from the range $0 \leq |\psi| < 1$. It follows that the limiting damage curve ψ is described by the equation of the circle and is located in the first quarter of the plane coordinate system $\lambda 0 \omega$. Figure 4 illustrates this conclusion.

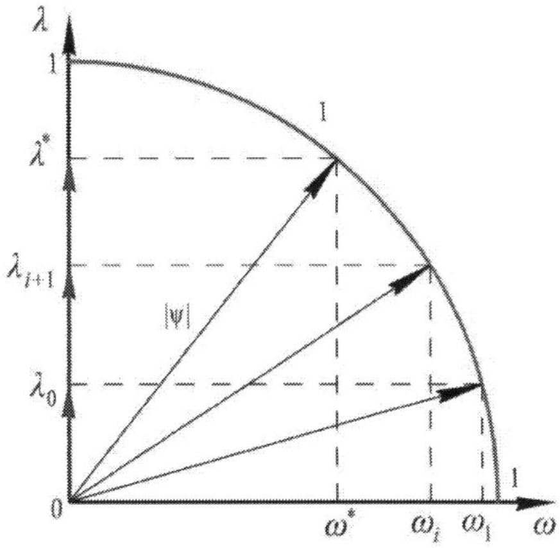

Fig. 4. Trace of the fracture curve 1 in the coordinate plane $\omega 0 \lambda$.

The destruction criterion takes the form:

$$|\psi| = 1 \tag{19}$$

We proceed to formulate the fracture mechanism based on equation (18). Analysis of service fractures and the results of bench tests for the long-term strength of full-scale models indicate a discrete nature of the growth of a creep crack. In all tested models, cracks originate from the top of the narrow structural gap at the root of the weld and develop almost along the axis of the weld, perpendicular to the pipe wall. All cracks are intergranular in nature. Figure 5a, 5b and 5c show the results of metallographic studies of a particular model, prematurely withdrawn from testing after 5762 hours.

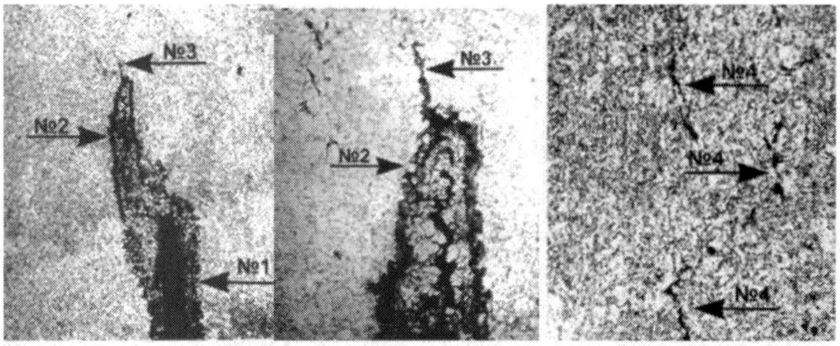

a) b) c)

Fig. 5. a) the structure of the weld metal near the initial defect No. 1, the growth of yield crack No. 2, the formation of microcracks No. 3 (st.12X1MF, temperature 699 °C, test time 5762 hours, x100); b) the structure of the weld metal near the creep crack No. 2, the formation of microcracks No. 3 (x500); c) the structure of the weld metal near the crack, the pore fusion in a microcrack No. 4 (x500).

Figure 5a illustrates the evolution of the growth of a creep crack: from the crack to which arrow No. 1 is assigned, and then to cracks No. 2 and No. 3. Such a description of the process at the crack tip with steady creep shows that at the stage of a continuous increase in the parameter ω from ω_i (current value $|\psi_I|<1$) to ω_{i+1}, at which $|\psi_{I+1}|= 1$, the growth rate of relative length crack equals to zero ($d\lambda/dt=0$). This circumstance makes it possible to explicitly integrate the modified kinetic equation (18) and to obtain an expression for finding the time interval between the current increments of the relative crack length. As a result of integration, we get:

$$t_{i+1} - t_i = \frac{\left[1 - \sqrt{\omega_i^2 + \lambda_i^2}\right]^{1+n}}{(1+n)\cdot A \cdot \sigma_{\max}^n} - \frac{\left[1 - \sqrt{\omega_{i+1}^2 + \lambda_{i+1}^2}\right]^{1+n}}{(1+n)\cdot A \cdot \sigma_{\max}^n}. \quad (20)$$

To estimate the residual life using expression (20), we assume the nature of the distribution of ω along the crack propagation plane.

First, note that in smooth bodies with a constant cross-sectional area, tensile stresses σ_y are uniformly distributed in the section plane orthogonal to the applied force and maximum stresses $\sigma_{max}=\sigma_y$.

4 Conclusion

The process of pore nucleation and coalescence, as well as the value of the parameter ω at each fixed time t_i, can also be taken uniformly distributed along the cross section of the part, which, in turn, leads to a similarity of the stress diagrams σ_{max} and parameter ω as plane figures.

Following these assumptions, we assume that the parameter ω varies along the horizontal axis *or* (see Fig. 1) in the form of a plane figure similar to the stress diagram σ_y, i.e. $\omega(r)\sim\sigma_y(r)$. On the $[or_c]$ segment in the "plastic core" zone of action, as the damage accumulates, the parameter ω reaches a critical value of 1.

As a result, a discrete fracture of the material occurs on a segment of r_c length, and the initial crack is incremented by this length. After that, a "plastic core" of r_c length different from the previous one is formed in the newly formed crack tip, additional damage is accumulated and the crack development process continues until the stress level in the nett section weakened by the crack reaches the critical value. The time accumulated by this moment is calculated by the formula (20) and characterizes the crack growth time from the initial crack-like defect to its critical length.

References

1. R.V. Gol'dshtejn, M.N. Perel'muter, Continuous Computational Mechanics, **2,** 22-39 (2009)
2. R. Goldstein, E. Shifrin, P. Shushpannikov, Comptes Rendus Acad. Sciences. Paris, **336(1),** 108-117 (2008)
3. F.D. Corso, D. Bigoni, J. Mech. Phys.Solids, **56,** 815-838 (2008)
4. H.B.H. Hamouda, L. Laiarinandrasana, R. Piqiues, Int. J. of Pressure Vessels and Piping, **86(2),** 228-238 (2009).
5. M. Rahman, J.W. Hancock, Int. J. Solids and Structures, **43,** 3692-3704 (2006)
6. H. Liu, B. Rui, J. Zhang, B. Fei, International Journal of Fatigue, **59,** 34-42 (2014)
7. S.Ju. Googe, I.S. Tabolin, E.I. Shirjaev, L.B. Shron, Bulletin of the Kuzbass State Technical University **1,** 137-140 (2012)
8. Parton V.Z., Morozov E.A. *Mehanika uprugoplasticheskogo razrushenija* (M.: LKI, (2008)
9. G. R. Irwin, In Proceeding of the Seventh Sagamore Ordnance Materials Conference, **4**, 63-78 (1960)
10. J.W. Hutchinson, J. Mech. Phys. Solids, **16,** 13-31 (1968)
11. J. R. Rice, G.F. Rosengren, J. Mech. Phys. Solids, **16,** 1-12 (1968)
12. L.V. Stepanova, Proceedings CanCNSM Vancouver, **2,** 523-532 (2002)
13. K.A. Agahi, Yy.G. Basalov, V.N. Kuznecov, L.V. Fomin, Bulletin of the Samara State Technical University, **2(19),** 243-247 (2009)
14. L.V. Stepanova, S.A. Igonin, Bulletin of the Samara State Technical University, **3(114),** 97-114 (2014)
15. A. Saxena, Int J Fract, **191**, 31-51 (2015)
16. V. Kalyanasundaram, A. Saxena, S. Narasimhachary, B. Dogan, ASTM STP, **1539** 23-40 (2011)
17. H. Riedel, Encyclopedia of Materials: Science and Technology, 1767-1773 (2008)
18. J. Pan, P.C. Lin, Engng. Fracture Mechanics, **73,** 1797-1813 (2006)
19. G.C. Sih, Theoretical and Applied Fracture Mechanics, **51,** 11-32 (2009)
20. H.Y. Subramanya, S. Viswanatan, R. Narasimhan, Intern. J. of Fracture, **136,** 167-185 (2005)

The study of the regularities of structure formation and properties of the L-PBF metal as a set of processes on the way to create a controlled structure

Ivan Shakirov[1,*], *Pavel* Kuznetsov[1], *Mikhail* Staritsyn[1], *Anton* Zhukov[1] and *Vitaliy* Bobyr[1]

[1]NRC "Kurchatov institute"– CRISM "Prometey", Shpalernaya str., StPetersburg 191017, Russia

Abstract. In this work, to study the effect of laser powder bed fusion (L-PBF) parameters on the microstructure and mechanical properties of 321 austenitic steel, a series of samples were created combining various combinations of L-PBF technological modes, such as: laser spot diameter, scanning speed, laser power, scanning strategy. The possibility of controlling the structure formation of steel in the L-PBF process with the aim of obtaining a given crystallographic texture, grain size and morphology is estimated. The relationship between the resulting anisotropic structure and mechanical properties is investigated.

1 Introduction

Layer-by-layer selective laser fusion of the powder layer already finds practical application for the manufacture of products of complex shape. Various studies demonstrate the possibility of growing samples with a geometric shape and internal microstructure, that cannot be obtained using classical technologies based on the mechanical processing of the initial workpiece [1].

This serves as a good signal for designers that they have another kind of "degree of freedom" when designing products of almost any shape. However, the process of fusion of the powder layer occurs when it is sequentially scanned by a laser beam. With this approach, the fused material consists of a large number of microwelded joints in which formed molten bath solidifies under nonequilibrium conditions [2]. Therefore, when changing the geometry of the melt pool, the thermodynamic conditions of crystallization will change, that will lead to the formation of various structures. Based on this, we can say that when changing the scanning parameters, it may be possible to control the structure, that may be another additional "degree of freedom" when creating the product. Thus, it becomes possible to create functional gradient products with a given distribution of properties by volume. For example, you can create an isotropic or anisotropic material, reproduce a predicted structure in a separate volume of a part. This, in turn, creates the prerequisites for the creation of functional products with a heterogeneous structure and untypical properties.

* Corresponding author: i.v.shakirov@yandex.ru

Currently, studies of the influence building parameters on the structure of fused steels are already being actively conducted. For example, it was shown in [3, 4] that, using various scanning strategies, a clear epitaxial growth along the heat flux with an orientation of <100> can be obtained, that can enhance isotropy. It was also shown that the change in yield strength is mainly associated with different grain sizes, and improved ductility was associated with a changed grain structure, based on the use of different scans [4]. In [5], it was proposed to add to the standard SLM process batch processing using laser shock peening (LSP) in order to increase fatigue properties. Thus, it can be said that despite the accumulated experience in the direction of numerical modeling, the task of assessing the effect of L-PBF regimes on the structure being formed in practice seems relevant. Since, computer modeling simulation results are unable to describe accurately the structure actually obtained on a specific sample at the micro level.

The aim of this work is to study the structure and properties of samples made of stainless austenitic steel powder at various values of the diameter of the focal spot and laser radiation power, scanning speed and the number of repeated scans.

2 Materials and methods

Powder of austenitic stainless 321 steel with a dispersion of 0-60 μm, made by gas spraying of a melt at Polema JSC, was used as a starting material. The prototypes were made from 321 steel powder by the method L-PBF using the EOSINT M270 setup in which Beam Expander (BE) was installed to change the diameter of the focal spot of laser radiation.

To study the effect of L-PBF parameters on the microstructure and mechanical properties of 321 steel, a series of samples were built combining various combinations of L-PBF technological modes, such as laser spot diameter, power, speed, and number of scans. The generalized energy input parameter is calculated as the ratio of the power of the laser source to the speed of movement of the laser spot along the irradiated surface. Three batches of samples were made at various Beam Expander values corresponding to the diameter of the laser spot in the focal plane: 1, 2, and 4.5. Within each batch, the samples correspond to the modes presented in the Table 1.

Table 1. L-PBF Parameters.

Sample no.	Power, W	Speed, mm/s	%	Energy input, W·s/mm	Scanning
1	150.0	800	100	0.188	Single scan
2	170.0	906	100	0.188	
3	190.0	1013	100	0.188	
4	75.0	800	50	0.094	Dual scan
5	85.0	906	50	0.094	
6	95.0	1013	50	0.094	
7	112.5	800	75	0.141	
8	127.5	906	75	0.141	
9	142.5	1013	75	0.141	

The modes were chosen in such a way that samples 1–3 were obtained with an energy input of 0.188 W·s/mm taken as 100%, samples 4–6 with an energy input of 0.141 W·s/mm (50%), samples 7–9 with an energy input of 0.094 W·s/mm (75%), with each sample differing in power and scanning speed. In addition, various scanning modes were used: samples 1-3 single scan, samples 4-9 dual scan.

The structure was studied using a scanning electron microscope (SEM) VEGA 3 TESCAN. EBSD maps were obtained on a Tescan LYRA 3 microscope with a step of 3 μm (grid 400x300 pixels). The density of the samples was determined by hydrostatic weighing on a Mettler Toledo XS105 Excellence XS Analytical scales. Hardness measurements were performed on a Zwick / Roell ZHU 750 top with a load of 10N. On each sample, 4 measurements were performed.

3 Results and discussions

3.1 Sample density

The results of density measurement depending on the L-PBF modes are presented in Fig. 1. The figure shows that with the values of Beam Expander 1, 2 and energy inputs of 100 and 75% for single or dual scanning, respectively, a density of 7.8 g/cm^3 is achieved, that corresponds to a porosity of 1% or less. With an energy input of 50%, the density of the samples decreases to 7.2 g/cm^3, that corresponds to a porosity of about 10%. For the Beam Expander 4.5 value, it is not possible to achieve a density of more than 7.6 g/cm^3 for any energy input. The lowest density value is 6.7 g/cm^3, that corresponds to a porosity of 15%. The data obtained can be used to create samples with a given value of porosity and more than 15%.

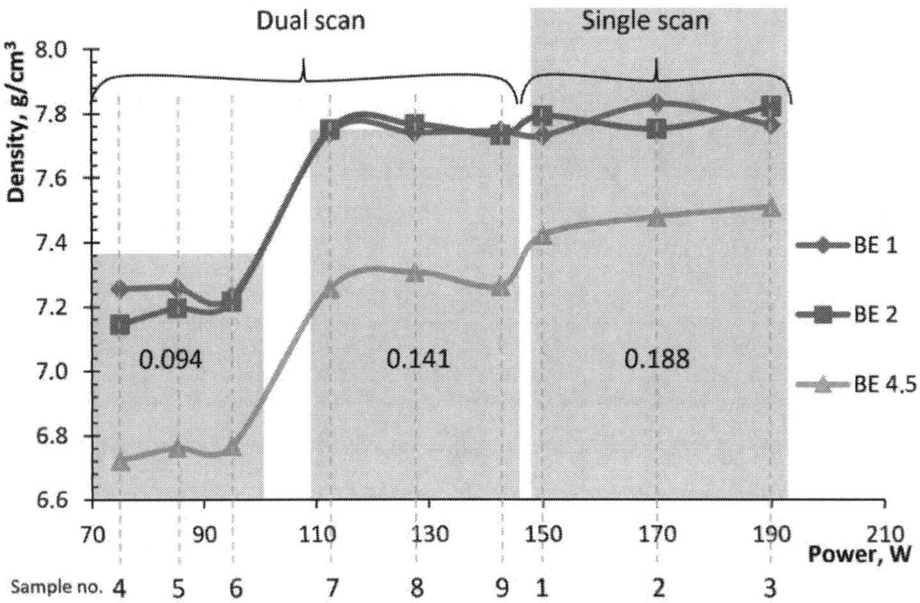

Fig. 1. Dependences of sample density on laser power for various Beam Expander positions.

Thus, we can conclude that the energy input at the level of 100 and 75% with a single or dual scan and a Beam Expander value of 1 or 2 is sufficient to melt the powder layer and form a high-quality melt bath, which allows to achieve a high density. As the diameter of the laser beam spot increases, an increase in energy input is necessary to achieve the desired density level.

3.2 Vickers hardness

The measured values of the hardness of the samples in a plane parallel to the building direction (BD) are presented in Fig. 2. It can be seen that with a single scan of 100% energy input, the hardness values are slightly less than with a dual scan at 75% of the energy input and are 230 and 250 HV, respectively. Lower hardness values at the Beam Expander 4.5 position or when scanning at 50% power are apparently associated with high porosity of the samples.

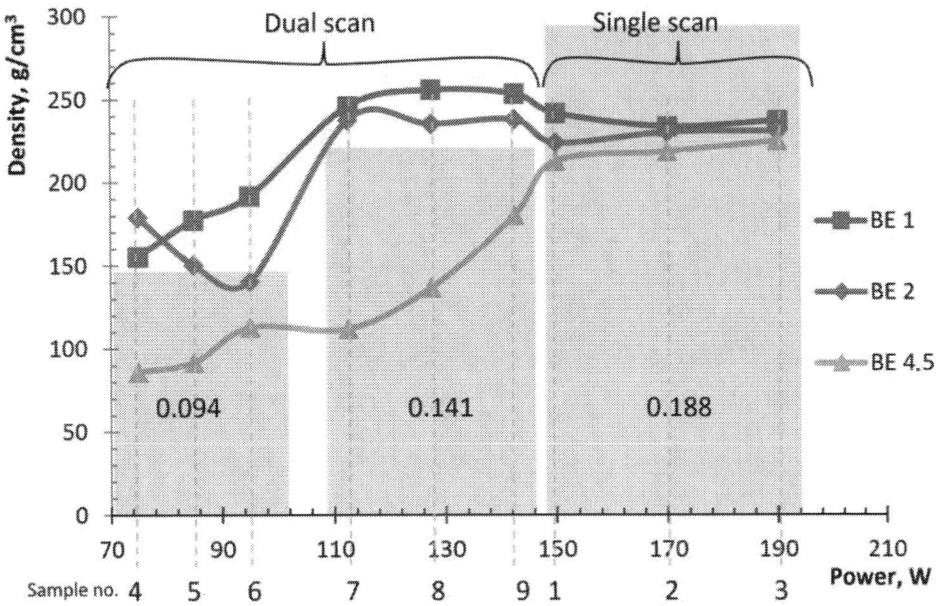

Fig. 2. Dependences of the hardness of the samples on the laser power for different positions of Beam Expander.

The measured hardness values make it possible to evaluate the strength properties of the samples. So, for samples obtained at 100% energy input, UTS is 800 MPa, and for samples obtained at 75% energy position - 750 MPa. This indicates a significant hardening of the samples, which is consistent with the work of other authors [7], in which hardening factors such as dislocation, size, and dispersed particles are noted.

3.3 Microstructure

Analysis of the microstructure of the samples by optical metallography, obtained at various Beam Expander values, shows that with an increase in the diameter of the focal spot, the width of the melt pool increases and its depth decreases. For example, Fig. 3 shows the microstructure of the samples obtained in regime№3 at Beam Expander values 1, 2, and 4.5 (see Table 1).

Samples 4–6 show a large number of pores, that corresponds to lower density values (Fig. 1).

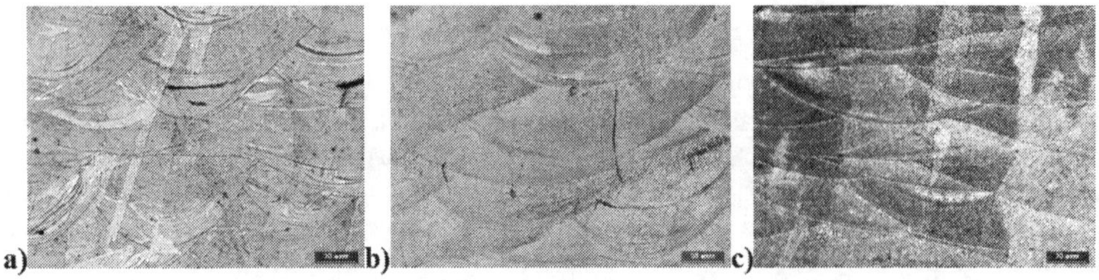

Fig. 3. Optical microscopy. Sample №3, Beam Expander: (a) 1, (b) 2, (c) 4.5.

Micrographs in Fig. 3 show the effect of the focal spot diameter on the solidification structure. The merge lines are clearly visible. The grains consisted of colonies of hardened cells with the same orientation. The grains are drawn parallel to BD and pass through several layers melted by the laser. The average size of the melt bath increases with the increase in the diameter of the laser focal spot with an increase in BE from 1 to 4.5 [6].

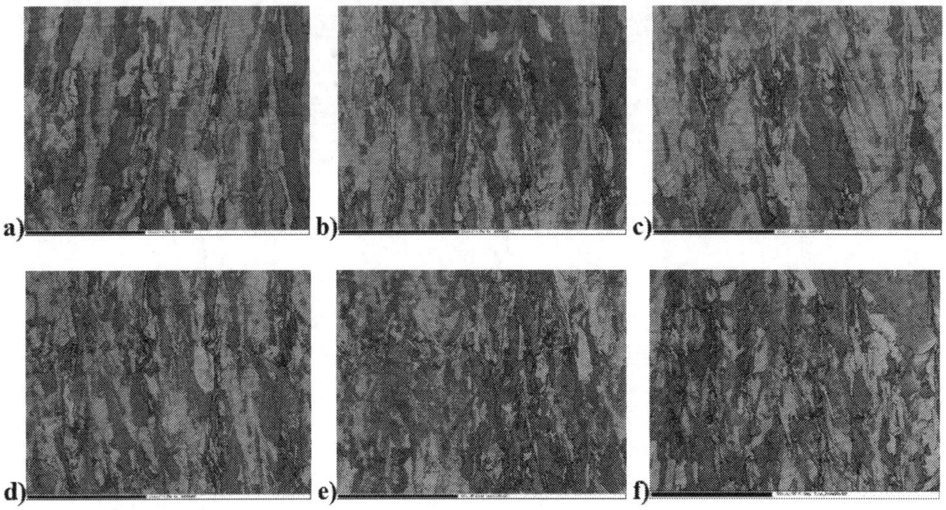

Fig. 4. EBSD maps for sample numbers: (a) 1, (b) 2, (c) 3, (d) 7, (e) 8, (f) 9 and Beam Expander 2.

The structure of samples 1-3 is represented by grains elongated in the sample building direction, spreading through several layers, that indicates their epitaxial growth (see Fig. 4 a, b, c). From the first to the third sample, a proportional increase in power and speed with the constant heat input occurred. In general, this lead to the acceleration of the thermokinetic process of structure formation. Hence, a change occurs from large columnar grains, similar to those formed during casting with directional solidification, to grains with a morphology repeating the shape of the melt baths. The lower rates of metal solidification contribute to the growth of crystallites with the formation of larger grains. With increasing speed, the grain structure is crushed, and crystallization conditions do not favor epitaxial growth. The grains are mainly oriented along the building axis (Z axis), and the length of these grains approaches 1 mm, which is much larger than the thickness of the layer used in the construction (40 µm).

The use of dual scanning leads to a refinement of the structure. Samples 7–9 with a Beam Expander 2 (see Fig. 4 d, e, f), in comparison with the structure of samples 1–3, have a structure with smaller grains of the order of 100 µm in length, whose morphology represented by a drop-shaped form. In the future, dual scanning can be applied in order to refine the structure, to obtain a favorable structure for imparting functional-gradient properties to the whole part or in a separate area of it.

4 Conclusion

In this work, the change in the parameters of the L-PBF process, that is, the diameter of the focal spot, laser radiation power, scanning speed and the number of repeated scans, was evaluated from the point of view of the effect on density, hardness and microstructure. The application of experimental methods contributed to a better fundamental understanding of the correlation between process, microstructure, and properties. The following conclusions can be made:

1) The influence of power and speed was estimated. Increasing the scanning velocity from 800 to 1013 mm/s with a proportional increase in power leads to a change in the thermokinetic process of structure formation. A change is observed from large columnar grains, similar to those formed during casting with directional solidification, to grains with morphology of the repeating shape of the melt baths.

2) Dual scanning does not have a summing effect, which affects the physical and mechanical properties of the resulting samples. The use of dual scanning leads to a refinement of the structure. In the future, dual scanning can be applied in order to refine the structure, to obtain a favorable structure for imparting functional-gradient properties to the whole part or in a separate area of it.

3) The diameter of the laser spot has a strong effect on the structure and properties of the resulting sample:

- the low penetration depth of the melt bath, the low ratio between the thermal gradient and the solidification rate, as well as the high cooling rate favor the refinement of the microstructure and a decrease in the texture intensity due to a partial change in the solidification regime from columnar to equiaxial;

- the density directly depends on the energy input – at the level of 100 and 75% with a single or dual scan and the Beam Expander value of 1 or 2 is enough to melt the powder layer and form a high-quality melt bath. As the diameter of the laser beam spot increases, an increase in energy input is necessary to achieve the desired density level;

The correlations revealed in this work can be effectively used to "tune" the mechanical properties of 321 steel and other austenitic steels on the way to creating functional products with a heterogeneous structure and untypical properties.

This work was supported by the Ministry of science and higher education of the Russian Federation, unique identifier of the agreement RFMEFI62519X0046.

References

1. P. Köhnen, M. Létang, M. Voshage, J.H. Schleifenbaum, C. Haase, Additive Manufacturing, *Understanding the process-microstructure correlations for tailoring the mechanical properties of L-PBF produced austenitic advanced high strength steel,* Volume **30**, (2019), 100914, ISSN 2214-8604, https://doi.org/10.1016/j.addma.2019.100914.
2. Y. Wang, C. Yu, L. Xing, K. Li, J. Chen, W. Liu, J. Ma, Z. Shen, Journal of Materials Processing Technology, *Grain structure and texture of the SLM single track,* Volume 281, 116591, (2020), ISSN 0924-0136, https://doi.org/10.1016/j.jmatprotec.2020.116591.
3. Jieren Guan, Yehua Jiang, Xiaowei Zhang, Xiaoyu Chong, Materials Characterization, *Microstructural evolution and EBSD analysis of AlSi10Mg alloy fabricated by selective laser remelting,* Volume 161, 110079, (2020), ISSN 1044-5803, https://doi.org/10.1016/j.matchar.2019.110079.

4. C.Y. Liu, J.D. Tong, M.G. Jiang, Z.W. Chen, G. Xu, H.B. Liao, P. Wang, X.Y. Wang, M. Xu, C.S. Lao, Materials Science and Engineering: A, *Effect of scanning strategy on microstructure and mechanical properties of selective laser melted reduced activation ferritic/martensitic steel,* Volume **766**, 138364, (2019), ISSN 0921-5093, https://doi.org/10.1016/j.msea.2019.138364.
5. Nikola Kalentics, Manuel Ortega Varela de Seijas, Seth Griffiths, Christian Leinenbach, Roland E. Logé, Additive Manufacturing, *3D laser shock peening – A new method for improving fatigue properties of selective laser melted parts,* Volume **33**, 101112, (2020), ISSN 2214-8604, https://doi.org/10.1016/j.addma.2020.101112.
6. M. Pham, B. Dovgyy, P.A. Hooper et al. *The role of side-branching in microstructure development in laser powder-bed fusion.* Nat Commun 11, **749** (2020). https://doi.org/10.1038/s41467-020-14453-3
7. M.V. Staritsyn, P. A. Kuznetsov, S. N. Petrov, M. S. Mikhailov, Physics of Metals and Metallurgy, *Composite structure as a strengthening factor for stainless austenitic chromium-nickel additive steel.* Volume 121, No. **4**, p. 1–7, (2020) (to be published)

Single-Stage Polystyrene Concrete 3D Printing Technology

Victor Molodin[1,*], *Ivan* Gasenko[2] , and *Pavel* Timin[3]

[1] Doctor of Technical Sciences, Professor, Head of the Department of Construction Technology and Organization at Novosibirsk State University Architectural and Civil engineering (Sibstrin), Novosibirsk, 630008, Russian Federation
[2] Student of the Institute of Civil Engineering, at Novosibirsk State University Architectural and Civil engineering (Sibstrin), Novosibirsk, 630008, Russian Federation
[3] Student of the Institute of Civil Engineering, at Novosibirsk State University Architectural and Civil engineering (Sibstrin), Novosibirsk, 630008, Russian Federation

Abstract. The 3-D printers offered to the construction market can perform retained form layer-by-layer for subsequent filling with reinforced structural concrete and insulation. The use of a single-stage polystyrene concrete with polystyrene granules bloated directly in the printer head allows to form layer-by-layer the insulated wall for its entire width and reduce the erection operation to one technological conversion.

1 Introduction

Within the implementation framework of the National Project “Housing and Urban Habitat”, the builders of Russia have serious tasks to increase the commissioning of housing and social infrastructures. In 2019, only in the Siberian Federal District about 7 million sq. m of housing was commissioned that is 8.75% of the housing commissioning in Russia. The planned figures for 2020 are even higher. They are 8 million 400 thousand sq. m that is the increase should be up to 21% of the achieved result. It is a difficult task, and to solve it, we have to mobilize all available resources, and introduce new technologies and materials widely.

One of the perspectives for development of modern construction is additive technologies that are layer-by-layer growth and synthesis of three-dimensional objects by adding material to the base. The first proposal to use 3D printing for the needs of the construction industry is recorded by a patent issued in the United States to Behrokh Khoshnevis in 2004 [1]. The proposed robotic system was a portal printer with a working head in the form of an extruder (Fig.1) that firstly extrudes layer-by-layer two side walls out of structural concrete, and then, during the second run, fills the cavity between them with heat insulating or other binding material (Fig. 2).

* Corresponding author: molodin@sibstrin.ru

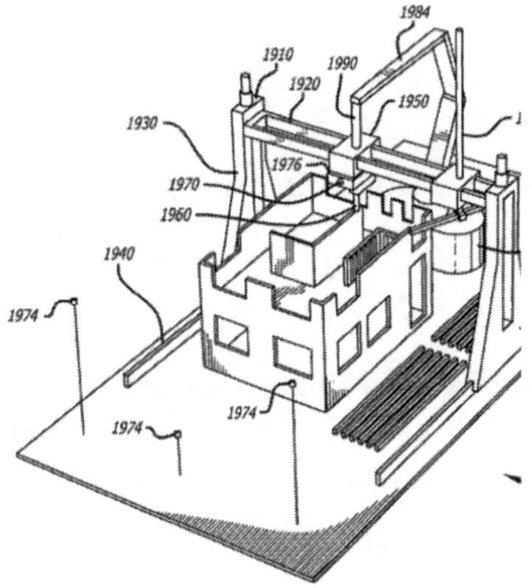

Fig.1. General view of the portal 3D printer printer working hea on the USA patent US7641461B2

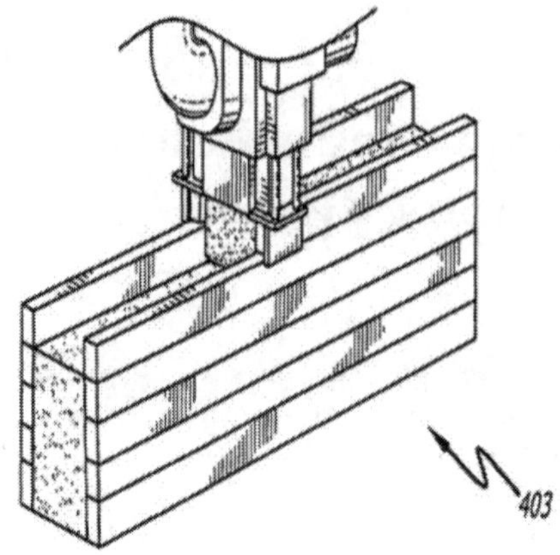

Fig.2. General view of the portal 3D on the USA patent US7641461B2.

The proposed technical solution has been the basic for long 12 years. It has been extensively improved both in terms of the printer itself and in terms of the material to be extruded [2, 3, 4, 5, 6]. During this period, more than 300 patents have been issued that improved the system. Since 2014 the technology has experimentally reached the construction sites. The Chinese company "WinSun" moved from semi-reproduction experiments to the printing of real one-storey and two-storey houses [7], using a bulky portal printer. The attempts to overcome the shortcomings of the predecessors have been made by "DCP" (USA) that offered to the market a hydraulic crane with a single-stage manipulator with six degrees of freedom and a set of working heads, including extruders for work with concrete, thermoplastic, soil and foamed polyurethane.

In 2016, a radial 3D construction printer, the development of "Apis Cor", appeared on the market [8]. It allowed simplifying the structure greatly while increasing the technological capabilities of the device. The application experience showed the advantages of a radial printer over a portal printer, but the extrusion technology – firstly layer-by-layer extrusion of the side walls and then filling the cavity between them – remained unchanged. Approximately at the same time, the 3D printer forming sidewalls not from cement-sand mixture, but from polyurethane foam has been developed at the University of Nantes together with the "Nantes Digital Sciences Laboratory (LS2N)" [9]. Then the cavity between the walls is filled with structural concrete. Such a wall has a structural part inside and insulation from outside. The use of polyurethane foam allowed accelerating the process of rigid side walls formation of the structure and at the same time to insulate them. However, the high cost of the materials and polyurethane surface of the wall, that require a lot of additional expenses for their protection from sunlight and finishing, make the project of little promise despite its attractiveness.

In recent years, there have been some attempts at the construction market to implement 3D technologies using clay to make wall blocks [10], sand that is impregnated with adhesive solution [11]. However, there is no serious market promoting of 3D printer yet.

It should be noted that all attempts to improve the 3D printer and extruding binders in recent years have not changed the fundamental solution of the basic patent [1]. First, the composition made of cement, clay or polyurethane form the wall frame or a retained form, and then the formed cavities are filled with insulation or structural concrete (fig.3)

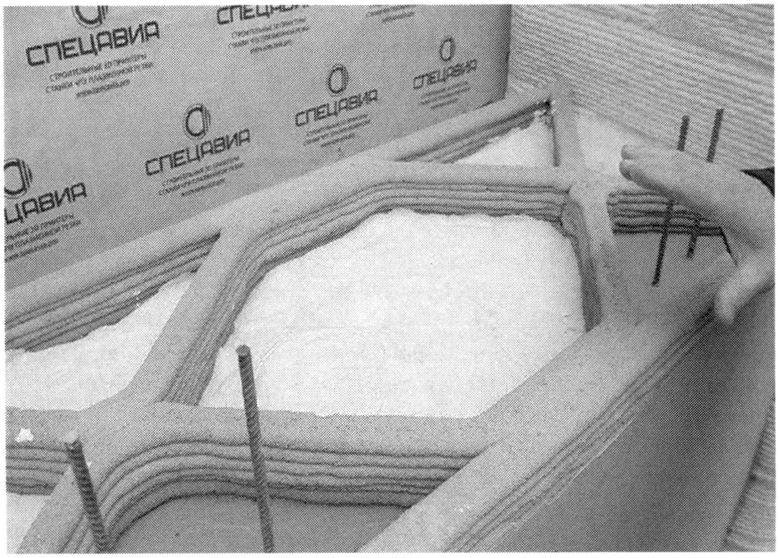

Fig. 2. A wall structure made with a 3D printer according to the traditional technology.

The developed technology has a number of disadvantages that delay its wide implementation into the construction.

1. A high labour intensity of the process to get a final product. First, a wall frame or a retained form is formed with a single binder by the 3D printer. Then, having waited for the loss of mobility of the extruded material, the formed cavities are partially reinforced and then filled with structural concrete and partially filled with insulation. It should be noted that all operations except the printing of a retained form are performed manually.
2. It is possible to combine the wall frame printing and a simultaneous filling of the formed cavity with light concrete [1]. At the same time, the structure of the printing head and its operation are considerably complicated by the need to include into the structure two systems to supply different binding materials.
3. The necessity for sufficiently rapid curing of the wall frame binder to carry the weight of the superimposed binder layers and the lateral pressure of the binder that fill the cavities. At the same time it should not cure too quickly because it can lead to its curing within material storage and delivery systems.
4. In order to avoid the bleeding of the lower, previously extruded layers under the weight of the upper layers, it is necessary to make a technological pause that significantly complicates the construction process and increases its duration.
5. A successful operation of a construction 3D printer using cement-based binder requires positive temperatures or a greenhouse arrangement during the cold season.

The objective of the present work is to develop additive technology to manufacture the enclosing structures of buildings that will allow to overcome disadvantages mentioned above and will be able to provide automated (without people) construction of insulated walls with a minimum number of technological changes.

The task of the present research is to study the workability of a single-stage polystyrene concrete as a binder.

2 Materials and methods

Figures and tables, as originals of good quality and well contrasted, are to be in their final form, ready for reproduction, pasted in the appropriate place in the text. Try to ensure that the size of the text in your figures is approximately the same size as the main text (10 point). Try to ensure that lines are no thinner than 0.25 point. The single-stage polystyrene concrete technology developed in NSUACE (Sibstrin) allows significantly simplifying the process of building enclosing structures using the 3D printer [12]. The essence of the technology is that granules of non-foaming (beaded) polystyrene are added into the mixture of cement and sand during its processing.

The density of the beaded polystyrene coordinates the density of the cement and sand mixture. Therefore, while mixing the polystyrene granules are uniformly distributed throughout the mass of the material. The ready mixture is placed in the space between plate electrodes and the current of industrial frequency, of 220÷380 V is applied to the plates. In the course of the forced electro heating the temperature of the mixture quickly rises up to 80° C, at such a temperature, the granules of beaded polystyrene bloats in many times, increasing in volume. At the same time, if the mixture is in a closed space, by increasing the volume of polystyrene granules during the foaming, self-consolidation of the material takes place. Changing the amount of polystyrene in the mixture, the structural or heat-insulating properties of the material can be enhanced.

All the advantages of *single-stage polystyrene* are realized for additive technology of construction with insulated walls. The technology is implemented as follows (Fig. 3). Polystyrene concrete mixture with non-foaming granules of beaded polystyrene 8 through the feeding system of the binder 2 enters the extruder 1

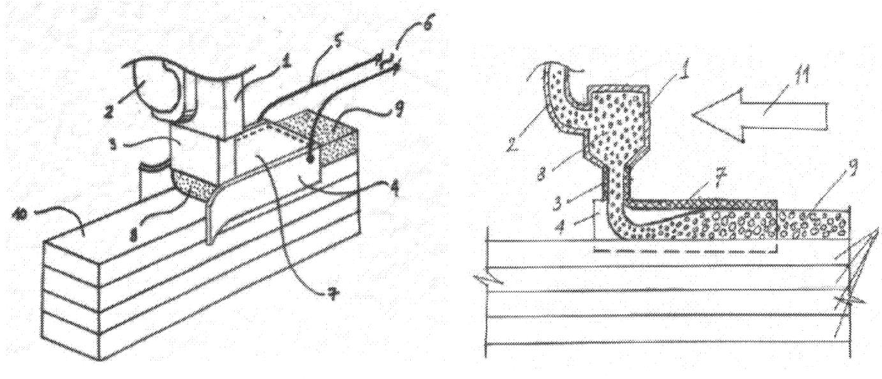

Fig. 3. Principal solution of 3D printer working head for extrusion of polystyrene concrete mixture that is laid according to one-stage technology.

and is heated there to the temperature close to the temperature of polystyrene foaming but it does not reach it. From that place, through the discharge outlet (nozzle) 3, the mixture is pressed onto the surface of the previously molded and solid layer of polystyrene concrete 10. The nozzle 3 has got a transverse dimension equal to the width of the erected wall 10. The polystyrene concrete mixture 8 fills the space between the spatula electrodes 4 those are being the part of the printing head of the 3D printer, move slowly and continuously 11 along the wall.

The electric current 220÷380 V that is coming from the electrical network 6 through the cables 5, passes through the polystyrene concrete mixture and heats it quickly. When the temperature reaches 80° C, the polystyrene granules soften, and under the influence of the boiling isopentane contained therein, the granules foamed and increased in volume many times over, changing the quality of the 9 mixture. While heating and polystyrene granules bloating, the volume of polystyrene concrete mixture 9 increases and occupies the whole space between spatula electrodes 4, the surface of previously laid layer 10 and pressing spatula 7, partially consolidates. Spatula electrodes 4 and pressure spatula 7 form another layer of the wall from polystyrene concrete 9 that was the result of electric thermal treatment of heat-insulating properties. Having a high temperature, polystyrene concrete begins to lose mobility very quickly and harden intensively still being in the space between the spatula electrodes. Having been uniformly distributed in volume the foamed granules of polystyrene, which are insulation as well, reduce heat loss from the structure into the environment.

The duration of maintaining high temperatures inside the material is increased; thereby it provides an intense increase in the strength of the polystyrene concrete. Changing the content of polystyrene in the binder formulation as required, it is possible to increase or decrease the thermal insulation or structural properties of the wall.

The full-scale testing of the 3D printer head for extrusion of polystyrene concrete mixture (fig. 4) that is laid according to one-stage technology showed that the working layer of the material formed according to the proposed technology retains cohesive and shows no signs of mobility. Laying the next working layers on the first one in 15 minutes after the first run of 3D printer does not practically affect the acquired shape and geometric dimensions of the first layer.

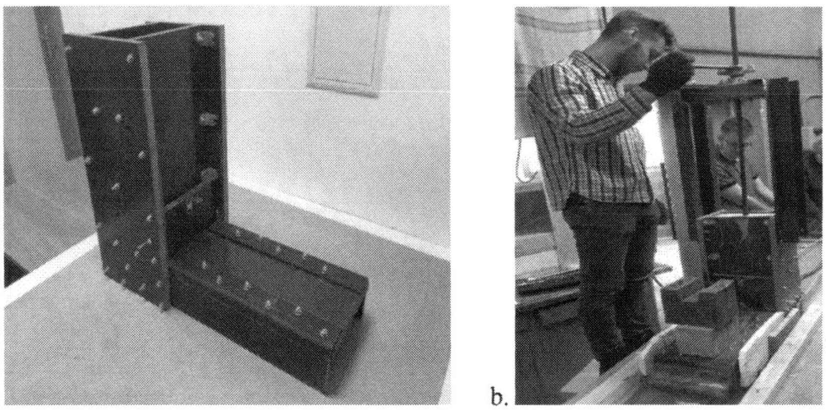

a. b.

Fig. 4. General view (a) and full-sale testing (b) of 3D printer working head for printing the enclosing structures with single-stage polystyrene concrete.

The obtained polystyrene concrete samples were tested for strength and distribution uniformity of the foamed polystyrene granules by the weight of the material.

The cube-strength tests of the samples were performed at 3, 7, 14 and 28 day period. Three standard samples were cut for the tests. Table 1 shows the average test values. The tests were carried out on a standard laboratory press (PSU-50).

Table 1 Dynamics of polystyrene concrete strength developing

Type of curing	3 days, R, MPa	7 days, R, MPa	14 days, R, MPa	28 days, R, MPa
Electro heating, D1200kg/m^3	3.83	7.71	9.38	12.73
Electro heating, D700 kg/m^3	2.72	5.05	5.87	7.46
Standard technology,D1200 kg/m^3	3.21	5.33	7.10	10.69

Analyzing the test results of the polystyrene concrete strength of the medium density D700 and D1200 after the heat treatment, it can be stated that the strength of the samples made according to the one-stage technology is higher than that of the natural hardening samples. The structure of the obtained material proved to be good. The beaded polystyrene foamed entirely and over the whole volume. The average strength compared to polystyrene of standard technology was 20% higher.

The uniformity of polystyrene granules distribution by the material mass was determined by means of measuring of flat samples thermal conductivity cut from the wall mass by sections perpendicular to the direction of possible surfacing of granules.

Before measuring the thermal conductivity, the samples were dried to a constant mass in the oven at 70° C within 24 hours.

The weight loss of the samples was about 5%. The geometric characteristics of the sample, the electrical power required to create heat current and the temperature of the opposite faces sides were measured.

In the course of the experiment, by means of thermal conductivity measuring device (PIT-2), the thermal conductivity of samples cut from a wall made of medium density polystyrene D700 and D1200, a working head of the 3D printer according to *one-stage technology*, as well as the sample made according to a standard technology of medium density D1200 (Table 2) were determined.

Table 2 Layer-by-layer thermal conductivity of polystyrene concrete samples of different density.

Layer	Thermal conductivity (W/m·K)		
	Sample 1 (Electro heating, 1200 kg/m^3)	Sample 2 (Electro heating, 700 kg/m ^{3})	Sample 3 (Standard technology, 1200 kg/m ^{3})
1 layer	0.2007	0.1283	0.2136
2 layer	0.1961	0.1274	0.2090
3 layer	0.1944	0.1268	0.2012
4 layer	0.1933	0.1261	0.1878
Segregation	3.68%	1.56%	12.08%

The non-uniformity of the sample made according to the proposed technology is 3.68% against 12.08% of the sample made according to the standard technology.

The significant difference in thermal conductivity of the upper and lower layers of the polystyrene concrete block manufactured according to conventional technology is due to the flotation of the foamed polystyrene granules and sedimentation of the mortar part.

3 Conclusion

Practically all additive technologies offered to the construction market propose the use the 3D printer to print retained form made of cement and sand mortar of sand, clay or foamed polyurethane.

Then the standard (manual) operations follow to reinforce, place concrete and insulate the future enclosure structure. It is known that in the complex of building structures concreting the labour consumption of formwork does not exceed 40%.

Consequently, the 3D printer will take over only this part of the labor intensity. The rest has to be done with the hands of the workers.

The use of the 3D printer that prints a full-width polystyrene wall allows to:

- form the working layer over the entire width of the wall simplifying the binder supply system;
- eliminate the mobility of binding material as soon as it leaves the working head and accelerate hardening due to high temperatures of heat treatment;
- provide self-consolidation of polystyrene concrete while forming the working layer by means of increasing the volume of granules during their bloating;
- ensure the uniform distribution of foamed polystyrene granules over the volume of the wall material, thus increasing its high-quality thermal resistance;
- guarantee the specified strength or heat insulation characteristics of the material of the erected wall.

References

1. Patent US7641461B2 Robotic systems for automated construction B29C64/106/
2. V.P. Grakhov, S.A. Mokhnachev, O.V. Borozdov . Fundamental researches, **11-12**, 2673-2676 (2014)
3. S.A. Udodov, *ICSC Science and Enlightment* 58-61 (Penza, 2017)
4. D.A. Luneva, E.O. Kozhevnikova, S.V. Kaloshina, Civil Engineering and Aarchitecture, **8** (**1**), 90–101 (2017)
5. K.A. Rudyak, Yu.O. Chernyshev, *Materials of International scientific and practical conference*, 147-151 (Kazan, 2016)
6. Ye.Yu. Stepanova, G.V. Barsukov, Yu.S. Stepanov, *Technical Science*, **8-2,** 243-249 (Tula: Tula State University, 2016)
7. Winsun Company. 3D housing design (Electronic resource – URL:http://www.yhbm.com/index.php?a=lists&c=index&catid=67&m=content) (date 14.12.2016).
8. Construction 3D: Nikita Chen-yun-tya was the first who printed the house in Russia. (Electronic resource - https://www.rbc.ru/magazine/2017/06/592567559a7947e1bb4b7ea9)
9. French project BATIPRINT for 3D house construction "inside out" (Electronic resource – URL:http://www3dprintingindustry.com/tags/ls2n/)
10. Institute for advanced architecture of catalonia takes a natural approach to on site 3d printing (Electronic resource – URL: http://www 3dprintingindustry.com/tags/terra-performa)
11. The Method and the Device (Electronic resource – URL: http://www d-shape.com/what-is-it/the-process/the-concept/)

AUTHOR INDEX

AUTHOR INDEX